Phosphodiesterase Methods and Protocols

METHODS IN MOLECULAR BIOLOGY™

John M. Walker, SERIES EDITOR

METHODS IN MOLECULAR BIOLOGY™

Phosphodiesterase Methods and Protocols

Edited by

Claire Lugnier

Pharmacologie et Physico-Chimie des Interactions Cellulaires et Moléculaires, Centre National de la Recherche Scientifique, Université Louis Pasteur de Strasbourg, Faculté de Pharmacie, Illkirch, France

HUMANA PRESS ✳ TOTOWA, NEW JERSEY

This publication is printed on acid-free paper. ∞
ANSI Z39.48-1984 (American Standards Institute)

Permanence of Paper for Printed Library Materials.

Cover illustration from Figure 3, Chapter 14, "Crystallization of Cyclic Nucleotide Phosphodiesterases," by Hengming Ke, Qing Huai, and Robert X. Xu.

Cover design by Patricia F. Cleary

For additional copies, pricing for bulk purchases, and/or information about other Humana titles, contact Humana at the above address or at any of the following numbers: Tel: 973-256-1699; Fax: 973-256-8341; E-mail: orders@humanapr.com; or visit our Website: www.humanapress.com

Printed in the United States of America. 10 9 8 7 6 5 4 3 2 1

ISSN 1064-3745

E-ISBN 1-59259-901-X

Library of Congress Cataloging-in-Publication Data

Phosphodiesterase methods and protocols / edited by Claire Lugnier.

    p. ; cm. — (Methods in molecular biology, ISSN 1064-3745 ; v. 307)

Includes bibliographical references and index. ISBN 1-58829-314-9 (alk. paper)

1. Phosphodiesterases–Laboratory manuals. I. Lugnier, Claire.

II. series: Methods in molecular biology (Clifton, N.J.) ; v. 307.

[DNLM: 1. Phosphoric Diester Hydrolases–analysis–Laboratory Manuals. 2. Phosphoric Diester Hydrolases–physiology–Laboratory Manuals. 3. Biological Assay–methods–Laboratory Manuals. 4. Protein Engineering–methods–Laboratory Manuals.   QU 25 P5746 2005]

QP609.P53P467 2005

612'.01519--dc22

                                                                            2005006204

Preface

Adenosine $3',5'$-cyclic monophosphate (cAMP) and guanosine $3',5'$-cyclic monophosphate are ubiquitous nucleotides that have been described as the first and second messengers. In concert with intracellular calcium and IP3, they play a major role in the control of intracellular signaling, which orchestrates normal and pathophysiological responses.

Downstream from the cyclic nucleotide synthesis by adenylyl and guanylyl cyclases, the multigenic family of cyclic nucleotide phosphodiesterases (PDEs), by specifically hydrolyzing cyclic nucleotides, controls cAMP and cGMP levels to maintain a basal state. Their critical role in intracellular signaling has recently designated them as new therapeutical targets. Several leading pharmaceutical companies are searching and developing new therapeutic agents that would potently and selectively inhibit PDE isozymes, notably PDE4 and PDE5. Nevertheless, the precise mechanism and the contribution of the various PDE isozymes in modulating intracellular signaling remain to be established.

The aim of *Phosphodiesterase Methods and Protocols* is to provide a palette of a variety of conceptual and technical approaches designed to solve questions concerning the role of PDEs, and ultimately of their different variants, in physiological functions as well as their implications in several pathologies.

During the four research decades spent characterizing cyclic nucleotide phosphodiesterases, PDE nomenclature (PDE1 to PDE11) was recently established according to their genes, biochemical properties, regulations, and sensitivities to pharmacological agents. Although PDE1 to PDE6 were first well characterized because of their predominance in various tissues, their specific contribution to tissue function and their regulatory rules in pathophysiology remain open research fields. Molecular biology as well as fluorescent cell imaging provide further insight into the knowledge of PDE implication in intracellular and subcellular signaling. This is particularly necessary for the PDE7 to PDE11 families, for which roles are not yet well established.

Many of the newest biotechnologies are reported by leader teams in PDE field in this book.

Chapters 1–4 deal with biosensors that allow the measurement of local variations of cyclic nucleotides in living cells as well as their visualization in a spatiotemporal manner. This approach is very helpful for analyzing the contribution of the various PDEs in cyclic nucleotide compartmentalization. Chapters 4–7, devoted to the localization and characterization of PDE activities in tissues and living cells, shed light on critical PDEs and their implications in cellular functions, thus indicating them as targets for specific pathologies. Chapters 8–14, which deal with PDE overexpression, promoter identification, purification, and biochemical and structural studies, describe several approaches for assessment of the potential role of targeted PDEs in the rational development of specific tools and drugs. Chapter 15 describes how to generate PDE4 knockout mice. If no compensatory mechanisms take place, this transgenic approach, which may be extended to various PDEs, is necessary to demonstrate the potential role of targeted PDEs. Chapters 16–21 mainly focus on PDE regulation, by phosphorylation, dimerization, or protein interactions, giving some starting points for further studies on the central role of PDEs in intracellular signaling control.

Phosphodiesterase Methods and Protocols is intended for biochemists, molecular biologists, cell biologists, and pharmacologists who wish to initiate or deepen studies in the PDE field. It also provides a basis for new approaches in drug design for medicinal chemists and pharmaceutical companies. Furthermore, our work will point out a new way for clinicians to find and test novel therapies for numerous pathologies where the molecular origin remains unknown and the treatment is principally symptomatic. In many pathologies, such as inflammation, neurodegeneration, and cancer, alterations of intracellular signalingrelated to PDE deregulation may explain the difficulties observed in their prevention and treatment. By specifically inhibiting the deregulated PDE isozyme(s) with newly identified selective PDE inhibitors, one could imagine the potential restoration of normal intracellular signaling.

We are grateful to all authors for their excellent contributions, which make this book a useful aid not only for scientists wishing to work in the PDE field, but also clinicians working to develop new therapeutic proteins.

1. Butcher, R. W. and Sutherland, E. W. (1962). Adenosine 3′,5′-phosphate in biological materials. 1. Purification and properties of cyclic 3′,5′-nucleotide phosphodiesterase and use of this enzyme to characterize adenosine 3′,5′-phosphate in human urine. J. Biol. Chem. 237, 1244–1250.

2. Beavo, J. A. and Brunton, L. L. (2002) Cyclic nucleotide research-still expanding after half a century. Nature Reviews, 3, 710–718.

Claire Lugnier

Contents

Contributors

FAIYAZ AHMAD • *Pulmonary Critical Care Medicine Branch, NHLBI, NIH, Bethesda, MD*

NIKOLAI O. ARTEMYEV • *Department of Physiology and Biophysics, University of Iowa College of Medicine, Iowa City, IA*

GEORGE S. BAILLIE • *Molecular Pharmacology Group, Division of Biochemistry and Molecular Biology, Davidson and Wolfson Buildings, IBLS, University of Glasgow, UK*

ALFREDA BEASLEY • *Department of Molecular Physiology and Biophysics, Vanderbilt University School of Medicine, Nashville, TN*

J. KELLEY BENTLEY • *Department of Pediatrics and Pulmonary Medicine, The University of Michigan, Ann Arbor, MI*

MITSI A. BLOUNT • *Department of Molecular Physiology and Biophysics, Vanderbilt University School of Medicine, Nashville, TN*

KIMBERLY K. BOYD • *Department of Physiology and Biophysics, University of Iowa College of Medicine, Iowa City, IA*

SHARON M. CAWLEY • *Department of Pharmacology, University of Vermont, College of Medicine, Burlington, VT*

RACHEL A. COLLUPY • *Department of Biochemistry and Molecular Biology, University of New Hampshire, Durham, NH*

MARCO CONTI • *Department of Obstetrics and Gynecology, Stanford University Medical Center Stanford, CA*

JACKIE D. CORBIN • *Department of Molecular Physiology and Biophysics, Vanderbilt University School of Medicine, Nashville, TN*

RICK H. COTE • *Department of Biochemistry and Molecular Biology, University of New Hampshire, Durham, NH*

EVA DEGERMAN • *Section for Molecular Signaling, Department of Cell and Molecular Biology, Lund University, Sweden*

DIETRICH DETTMER • *Institute of Biochemistry, Medical Faculty, University of Leipzig, Leipzig, Germany*

WOLFGANG R. G. DOSTMANN • *Department of Pharmacology, University of Vermont, College of Medicine, Burlington, VT*

SANDRINE EVELLIN • *Dulbecco Telethon Institute and Venetian Institute of Molecular Medicine, Padova, Italy*

SHARRON H. FRANCIS • *Department of Molecular Physiology and Biophysics, Vanderbilt University School of Medicine, Nashville, TN*

LINDA HÄRNDAHL • *Section for Molecular Signaling, Department of Cell and Molecular Biology, Lund University, Biomedical Center, Lund, Sweden*

FENG HE • *Verna and Marrs McLean Department of Biology and Molecular Biology, Baylor College of Medicine, Houston, TX*

THOMAS HERMSDORF • *Institute of Biochemistry, Medical Faculty, University of Leipzig, Leipzig, Germany*

ELAINE V. HILL • *Molecular Pharmacology Group, Division of Biochemistry and Molecular Biology, Davidson and Wolfson Buildings, IBLS, University of Glasgow, UK*

LENA STENSON HOLST • *Section for Molecular Signaling, Department of Cell and Molecular Biology, Lund University, Biomedical Center, Lund, Sweden*

AKIRA HONDA • *Department of Pharmacology, University of Vermont, College of Medicine, Burlington, VT*

SUZANNE HOSIER • *Department of Biochemistry and Molecular Biology, University of New Hampshire, Durham, NH*

MILES D. HOUSLAY • *Molecular Pharmacology Group, Division of Biochemistry and Molecular Biology, Davidson and Wolfson Buildings, IBLS, University of Glasgow, UK*

QING HUAI • *Department of Biochemistry and Biophysics and Lineberger Comprehensive Cancer Center, University of North Carolina, Chapel Hill, NC*

S.-L. CATHERINE JIN • *Division of Reproductive Biology, Department of Obstetrics and Gynecology, Stanford University Medical Center Stanford, CA*

JEFFREY W. KARPEN • *Department of Physiology and Pharmacology, Oregon Health and Science University, Portland, OR*

HENGMING KE • *Department of Biochemistry and Biophysics and Lineberger Comprehensive Cancer Center, University of North Carolina, Chapel Hill, NC*

THÉRÈSE KERAVIS • *Pharmacologie et Physico Chimie des Interactions Cellulaires et Moléculaires, Centre National de la Researche Scientifique, Université Louis Pasteur de Strasbourg, Illkirch, France*

ANNE M. LATOUR • *Department of Medicine, University of North Carolina, Chapel Hill, NC*

VALENTINA LISSANDRON • *Dulbecco Telethon Institute and Venetian Institute of Molecular Medicine, Padova, Italy*

HANGUAN LIU • *Pulmonary Critical Care Medicine Branch, NHLBI, NIH, Bethesda, MD*

CLAIRE LUGNIER • *Pharmacologie et Physico-Chimie des Interactions Cellulaires et Moléculaires, Centre National de la Researche Scientifique, Université Louis Pasteur de Strasbourg, Illkirch, France*

JUSTINE A. MALINSKI • *ATG Laboratories Inc., Eden Prairie, MN*

VINCENT C. MANGANIELLO • *Pulmonary Critical Care Medicine Branch, NHLBI, NIH, Bethesda, MD*

MARJANNE MARKERINK-VAN ITTERSUM • *Department of Psychiatry and Neuropsychology, Maastricht University, Maastricht, Netherlands*

MARCO MONGILLO • *Dulbecco Telethon Institute and Venetian Institute of Molecular Medicine, Padova, Italy*

KHAKIM G. MURADOV • *Department of Physiology and Biophysics, University of Iowa College of Medicine, Iowa City, IA*

DANA C. PENTIA • *Department of Biochemistry and Molecular Biology, University of New Hampshire, Durham, NH*

THOMAS C. RICH • *Department of Integrative Biology and Pharmacology, University of Texas Health and Science Center at Houston, Houston, TX*

WITO RICHTER • *Department of Obstetrics and Gynecology, Stanford University School of Medicine, Stanford, CA*

CAROLYN L. SAWYER • *Department of Pharmacology, University of Vermont, College of Medicine, Burlington, VT*

KONJETI RAJA SEKHAR • *Department of Molecular Physiology and Biophysics, Vanderbilt University School of Medicine, Nashville, TN*

JING RONG TANG • *Pulmonary Critical Care Medicine Branch, NHLBI, NIH, Bethesda, MD*

YAN TANG • *Pulmonary Critical Care Branch, NHLBI, NIH, Bethesda, MD*

ANNA TERRIN • *Dulbecco Telethon Institute and Venetian Institute of Molecular Medicine, Padova, Italy*

RIMA THASELDAR-ROUMIÉ • *Pharmacologie et PhysicoChimie des Interactions Cellulaires et Moléculaires, Centre National de la Recherche Scientifique, Université Louis Pasteur de Strasbourg, Illkirch, France*

MELISSA K. THOMAS • *Department of Molecular Physiology and Biophysics, Vanderbilt University School of Medicine, Nashville, TN*

BEVERLY A. VALERIANI • *Department of Biochemistry and Molecular Biology, University of New Hampshire, Durham, NH*

WILMA C. G. VAN STAVEREN • *Department of Psychiatry and Neuropyschology, Maastricht University, Maastricht, Netherlands*

JAMES L. WEEKS, II • *Department of Molecular Physiology and Biophysics, Vanderbilt University School of Medicine, Nashville, TN*

THEODORE G. WENSEL • *Verna and Marrs McLean Department of Biology and Molecular Biology, Baylor College of Medicine, Houston, TX*

ROBERT X. XU • *Department of Structural Science, GlaxoSmithKline, Research Triangle Park, NC*

CHEN YAN • *Center for Cardiovascular Research, University of Rochester School of Medicine, Rochester, NY*

MANUELA ZACCOLO • *Dulbecco Telethon Institute and Venetian Institute of Molecular Medicine, Padova, Italy*

ROYA ZORAGHI • *Department of Molecular Physiology and Biophysics, Vanderbilt University School of Medicine, Nashville, TN*

1

Study of Cyclic Adenosine Monophosphate Microdomains in Cells

Marco Mongillo, Anna Terrin, Sandrine Evellin, Valentina Lissandron, and Manuela Zaccolo

Summary

Cyclic adenosine monophosphate (cAMP) controls the physiological response to many diverse extracellular stimuli. To maintain signal specificity, cAMP-mediated signaling is finely tuned by means of a complex array of proteins that control the spatial and temporal dynamics of the second messenger within the cell. To unravel the way a cell encodes cAMP signals, new biosensors have recently been introduced that allow imaging of the second messenger in living cells with high spatial resolution. The more recent generation of such biosensors exploits the phenomenon of fluorescence resonance energy transfer between the green fluorescent protein-tagged subunits of a chimeric protein kinase A, as the way to visualize and measure the dynamic fluctuations of cAMP. This chapter describes the molecular basis on which such a genetically encoded cAMP sensor relies and the tools and methods required to perform cAMP measurements in living samples.

Key Words

Cyclic adenosine monophosphate (cAMP); protein kinase A; fluorescence resonance energy transfer; green fluorescent protein; imaging; biosensor; fluorescence microscopy; phosphodiesterase.

1. Introduction

1.1. General Aspects

Soon after the discovery of cyclic adenosine monophosphate (cAMP), evidence rapidly accumulated on its involvement in the regulation of a large variety of physiological and pathophysiological cellular processes, and methods to assess directly cAMP were developed. The most commonly used assay for cAMP is a radioimmunoassay. Although its introduction in the early 1970s (1) allowed to define important aspects of cAMP signaling, such a method is

From: *Methods in Molecular Biology, vol. 307: Phosphodiesterase Methods and Protocols*
Edited by: C. Lugnier © Humana Press Inc., Totowa, NJ

hampered by the necessity of fixing the sample and extracting the second messenger from the cells. As a result, cAMP studies have been restricted to assessing its concentration in steady-state conditions or concentration changes in a timescale of minutes. Moreover, lysis of the sample, which is a necessary step in such an assay, determines unavoidably the modification of the cellular environment, and, as a consequence, any information on the local modulation of cAMP is lost. This is an obvious limitation given the increasing evidence that intracellular cAMP concentration can vary rapidly and transiently on hormone stimulation, and that such changes can occur nonuniformly throughout the cell *(2)*. In addition, several experimental data indicate that the tuning of cAMP-mediated signal transduction relies on the specific localization and molecular interaction of diverse effectors and modulators within precisely structured signaling domains *(3)*.

For a detailed understanding of the way a cell computes cAMP signals, a method to assess dynamic fluctuations of cAMP in the living cell with a high spatial and temporal resolution is therefore needed. The first methodology to fulfill this requirement was based on the microinjection of fluorescein-labeled catalytic subunit and rhodamine-labeled regulatory subunit of protein kinase A (PKA), the main effector of cAMP *(4)*. Such an approach relies on the change in fluorescence resonance energy transfer (FRET) between fluorescein and rhodamine that occurs when the two labeled subunits separate on binding of cAMP to PKA. Although this methodology allowed, for the first time, imaging of cAMP fluctuations in a living sample, it was hindered by major limitations and technical difficulties (the necessity to microinject a large amount of probe, aggregation of the labeled subunits, nonspecific interactions with cellular structures) *(5)*.

More recently, such limitations have been overcome with the introduction of a new generation of PKA-based probes in which the regulatory and catalytic subunits are genetically fused to mutants of green fluorescent protein (GFP) with the appropriate spectral characteristics to generate FRET. Entirely genetically encoded, such a sensor is introduced into the cells by transfection of the cDNAs encoding for the two chimeric subunits *(6)*.

In this chapter, we describe the molecular basis on which the genetically encoded cAMP sensor relies and the tools and methods required to perform cAMP measurements in living samples.

1.2. Principles of the Method for cAMP Measurements in Living Cells

The genetically encoded cAMP sensor is a PKA whose catalytic (C)- and regulatory (R)-subunits are fused, respectively, to the yellow (YFP) and cyan (CFP) mutants of GFP. CFP and YFP act, respectively, as donor and acceptor

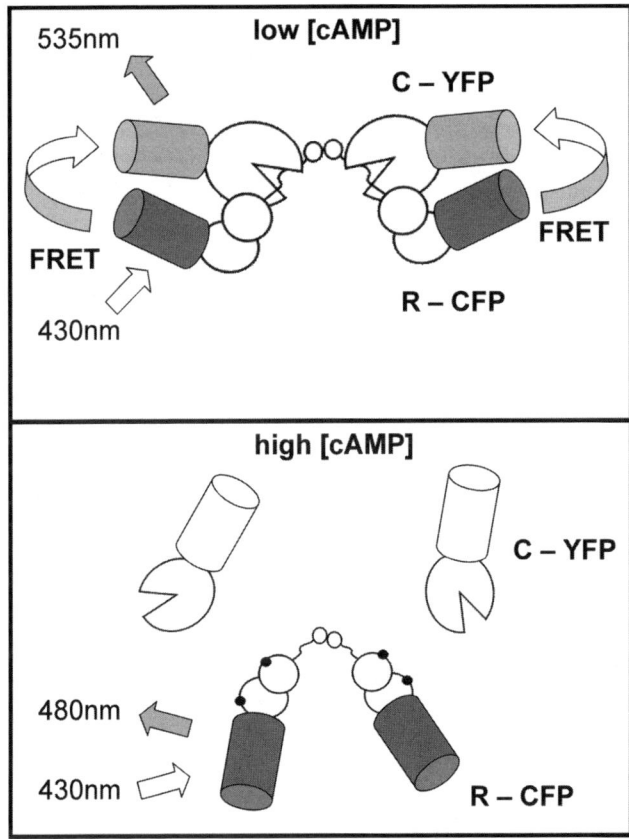

Fig. 1. Genetically encoded cAMP sensor. C and R subunits of PKA are fused, respectively, to the yellow (YFP) and cyan (CFP) mutants of GFP. CFP and YFP act, respectively, as donor and acceptor fluorophores for FRET. (**A**) At low [cAMP] most of the R and C subunits are in a complex, and the fused GFPs are close enough for FRET to occur. (**B**) When [cAMP] rises, leading to dissociation of the C and R subunits, FRET is abolished.

fluorophores for FRET (*see* **Fig. 1**). FRET is a quantum-mechanical event that occurs when two fluorophores are placed in close proximity (<100 Å) and the emission spectrum of the fluorophore that acts as the "donor" overlaps the excitation spectrum of the "acceptor" fluorophore. Under these circumstances, part of the vibrational energy of the excited state of the donor is transferred to the acceptor, which emits at its own wavelength (*7*).

In the absence of cAMP, PKA forms a holotetramer of two R-subunits and two C-subunits. The R-subunit responds to binding of cAMP with a conformational change that reduces its affinity for the C-subunit, leading to dissociation of the holotetramer. At the low concentration of cAMP of a resting cell, most of the R- and C-subunits are in a complex, and the fused GFPs are close enough for FRET to occur. When cAMP rises and the two subunits dissociate, CFP and YFP disengage and FRET is no longer possible. When FRET occurs, on excitation of CFP, part of CFP emission is taken up by YFP that emits, in turn, whereas at high cAMP, when FRET is abolished, only CFP emission can be detected. FRET can therefore be measured as a change in the emission spectrum of the probe on illumination at a wavelength that excites selectively the donor CFP. A convenient way to estimate FRET changes is to calculate the ratio between the emitted fluorescence of the probe within two spectral windows centered, respectively, on CFP and YFP emission peaks (*see* **Note 1**).

A rise in cAMP (and therefore a dissociation of the probe subunits) results in an increase in the ratio value when FRET is calculated as CFP emission/YFP emission (*see* **Notes 2** and **3**). Such a ratio can be computed in a series of images collected at different time points, and the kinetics of cAMP changes can thus be recorded.

2. Materials

2.1. Preparation of Samples

1. Borosilicate glass cover slips, 0.17 mm thick (cat. no. 406/0189/50; BDH) (*see* **Note 4**).
2. Cover slip holder (any device on which the cover slip can be mounted and sealed), i.e., gasket (*see* Molecular Probes catalog).
3. FuGene transfection reagent (Roche Molecular Biochemicals, Indianapolis, IN).
4. Serum-free culture medium.
5. Tris-EDTA: 1 M Tris-HCl, pH 8.0; 0.5 M EDTA.
6. Plasmid DNA (RII-CFP and Cat-YFP), cesium chloride grade.
7. Ringer's saline: 135 mM NaCl, 2.5 mM KCl, 1 mM CaCl$_2$, 1 mM MgCl$_2$, 1.25 mM NaH$_2$PO$_4$, 26 mM NaHCO$_3$, and 12 mM glucose and buffered to pH 7.4.

2.2. Imaging System

1. Olympus axiovert IX series microscope (ww.olympus.com).
2. Polychrome IV monochromator (Till Photonics GmbH, Martinsried, Germany, www.till-photonics.de).
3. Olympus 100X PlanApo, 1.3 numerical aperture (N.A.), oil-immersion objective.
4. 455DCLP Dichroic mirror.
5. CFP emission filter (480/30) (www.chroma.com).
6. YFP emission filter (535/40) (www.chroma.com).
7. Microimager beam-splitter device (Optical Insight, Santa Fe, NM).
8. Till Imago digital camera (Till Photonics) (*see* **Note 5**).

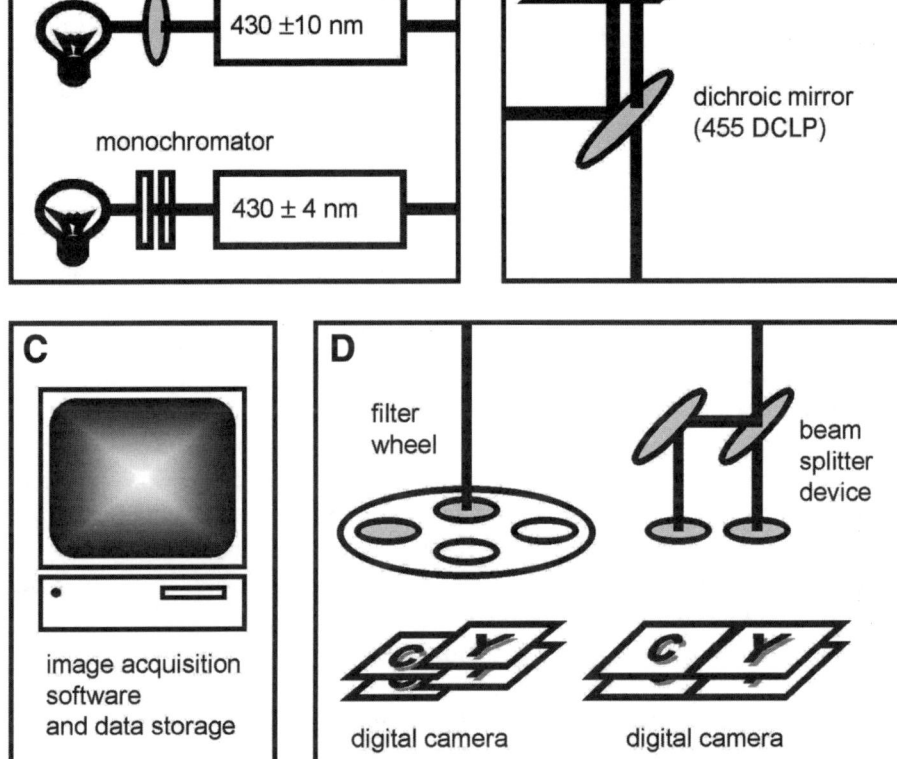

Fig. 2. Imaging system. Major components of the setup for monitoring cAMP in single live cells are shown. The system is based on a wide-field fluorescence microscope (**B**) equipped with a light source for excitation of CFP (**A**), a device for collection of CFP and YFP emission signals separately onto a digital camera (**C**), and a computer to store the acquired images (**D**).

9. Computer: PC Pentium 4 processor, 800 MHz, 512-Mb SDRAM, 40-Gb hard disk.
10. Till VisiON image acquisition and analysis software (Till Photonics, www.till-photonics.de).

3. Methods

3.1. Imaging System Setup

A system for FRET imaging of cAMP in single live cells is, in its simplest form, based on a wide-field fluorescence microscope, a light source for excitation of CFP, a device for collection of CFP and YFP emission signals separately onto a digital camera, and a computer to store the acquired images (**Fig. 2**). On this basis, a wide spectrum of choices (and prices) is possible. We therefore discuss general guidelines on what we consider are the advantages and disadvantages of the more common solution offered on the market to assemble a setup for FRET imaging.

In principle, any fluorescence microscope can be used/adapted for FRET imaging. Both inverted and upright microscopes can be used. An inverted microscope is generally more flexible, because the addition of compounds is easier as is access to the sample with manipulators or microelectrodes. Regardless of the magnification of the objective, high-quality high-N.A. optics must be used, in order to collect efficiently emission light with high resolution.

On the excitation side, the microscope must be equipped with an appropriate light source. Excitation light must fall into the spectral window where CFP is excited selectively. For the CFP/YFP couple an appropriate spectral range of excitation is between 430 and 440 nm. Such an excitation can be obtained with both a monochromatic source and a conventional mercury or xenon bulb coupled to an appropriate bandpass filter in excitation (**Fig. 2A**). Filter manufacturers (Chroma, Omega optical) provide filter sets with exciter, dichroic, and emitter that are optimized for FRET measures with CFP and YFP (*see* **Fig. 3**). Monochromators are normally tunable to the appropriate excitation wavelength. To avoid continuous illumination of the sample during idle periods, and the consequent unnecessary photodestruction of fluorescence and damage to the cells, a device to shutter excitation light is needed. In case a preexisting microscope is adapted for imaging, it must be equipped with a software-controlled shutter.

On the emission side, different solutions can be adopted to collect CFP and YFP emission separately (**Fig. 2D**). One possibility is to use a motorized, software-controlled filter wheel that mounts emission filters for CFP and YFP (**Fig. 2D**). In this case, each time point of the FRET measure is composed of two sequential pulses of excitation and acquisition of two sequential images, each with a different emission filter. Such a solution is flexible, because the same setup can mount several filters and can be used to image other fluorescent indicators and probes. On the other hand, CFP and YFP emission images will necessarily be collected with a delay, the duration of which depends on the time required for the wheel to switch between the two filters (*see* **Note 6**). This problem is often negligible, but it might introduce artifacts in case the sample or the focal plane moves during the lag time between the acquisition of the CFP and YFP images. In addition, the total time required for the acquisition of each time point is doubled, leading to increased photodamage and hindering imaging at fast acquisition rates (*see* **Note 7**).

	CFP imaging	YFP imaging	FRET imaging
Excitation (monochromator)	D436/20x (430nm)	HQ500/20x (514nm)	D436/20x (430nm)
Dichroic	455DCLP	Q515LP	DCXR500
Emission	D480/30m	HQ535/30m	HQ535/30m

Fig. 3. Filter set for CFP, YFP, and FRET imaging. Filter specifications for CFP, YFP, and FRET. Product numbers from Chroma catalog are indicated.

To overcome such constraints, two strategies can be implemented. The first (very expensive) is to equip the microscope with a dichroic mirror to separate CFP from YFP emission channels that are then collected with two separate digital cameras. Alternatively, an optical device can be used that allows simultaneous collection of CFP and YFP emission images on the two halves of the chip of the digital camera. Such a device mounts on a dichroic mirror and two bandpass filters within a single apparatus and is commercially available (Microimager; Optical Insight).

For image acquisition, a high-sensitivity cooled charge-coupled device camera is usually the preferred option. The market offers a various range of price/performance cameras; some of the newer models have a dual chip that is fitted to couple with a beam-splitter device.

All the devices of the imaging system—the shutter on the excitation beam path, the motorized filter wheel (if present), and the digital camera—must be software controlled. Several commercial softwares exist that generally allow image acquisition and offline image analysis and processing. Image analysis can also be optimally performed with freeware softwares that are downloadable from the Internet (i.e., ImageJ; Wayne Rasband, Bethesda, MD).

3.2. cAMP Imaging

For clarity we divide a cAMP-imaging experiment in three steps. The first step consists of preparation of the sample and is normally carried out in the cell culture laboratory. The second step is performed at the microscope, where a series of images are acquired that correspond to CFP and YFP emission intensities on illumination of the sample at 430 nm (CFP excitation wavelength). The third step consists of processing and analysis of the acquired images. In the following sections we provide a more detailed description of the three stages of a cAMP-imaging experiment.

3.2.1. Phase 1: Culture and Transfection of Cells

Expression of the probe in the specimen is achieved with transfection of the cDNA encoding for the two GFP-PKA chimeric subunits. Transcription level is under control of a cytomegalovirus promoter, which ensures a sufficiently high expression level to perform routine experiments in most of the cell types that we have tested, both primary cultured cells and cell lines.

The chosen transfection technique depends on the cell type. We routinely use Fugene transfection reagent (Roche molecular biochemicals), and transfection efficiency ranges from 10 to 60%, depending on the cell type (*see* **Note 8**).

1. Seed cells onto 24-mm glass cover slips.
2. When the cells are 50–60% confluent, transfect them by mixing 1.5 µg of RII-CFP with 1.5 µg of C-YFP, 6 µL of FuGene, and 100 µL of serum-free medium. Incubate the mixture for 15 min and then add to the cells.
3. After 24–36 h inspect the cells at the microscope (*see* **Note 9**).

3.2.2. Phase 2: Image Acquisition

Three parameters have to be set before starting the acquisition protocol: exposure time, frequency of acquisition, and number of acquisitions.

1. Before starting a FRET-imaging experiment, mount a cover slip on a sealed cover slip holder containing experimental medium. Both saline solutions and culture media can be used provided that they are HEPES buffered for pH control (*see* **Note 10**).
2. Set the exposure time: This determines how long the cell is illuminated and how long the camera will integrate emitted photons at each frame. Exposure time depends on the fluorophore content and on the characteristics of the system (fluorescence lamp, optics, and camera). As a rule of thumb, average signal on the cell should be at least threefold higher than background signal. To achieve an adequate signal-to-noise ratio, it is possible to increase either the exposure time or the pixel binning on the camera. This second operation increases sensitivity of the camera at the expense of image resolution (*see* **Note 11**).

   One should consider that increasing exposure time leads to increased photodestruction of CFP and YFP, resulting in bleaching of the sample fluorescence (*see* **Note 7**). In our experience, illumination times range from 50 to 400 ms, depending on the brightness of the cell and on the setup.
3. Set the frequency of image acquisition: This parameter must be determined empirically based on the kinetics of the cAMP changes expected in the specific cell model studied. Although, in principle, the probe allows real-time imaging, high rates of illumination cycles can result in photoisomerization of CFP and YFP and

consequent artifactual changes in the CFP-to-YFP ratio (*see* **Note 12**). In our experience, sampling at more than 4 Hz is difficult to obtain.

3.2.3. Phase 3: Image Processing and Analysis

Once the experimental protocol is completed, the raw data stored in the computer memory consist of two series, or stacks, of images of the analyzed cell along time. The two stacks represent the fluorescence intensity of CFP and YFP emission at each time point. In case the system is equipped with a beam splitter device only one stack will be collected, with each frame split into two halves corresponding to CFP and YFP intensity emissions (**Fig. 4A**). The image processing is a multistep procedure, the aim of which is to obtain a new stack of images in which the value of each pixel is equivalent to the ratio between CFP and YFP values in the corresponding pixel of the raw CFP and YFP images. The image-processing software of choice must allow the following operation:

1. If using a beam-splitter device, split, superimpose, and align the two halves of each image; a perfect alignment is very important in order to avoid artifacts in calculation of the ratio (**Fig. 4B**).
2. Subtract background noise from each image (**Fig. 4D**). Background noise results from the intrinsic noise of the camera, from autofluorescence of the cells and of the media in which the cells are immersed. Since background can vary along time with illumination of the sample or movement of the focal plane during the experiment, background should be subtracted from each frame by calculating the average intensity in an area outside the analyzed cell.
3. Calculate the pixel-by-pixel ratio image for each time point (**Fig. 4F**). Image analysis software normally has an in-built function to perform such an operation. In the "ratio" stack obtained, the numerical values of each pixel result from CFP/YFP intensity emission and correlate with the dissociation state of the probe and, therefore, with the amount of cAMP in the area of the cell that corresponds to that pixel.
4. Plot the CFP-to-YFP ratio value along time by averaging pixel intensities within a defined area of the cell (**Fig. 4E**). By drawing a region of interest in correspondence to specific areas, the kinetics of changes in cAMP that take place in spatially resolved compartments of the cell can be assessed *(8)* (**Fig. 5**).

4. Notes

1. As a result of spectral overlap, the FRET signal is always contaminated by donor emission into the acceptor channel and by the excitation of acceptor molecules by the donor excitation wavelength. The component of the FRET signal owing to such bleedthrough can be estimated and subtracted by image processing, in order

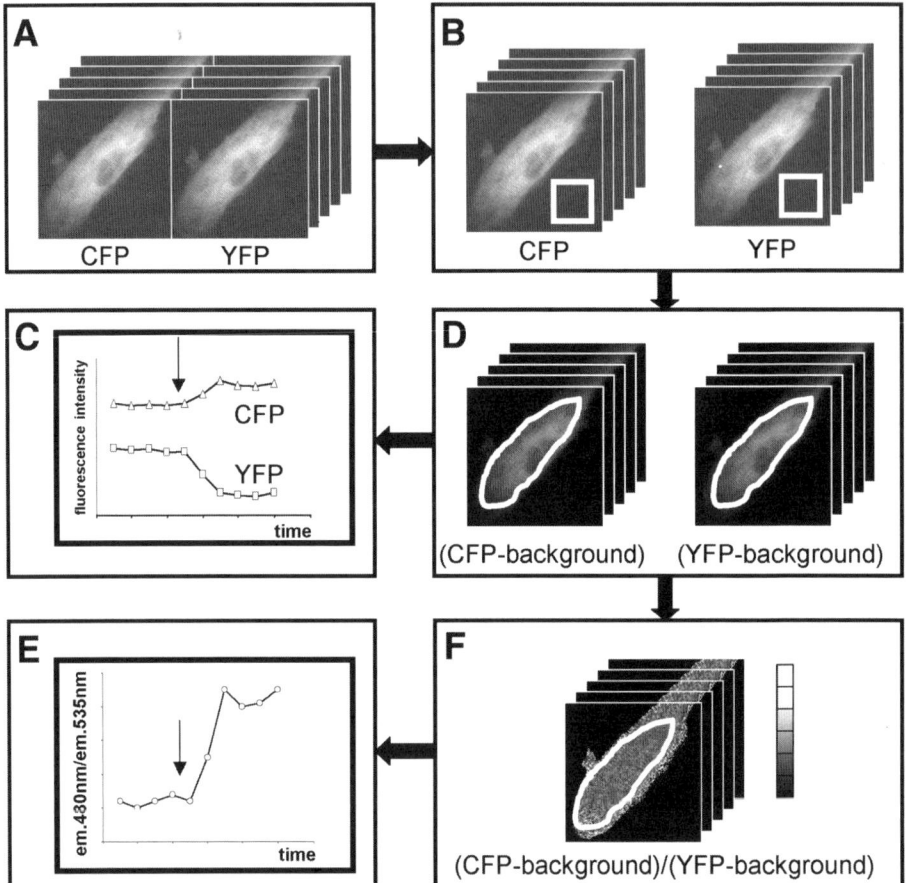

Fig. 4. Image analysis. (**A**) The raw images of a transfected cell as acquired by a digital camera are shown. CFP and YFP emission images are recorded in the two halves of each frame. (**B**) CFP and YFP images are separated into two different stacks. Background is estimated by averaging intensity within an area out of the cell (white frame) and is then subtracted from every frame, resulting in a new set of images. A CFP-to-YFP ratio image is computed (**D**), and the average pixel value within an area of interest is calculated. (**E**) Changes in such value along the stack correspond to changes in [cAMP] along time, as shown in the plot. (**F**) A plot of the corresponding changes (**C**) in emission intensity of CFP and YFP within the same area of interest is shown.

to increase the dynamic change in the measure. CFP bleedthrough in the YFP channel can be measured as the ratio between YFP and CFP channel intensity in cells expressing only CFP on excitation of CFP. In our system, we extimated that about 50% of the CFP signal bleeds into the YFP channel.

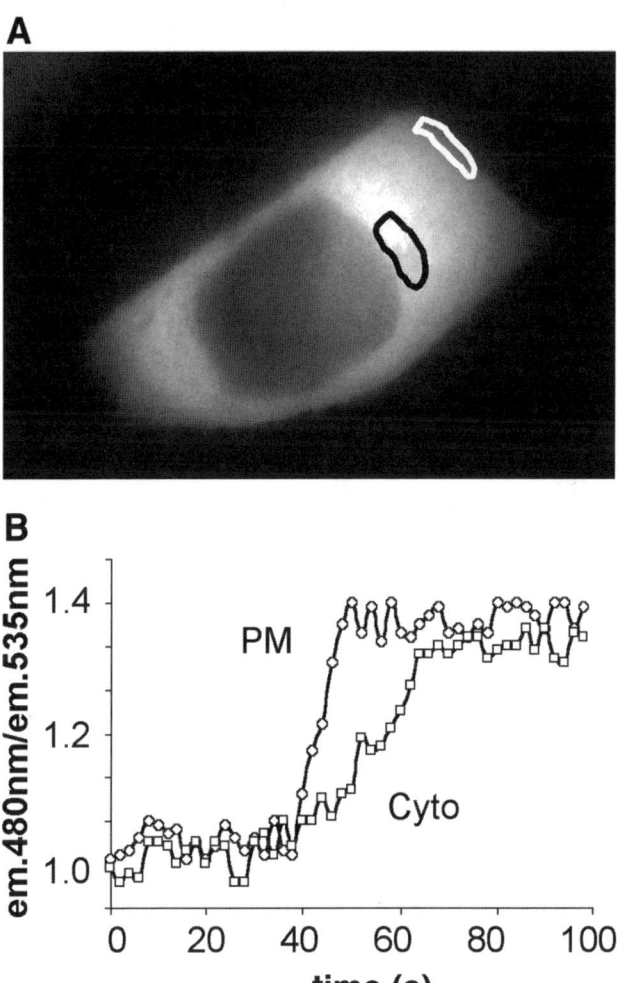

Fig. 5. Analysis of cAMP fluctuations in subcellular compartments. Changes in cAMP assessed in Chinese hamster ovary cells expressing the cAMP sensor and challenged with the adenylate cyclase activator forskolin (50 m*M*) are shown. (**A**) Average 480 nm/535 nm emission ratio was calculated within two spatially resolved compartments, in areas corresponding to the subplasma membrane (PM) region and to the cytoplasm. (**B**) The kinetics of the rise in cAMP is faster under the PM and slower in the deep cytosol (Cyto).

2. Because the output of such measurement expressed as the ratio of two intensities is a pure number, the absolute value depends strictly on the experimental conditions and setup.
3. Even if a "ratiometric" measurement corrects intrinsically for the unequal distribution of the probe and, within a certain limit, also for bleaching occurring

during the experiments and movement of the probe, care must be taken when interpreting changes in ratio value. A rule of thumb is to score as a change in cAMP concentration only changes in the CFP emission/YFP emission ratio value that result from a drop in YFP emission and a concomitant rise in CFP emission (**Fig. 3C**).

4. Good-quality glass cover slips should be used.
5. Chip size is 640×480. The fastest acquisition rate at full frame is up to 30 frames/s. This camera is sold with a frame grabber that fits the PCI bus on the motherboard of the computer.
6. The filter wheel must be set to change regularly emission filters after each frame. To reduce delay between the acquisition of CFP and YFP emission images at each time point, mount the emission filters in two neighboring slots in the filter wheel.
7. Photobleaching is the irreversible destruction of fluorescence that occurs on illumination of a fluorophore. Because photodestruction affects CFP and YFP differently, an artifactual change in CFP-to-YFP emission ratio can result. To reduce photobleaching, try to decrease exposure times.
8. Although having a higher number of transfected cells is helpful, transfection efficiency is not a crucial parameter because the experiment is made on single cells.
9. The time required for the synthesis of a sufficiently high amount of the probe for imaging depends on the cell type. More sensitive systems allow imaging of dimmer cells.
10. If culture media have to be used, it is preferable to avoid phenol red, which is slightly autofluorescent.
11. A pixel of the digital image corresponds to the smallest photosensitive unit on the chip of the digital camera, and its size determines the resolution of the image. In some digital cameras, groups of photosensitive units of the chip can be "binned" in one pixel, in order to increase intensity of the image. The camera's binning function can be used to increase brightness, but this results in reduced resolution.
12. Photoisomerization is the reversible conformational change that occurs on illumination of a fluorophore. When photoisomerization occurs, CFP and YFP do not fluoresce on further illumination. Because time to recovery from photoisomerization differs for CFP and YFP, such a phenomenon can introduce artifacts in CFP-to-YFP emission ratio. To avoid photoisomerization, the minimum delay between the time points of the measure should be about 500 ms.

Acknowledgments

Data included in this chapter are drawn from research funded by Telethon Italy and the European Commission (project QLK3-CT-2002-02149).

References

1. Steiner, A. L., Parker, C. W., and Kipnis, D. M. (1970) The measurement of cyclic nucleotides by radioimmunoassay. *Adv. Biochem. Psychopharmacol.* **3,** 89–111.
2. Beavo, J. A. and Brunton, L. L. (2002) Cyclic nucleotide research—still expanding after half a century. *Nat. Rev. Mol. Cell. Biol.* **3,** 710–718.

3. Zaccolo, M., Magalhaes, P., and Pozzan, T. (2002) Compartmentalisation of cAMP and Ca(2+) signals. *Curr. Opin. Cell. Biol.* **14,** 160–166.
4. Adams, S. R., Harootunian, A. T., Buechler, Y. J., Taylor, S. S., and Tsien, R. Y. (1991) Fluorescence ratio imaging of cyclic AMP in single cells. *Nature* **349,** 694–697.
5. Goaillard, J. M., Vincent, P. V., and Fischmeister, R. (2001) Simultaneous measurements of intracellular cAMP and L-type Ca2+ current in single frog ventricular myocytes. *J. Physiol.* **530,** 79–91.
6. Zaccolo, M., De Giorgi, F., Cho, C. Y., Feng, L., Knapp, T., Negulescu, P. A., Taylor, S. S., Tsien, R. Y., and Pozzan, T. (2000) A genetically encoded, fluorescent indicator for cyclic AMP in living cells. *Nat. Cell. Biol.* **1,** 25–29.
7. Miyawaky, A. and Tsien, R. Y. (2000) Monitoring protein conformations and interactions by fluorescence resonance energy transfer between mutants of green fluorescent protein. *Methods Enzymol.* **327,** 472–500.
8. Zaccolo, M. and Pozzan, T. (2002) Discrete microdomains with high concentration of cAMP in stimulated rat neonatal cardiac myocytes. *Science* **295,** 1711–1715.

2

High-Resolution Measurements of Cyclic Adenosine Monophosphate Signals in 3D Microdomains

Jeffrey W. Karpen and Thomas C. Rich

Summary

A large number of hormones, neurotransmitters, and odorants exert their effects on cells by triggering changes in intracellular levels of cyclic adenosine monophosphate (cAMP). Although the effector proteins that bind cAMP have been identified, it is not known how this single messenger can differentially regulate the activities of hundreds of cellular proteins. It has been clear, for some time, that compartmentation of cAMP signals must be taking place, but the physical basis for compartmentation and the nature of local cAMP signals are mostly unknown. We present here a high-resolution method for measuring cAMP signals near the membrane in single cells. Cyclic nucleotide-gated (CNG) ion channels from olfactory receptor neurons have been genetically modified to improve their cAMP-sensing properties. We outline how these channels can be used in electrophysiological experiments to measure accurately changes in cAMP concentration near the membrane, where most adenylyl cyclases reside. We also describe how the method has been employed to dissect the roles of diffusion barriers and differential phosphodiesterase activity in creating distinct cAMP signals. This approach has much greater spatial and temporal resolution than other methods for measuring cAMP and should help to unravel the complexities of signaling by this ubiquitous messenger.

Key Words

Phosphodiesterase; adenylyl cyclase; cyclic nucleotide-gated ion channels; G protein-coupled receptors; second messengers; cyclic adenosine monophosphate (cAMP); subcellular compartmentation; diffusion; permeability barrier; biosensors.

1. Introduction

Cyclic adenosine monophosphate (cAMP), the prototypical second messenger, regulates a wide variety of cellular processes. Changes in cAMP concentration transmit information to downstream effectors including protein kinase A (PKA), cyclic nucleotide-gated (CNG) channels, hyperpolarization-activated

From: *Methods in Molecular Biology, vol. 307: Phosphodiesterase Methods and Protocols*
Edited by: C. Lugnier © Humana Press Inc., Totowa, NJ

(I_hHCN) channels, and Epac *(1–4)*. However, it is largely unclear how differential regulation of cellular targets occurs. The concept of compartmentation emerged more than 20 yr ago in studies of cardiac myocytes, to help explain how a variety of extracellular stimuli that primarily act through cAMP can have very different downstream effects on the cell *(5,6)*. The basis for compartmentation and, indeed, the nature of cAMP signals themselves, have remained mysteries *(7,8)*. To understand how these signals function within the cell, it is important to answer the following questions: (1) How are cAMP signals localized? (2) What are the kinetics of cAMP signals in localized domains? and (3) What information is contained in the amplitude and frequency of cAMP signals? To address these questions, olfactory CNG channels (CNGA2) have been adapted to measure cAMP in single cells *(9–11)* (*see* **Note 1**). There are several advantages to using genetically engineered CNG channels as high-resolution cAMP sensors: First, CNG channels are expressed in the plasma membrane, where most types of adenylyl cyclase (AC) are localized. Second, several hundred channels provide a readily detectable readout of cAMP concentration without significantly buffering the cAMP signal being measured *(9,11,12)*. Third, CNG channels respond rapidly to changes in cyclic nucleotide concentration *(9,12–14)*. Finally, the sensor can be calibrated in different cell types *(9–11)*.

2. Materials

1. Adenovirus constructs for expression of CNG channels (pertinent characteristics of four CNG channel constructs used for monitoring cAMP levels are described in Chapter 4).
2. Primary or cultured cell lines in which measurements are to be conducted: The examples given here are in HEK-293 and C6-2B cells.
3. Minimal essential medium (MEM) supplemented with 26.2 m*M* NaHCO$_3$, 10% (v/v) fetal bovine serum (Gemini), penicillin (50 µg/mL), and streptomycin (50 µg/mL), pH 7.0, for HEK-293 cells, and F-10 medium supplemented with 26.2 m*M* NaHCO$_3$ and 10% bovine calf serum (Gemini), pH 7.0, for C6-2B cells.
4. 100 m*M* hydroxyurea stock solution: This is required only for cell lines in which adenovirus can readily replicate, including HEK-293 cells (*see* **Note 2**).
5. Electrophysiology setup for whole-cell and perforated patch-clamp experiments, and rapid perfusion system.
6. Extracellular buffer 1: 145 m*M* NaCl, 11 m*M* D-glucose, 10 m*M* HEPES, 4 m*M* KCl, and 0.1 m*M* MgCl$_2$, pH 7.4.
7. Extracellular buffer 2: 145 m*M* NaCl, 11 m*M* D-glucose, 10 m*M* HEPES, 4 m*M* KCl, and 10 m*M* MgCl$_2$, pH 7.4.
8. Perforated patch pipet buffer: 70 m*M* KCl, 70 m*M* potassium gluconate, 11 m*M* D-glucose, 10 m*M* HEPES, 4 m*M* NaCl, 0.5 m*M* MgCl$_2$, 1 m*M* cAMP, and 50–200 µg/mL of nystatin, pH 7.4 (*see* **Note 3**).

9. Whole-cell pipet buffer: 70 mM KCl, 70 mM potassium gluconate, 11 mM D-glucose, 10 mM HEPES, 4 mM NaCl, 0.5 mM MgCl$_2$, 5 mM K$_2$ATP, and 0.1 mM Na$_2$GTP, pH 7.4 (*see* **Note 4**).

10. Activators of adenylyl cyclase and inhibitors of phosphodiesterase (PDE). Add forskolin, prostaglandin E$_1$ (PGE$_1$), 3-isobutyl-1-methylxanthine (IBMX), and 4-(3-butoxy-4-methoxybenzyl)-2-imidazolidinone (RO-20-1724) to the control solution from concentrated dimethyl sulfoxide (DMSO) stocks, with the final concentrations as indicated (final DMSO concentrations were <0.5%).

11. Data analysis software: Appropriate software should be available from the manufacturer of the patch-clamp amplifier (e.g., Axon, Foster City, CA). We typically use the MATLAB software package (Mathworks, Natick, MA) to design custom analysis software in-house.

3. Methods

The following methods outline an approach to accurately measure cAMP concentrations near the surface membrane of living cells using modified olfactory CNG channels (*see* **Note 1**). This method offers unprecedented spatial and temporal resolution of cAMP signals. Among the findings with this approach are that cAMP in several cell types is produced in subcellular compartments under the plasma membrane with restricted diffusional access to the bulk cytosol, and that the amplitude and kinetics of cAMP signals within these compartments are distinct from those in the remainder of the cell (*see* **Note 5**).

3.1. Cell Culture and CNG Channel Expression

It is important to note that CNG channel expression in different cell types requires different multiplicity of infection (MOI) and incubation time. **Table 1** of Chapter 4 summarizes the conditions for optimal expression of CNG channels in several cell types.

1. Maintain HEK-293 and C6-2B cells in supplemented MEM and F-10 medium, respectively (*see* **Subheading 2.**, **item 3**), in a humidified atmosphere of 95% air and 5% CO$_2$.

2. Plate the cells at approx 60% confluence in 100-mm culture dishes 24 h prior to infection with the CNG channel-encoding adenovirus constructs (MOI = 10 plaque-forming units/cell for HEK-293 cells and 100 for C6-2B cells).

3. When using adenovirus to transfect HEK-293 cells, add 2 mM hydroxyurea (final concentration) to the cell media 2 h postinfection to partially inhibit viral replication (*see* **Note 2**).

4. Twenty-four (HEK-293) or 48 h (C6-2B) postinfection, detach the cells with phosphate-buffered saline containing 0.03% EDTA (PBS-EDTA), resuspend in serum-containing medium, allow to recover for 1 h, and assay within 12 h. We use PBS-EDTA rather than trypsin-containing solutions so as not to modify extracellular regions of the CNG channels.

3.2. Measurement of Local cAMP in Single Cells

Single-cell cAMP measurements are made using either the perforated patch or whole-cell patch-clamp technique. A detailed consideration of patch-clamp methods is beyond the scope of this chapter, but there are excellent monographs and articles that provide a clear description of these electrophysiological approaches *(15–17)*. Here, we focus on how patch-clamp techniques are used to measure the activity of CNG channel-based cAMP sensors, in order to measure cAMP concentrations near the surface membrane.

1. Pull patch pipets (electrodes) from borosilicate glass and heat polish. In our experiments, pipet resistance was limited to 5 MΩ and averaged 3.4 ± 0.5 MΩ (mean ± SD).
2. Lower the pipets onto the cells and form gigaohm seals (8.3 ± 3.3 GΩ).
3. Make recordings using a patch-clamp amplifier (e.g., Axopatch-200A from Axon).
4. Digitize the signals corresponding to ionic currents and sample at five times the low-pass filter setting. For example, in some of our experiments, records were low-pass filtered with a 12-Hz bandwidth and sampled at 60 Hz.
5. Store and analyze the digitized records on a computer. In most cases records are later corrected for errors owing to series resistance (a combination of pipet resistance and access resistance to the cell interior).

3.2.1. Whole-Cell Experiments

In the whole-cell configuration, the piece of membrane underneath the patch pipet is ruptured by applying light suction or a brief electrical pulse (the "zap" feature available on most patch-clamp amplifiers). This allows dialysis of solutions from the patch pipet into the bulk cytosol. After achieving the whole-cell configuration, capacitive transients are elicited by applying 20-mV steps from the holding potential and recorded at 40 kHz (filtered at 10 kHz) for calculation of access resistance, which is typically <8 MΩ.

The whole-cell configuration is particularly useful for (1) washing in known concentrations of compounds that affect signal transduction (e.g., PDE inhibitors) into the cell, and (2) measuring the rate at which small molecules such as Na^+ or cAMP wash into the cell and diffuse to their targets. The latter experiments can be used to estimate flux coefficients for the compartmental models described in **Subheading 3.4.**, or the effective diffusion coefficients of cAMP within the cell *(9)*. For example, in HEK-293 cells, the wash in of Na^+ from the patch pipet into the bulk cytosol is 90% complete in approx 22 s *(9,18)*.

3.2.2. Perforated Patch Experiments

In the perforated patch configuration, the pore-forming antibiotic nystatin is added to the pipet solution to gain electrical access to the cell's interior while retaining divalent cations and larger molecules such as cAMP in the cell (*see*

Note 3). A steady access resistance is obtained 5–15 min following seal formation. Capacitive transients are elicited by applying –30-mV steps from the holding potential of –20 mV for calculation of access resistance, typically less than 100 MΩ. These quantities are monitored throughout the experiments to ensure that stable electrical access is maintained. Solutions are applied using the SF-77B fast-step solution switcher (Warner, Hamden, CT) (*see* **Note 6**). In most of these experiments, 1 m*M* cAMP is included in the pipet solution; at the end of the experiment, the maximal cAMP-induced current can be measured by rupturing the cell membrane at the tip of the pipet with suction, allowing saturating cAMP to diffuse to the channels *(9)*.

Examples of this technique are given in **Figs. 1** and **2**. In **Fig. 1** a single C6-2B cell expressing wild-type (WT) CNG channels (CNGA2; *see* **Note 1**) was stimulated with 50 μ*M* forskolin. This triggered a steady increase in inward current (measured at –50 mV). Currents were measured in 0.1 m*M* external MgCl$_2$ (extracellular buffer 1; *see* **Subheading 2.**, **item 6**) and subsequently blocked by 10 m*M* external MgCl$_2$ (extracellular buffer 2; *see* **Subheading 2.**, **item 7**) in a voltage-dependent fashion, a signature of currents through CNG channels. The ratio of the maximal forskolin-induced current to the maximal CNG channel current indicates that 50 μ*M* cAMP accumulated near CNG channels (*see* **Note 5**). In **Fig. 2A,B,** the responses of two different HEK-293 cells expressing C460W/E583M channels to rapid application of 1 μM PGE$_1$ are shown. C460W/E583M channels can be used to measure cAMP signals at physiological concentrations (0.1–4 μM) without significantly buffering the signal being measured; this is discussed in Chapter 4 and **refs.** *11* and *12*. The inward currents through these channels were measured at a holding potential of –20 mV. Currents were converted to cAMP concentration based on the channel's dose-response relation as described in **Subheading 3.3.** Both the raw currents and the calibrated responses indicate that in response to PGE1, cAMP levels increased transiently. We demonstrated that this transient response was owing to an initial increase in adenylyl cyclase activity followed by a more profound increase in PDE type IV activity *(11)*. Interestingly, the total cellular cAMP levels rose to a steady plateau over the same time frame. To describe these results quantitatively, we used the framework of compartmental models discussed in **Subheading 3.4.**

3.3. Calibration of Single-Cell Measurements

The cyclic nucleotide sensitivity of WT and C460W/E583M CNG channels was assessed in excised, inside-out patches *(9,10)*. The cAMP dose responses of these channels are fit with the Hill equation, $I/I_{max} = [cAMP]^N/([cAMP]^N + K_{1/2}^N)$, in which I/I_{max} is the fraction of maximal current, $K_{1/2}$ is the cAMP concentration that gives a half-maximal current, and N is the Hill coefficient. We found

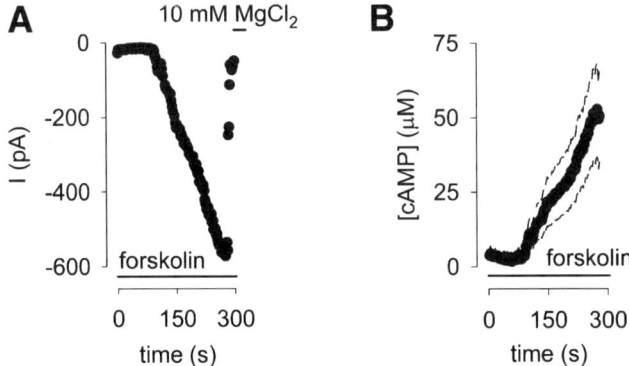

Fig. 1. cAMP measurements in a single C6-2B cell expressing WT CNG channels. (A) Response of a C6-2B cell to 50 μ*M* forskolin (applied at time = 0), V_m = –60 mV, I_{max} = 910 pA. The response reached a plateau in approx 270 s. No forskolin-induced currents were observed in uninfected cells (cells not expressing CNG channels). (B) cAMP concentrations calculated using Hill equation (•) ± SD, based on uncertainty in calibration. (Reproduced from **ref. 9** by copyright permission of Rockefeller University Press.)

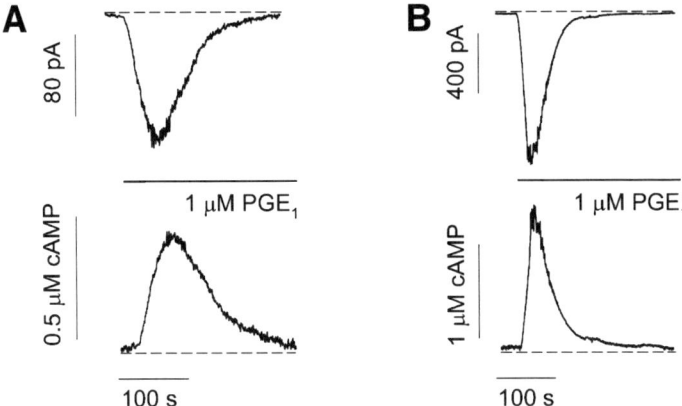

Fig. 2. Measurements of local cAMP signals in single HEK-293 cells. (A,B) At top, rapid application of PGE$_1$ triggered transient inward currents through C460/E583M CNG channels (–20 mV). (A) and (B) are the responses of two different cells. No PGE$_1$-induced currents were observed in cells not expressing CNG channels. At bottom, the corresponding cAMP signals were calibrated as described in the text. Dashed lines indicate either zero cyclic nucleotide–induced current (the current in 10 m*M* MgCl$_2$) or zero cAMP. To maximize the response and remove the possibility of Ca^{2+} feedback, these experiments were conducted in nominally Ca^{2+}-free solutions (extracellular buffer 1; *see* **Subheading 2., item 6 *[11]***).

the $K_{1/2}$ and N to be about 40 μM and 2.2 for the WT channels and 1.0 μM and 2.0 for the C460W/E583M channel *(10,11)*. With these values, the cAMP concentration can be calculated from currents measured in perforated patch experiments *(9,11)*. For example, if I/I_{max} were found to be 0.6 in an experiment using C460W/E583M channels, the estimated cAMP concentration would be 1.2 μM. Note that the low concentration of channels expressed in these cells (~1 nM) does not significantly buffer the measured cAMP signals.

We confirmed this calibration technique in the whole-cell setting in two ways, measuring the responses induced by photolysis of caged cAMP and washing in different cAMP concentrations from the patch pipet into the cell *(9)*.

3.4. Compartmental Models of Localized cAMP Signals

Several lines of evidence suggest that cAMP is produced in subcellular compartments near the surface membrane, and that diffusion between these domains and the bulk cytosol is significantly impeded (*see* **Note 5**) *(9,11)*.

To describe these data quantitatively, we have used the framework of compartmental models. The major assumptions in these models are that each compartment is well mixed (the concentration within the compartment is uniform), and that there is a permeability barrier between compartments. This assumption is justified based on how rapidly cAMP diffuses without restriction across the entire cell (<0.2 s in a 20-μm-diameter cell).

We have used a simple compartmental model (**Fig. 3A**, inset) to describe several independent experimental observations. This model contains two compartments: a small subcellular compartment beneath the plasma membrane (compartment 1) whose volume is 2% of the bulk cytosol (compartment 2). In this model, PGE$_1$ triggers an increase in AC activity in compartment 1. PGE1 also triggers an increase in PDE activity within compartment 1. This is consistent with our finding that the decline in the transient response (**Fig. 2**) is owing to a time-dependent upregulation of PDE type IV activity *(11)*. The flux of cAMP from the microdomain to the bulk cytosol (from compartment 1 to compartment 2) is hindered by a permeability barrier. In addition, there is a constitutively active PDE in compartment 2.

The system is described by the following equations:

$$\frac{dC_1}{dt} = E_{AC} + \frac{J_{12}}{V_1}(C_2 - C_1) - \frac{A \bullet E_1 \bullet C_1}{K_{M1} + C_1} \tag{1}$$

$$\frac{dC_2}{dt} = \frac{J_{12}}{V_2}(C_1 - C_2) - \frac{E_2 \bullet C_2}{K_{M2} + C_2} \tag{2}$$

$$\frac{dA}{dt} = k_A I - k_I A \tag{3}$$

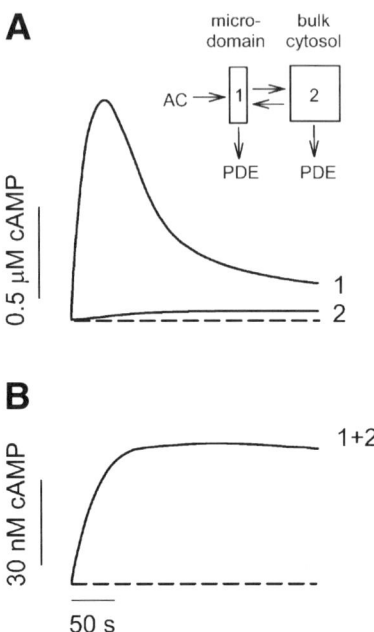

Fig. 3. Quantitative description of localized transient cAMP response and total cellular cAMP accumulation. The inset shows a two-compartment model of the cell with a diffusional restriction between the membrane-localized microdomain (compartment 1) and the bulk cytosol (compartment 2). See the text for details. **(A)** Rapid activation of AC and slower activation of PDE shape the transient signal in the microdomain. The slow flux of cAMP from the microdomain allows low levels to accumulate in the cytosol. Even in the small volume of the microdomain, the concentration of CNG channels is low (~40 n*M*) and will not significantly buffer the cAMP signal. **(B)** Total cAMP levels (microdomain and cytosol) reach a plateau. Dashed lines indicate zero cAMP *(11)*.

in which V_1 and V_2 are the volumes of compartments 1 and 2, respectively; C_1 and C_2 are the cAMP concentrations; J_{12} is the flux coefficient between compartments; E_{AC} is the synthesis rate of cAMP; E_1 and E_2 are the maximal cAMP hydrolysis rates; K_{M1} and K_{M2} are the Michaelis constants for PDE activity; A and I are the fraction of active and inactive PDE in compartment 1, respectively ($A + I = 1$); and k_A and k_I are the rate constants of PDE activation and inactivation, respectively. $J_{12} = 8.0 \times 10^{-16}$ L/s, $V_1 = 0.040$ pL, and $V_2 = 2.0$ pL. AC activity is considered constant, with $E_{AC} = 0.13$ μ*M*/s. K_{M1}, E_1, K_{M2}, and E_2 are 0.30 μ*M*, 0.83 μ*M*/s, 1.0 μ*M*, and 0.0020 μ*M*/s, respectively. The rate constants k_A and k_I are 0.0015 s^{-1} and 0.0010 s^{-1}, respectively. The initial and final (300 s) values of A are 0.10 and 0.36, respectively. The parameters used in this simulation reflect similar total PDE activities near the plasma membrane and throughout the cytosol. This is in broad agreement with the findings of experiments on PDE type IV in several cell types *(19)*.

Simulations of the model successfully reproduce the local transient change in cAMP, as well as the rise in total cAMP to a plateau (**Fig. 3**). A PGE_1-induced increase in PDE activity within compartment 1 is required to explain the data. Slow efflux of cAMP from the microdomain is ultimately balanced by low rates of hydrolysis within the bulk cytosol. Thus, different relative PDE activities within more than one diffusionally restricted compartment can explain the generation of distinct cAMP signals (*see* **Note 7**).

4. Notes

1. This approach was inspired by the field of retinal phototransduction, the best-studied second-messenger signaling system, in which elegant biochemical studies have been complemented by real-time measurements of cyclic guanosine 5′-monophosphate (cGMP) signals using endogenous CNG channels *(20–23)*. CNG channels are directly opened by the binding of cyclic nucleotides. They were discovered in retinal photoreceptor cells and olfactory receptor neurons, where they generate the electrical response to light and odorants. The native retinal channel is cGMP specific, whereas the native olfactory channel is equally sensitive to cAMP and cGMP. Native CNG channels consist of A- and B-subunits, both of which bind cyclic nucleotides, although most A-subunits form functional channels on their own. We have modified an olfactory channel A-subunit (CNGA2) to improve its sensitivity and selectivity for cAMP.

2. Higher hydroxyurea concentrations ($\geq$10 m*M*) are required to completely inhibit viral replication. However, such high concentrations affect cellular adenosine triphosphate (ATP) pools and can significantly alter second-messenger signaling. We have determined that 1 to 2 m*M* hydroxyurea sufficiently inhibits viral replication without significantly altering ATP pools in HEK-293 cells. As a further control for potential effects of hydroxyurea, the experiments described here can be repeated with CNG channels expressed using other transfection techniques such as $Ca_3(PO_4)_2$ precipitation *(24)*. We have done these controls and found that the lower hydroxyurea concentrations have little or no effect on cAMP signals *(9–11)*. It is important to note that we use adenovirus to transfect cells because it allows relatively uniform CNG channel expression in a majority ($\geq$70%) of cells.

3. In the perforated patch configuration, pipet solutions contained nystatin (diluted from 50 mg/mL stock in DMSO) to gain electrical access to the cell. These solutions were kept on ice and shielded from light until use. Using this approach, the solutions can be used for 4–8 h. We fill the entire pipet with nystatin-containing solution (*see* **Subheading 2., item 8**). In our hands this does not impede seal formation. In general, we start each day using low nystatin concentrations (50 µg/mL). A steady access resistance is typically obtained 5–15 min following seal formation. If low access resistances (<100 MΩ) are not readily achieved with this solution, we use a higher nystatin concentration (100 µg/mL). Nystatin did

not induce measurable currents up to 20 min after break-in in whole-cell experiments; this is of particular importance for the single-cell calibration of cAMP concentrations.

4. If one is studying only forskolin activation of adenylyl cyclase in the whole-cell configuration, it may be beneficial to omit Na_2GTP from the patch pipet solution. This will minimize/eliminate G protein-mediated signaling from the preparation.

5. Several lines of evidence using CNG channels as biosensors in HEK-293 and C6-2B glioma cells point to the existence of a barrier under the surface membrane that significantly hinders the diffusion of cAMP into the rest of the cell: First, stimulation of membrane adenylyl cyclase with forskolin causes cAMP to build to much higher concentrations near the surface membrane than throughout the cell (>12-fold difference). The proximity of the sensor to adenylyl cyclase cannot explain this observation. In effect, in the absence of a diffusion barrier, each cAMP would diffuse away faster than the next one is produced *(9)*. Second, rapid dialysis of the bulk cytosol in the whole-cell patch configuration (as opposed to the perforated patch configuration) does not measurably alter the magnitude or kinetics of cAMP accumulation near the membrane *(9)*. Third, the wash in of cAMP from the patch pipet to the CNG channels (in the whole-cell configuration) is considerably slower than would be expected compared to the wash in of other small molecules *(9,18)*. These observations led to the prediction that distinct cAMP signals could coexist within a cell. Such distinct signals have been resolved in PGE_1-stimulated HEK-293 cells (*see* **Figs. 2** and **3**) *(11)*. So far the evidence for a diffusional restriction is solely from functional measurements. The biochemical and morphological basis for the barrier(s) remains to be determined. This is a fascinating subject for future study.

6. The mechanical switch time was 1 ms for the SF-77B fast-step solution switcher (Warner). The time to exchange the extracellular solution was measured by applying a 140 m*M* KCl solution to a depolarized HEK-293 cell (+50 mV), and monitoring changes in current through endogenous voltage-gated K^+ channels; for each experiment, it was <60 ms.

7. Recently, the subcellular localization of cAMP signals has become a topic of much discussion *(5,7,8)*. The frequency of cAMP signals is also likely to be important in the regulation of downstream targets *(12,25,26)*. The downstream targets of cAMP signals have vastly different apparent cAMP affinities and activation/deactivation kinetics. For example, PKA has a high apparent affinity for cAMP and relatively slow reassociation rates. This indicates that PKA will remain active after cAMP levels decline *(12)*. By contrast, CNG channels have a relatively low apparent cAMP affinity and fast kinetics. Thus, CNG channel activity will follow the time course of cAMP signals at frequencies as high as 10 Hz. It is unlikely that the different properties of these enzymes are accidental. Rather, they may have evolved to respond to different amplitudes and frequencies of cAMP signals. Indeed, it is likely that differential regulation of distinct types of PDE by a variety of intracellular messengers will be required in order to encode

information in a frequency-dependent manner. Chapters 5, 10–14, and 16–21, in this volume outline a variety of approaches to examine the regulation of PDE activity in vitro; however, these approaches cannot determine whether frequency encoding of information occurs in living cells. Thus, high-resolution, single-cell measurements of cAMP signals are required to test this hypothesis.

References

1. Montminy, M. (1997) Transcriptional regulation by cyclic AMP. *Annu. Rev. Biochem.* **66**, 807–822.
2. Francis, S. and Corbin, J. D. (1999) Cyclic nucleotide-dependent protein kinases: intracellular receptors for cAMP and cGMP action. *Crit. Rev. Clin. Lab. Sci.* **36**, 275–328.
3. Finn, J. T., Grunwald, M. E., and Yau, K.-W. (1996) Cyclic nucleotide-gated ion channels: an extended family with diverse functions. *Annu. Rev. Physiol.* **58**, 395–426.
4. de Rooij, J., Zwartkruis, F. J., Verheijen, M. H., Cool, R. H., Nijman, S. M., Wittinghofer, A., and Bos, J. L. (1998) Epac is a Rap1 guanine-nucleotide-exchange factor directly activated by cyclic AMP. *Nature* **396**, 474–477.
5. Steinberg, S. F. and Brunton, L. L. (2001) Compartmentation of G protein–coupled signaling pathways in cardiac myocytes. *Annu. Rev. Pharmacol. Toxicol.* **41**, 751–773.
6. Jurevicius, J. and Fischmeister, R. (1996) cAMP compartmentation is responsible for a local activation of cardiac Ca^{2+} channels by β-adrenergic agonists. *Proc. Natl. Acad. Sci. USA* **93**, 295–299.
7. Karpen, J. W. and Rich, T. C. (2001) The fourth dimension in cellular signaling. *Science* **293**, 2204, 2205.
8. Hall, D. D. and Hell, J. W. (2001) The fourth dimension in cellular signaling— response. *Science* **293**, 2205.
9. Rich, T. C., Fagan, K. A., Nakata, H., Schaack, J., Cooper, D. M. F., and Karpen, J. W. (2000) Cyclic nucleotide-gated channels colocalize with adenylyl cyclase in regions of restricted cAMP diffusion. *J. Gen. Physiol.* **116**, 147–161.
10. Rich, T. C., Tse, T. E., Rohan, J. G., Schaack, J., and Karpen, J. W. (2001) In vivo assessment of local phosphodiesterase activity using tailored cyclic nucleotide–gated channels as cAMP sensors. *J. Gen. Physiol.* **118**, 63–77.
11. Rich, T. C., Fagan, K. A., Tse, T. E., Schaack, J., Cooper, D. M. F., and Karpen, J. W. (2001) A uniform extracellular stimulus triggers distinct cAMP signals in different compartments of a simple cell. *Proc. Natl. Acad. Sci. USA* **98**, 13,049–13,054.
12. Rich, T. C. and Karpen, J. W. (2002) Cyclic AMP sensors in living cells: what signals can they actually measure? *Ann. Biomed. Eng.* **30**, 1088–1099.
13. Karpen, J. W., Zimmerman, A. L., Stryer, L., and Baylor, D. A. (1988) Gating kinetics of the cyclic-GMP-activated channel of retinal rods: flash photolysis and voltage-jump studies. *Proc. Natl. Acad. Sci. USA* **85**, 1287–1291.
14. Hagen, V., Dzeja, C., Frings, S., Bendig, J., Krause, E., and Kaupp, U. B. (1996) Caged compounds of hydrolysis-resistant analogues of cAMP and cGMP: synthesis and application to cyclic nucleotide–gated channels. *Biochemistry* **35**, 7762–7771.

15. Sakmann, B. and Neher, E. (1995) *Single Channel Recording*, Plenum, New York.
16. Hille, B. (2001) *Ionic Channels of Excitable Membranes*, Sinauer, Sunderland, MA.
17. Horn, R. and Marty, A. (1988) Muscarinic activation of ionic currents measured by a new whole-cell recording method. *J. Gen. Physiol.* **92**, 145–159.
18. Pusch, M. and Neher, E. (1988) Rates of diffusional exchange between small cells and a measuring patch pipette. *Pflügers Arch.* **411**, 204–211.
19. Houslay, M. D., Sullivan, M., and Bolger, G. B. (1998) The multienzyme PDE4 cyclic adenosine monophosphate-specific phosphodiesterase family: intracellular targeting, regulation, and selective inhibition by compounds exerting anti-inflammatory and antidepressant actions. *Adv. Pharmacol.* **44**, 225–342.
20. Stryer, L. (1991) Visual excitation and recovery. *J. Biol. Chem.* **266**, 10,711–10,714.
21. Yau, K.-W. (1994) Phototransduction mechanism in retinal rods and cones: The Friedenwald Lecture. *Invest. Ophthalmol. Vis. Sci.* **35**, 9–32.
22. Molday, R. S. (1998) Photoreceptor membrane proteins, phototransduction, and retinal degenerative diseases: The Friedenwald Lecture. *Invest. Ophthalmol. Vis. Sci.* **39**, 2493–2513.
23. Kaupp, U. B. and Seifert, R. (2002) Cyclic nucleotide–gated ion channels. *Physiol. Rev.* **82**, 769–824.
24. Jordan, M., Schallhorn, A., and Wurm, F. M. (1996) Transfecting mammalian cells: optimization of critical parameters affecting calcium-phosphate precipitate formation. *Nucleic Acids Res.* **24**, 596–601.
25. Rapp, P. E. and Berridge, M. J. (1977) Oscillations in calcium-cyclic AMP control loops form the basis of pacemaker activity and other high frequency biological rhythms. *J. Theor. Biol.* **66**, 497–525.
26. Cooper, D. M. F., Mons, N., and Karpen, J. W. (1995) Adenylyl cyclases and the interaction between calcium and cAMP signalling. *Nature* **374**, 421–424.

3

Cygnets

In Vivo Characterization of Novel cGMP Indicators and In Vivo Imaging of Intracellular cGMP

Akira Honda, Carolyn L. Sawyer, Sharon M. Cawley, and Wolfgang R. G. Dostmann

Summary

The second messenger cyclic guanosine 5'-monophosphate (cGMP) plays a key role in the control and regulation of a steadily increasing number of diverse physiological processes. As the appreciation of the importance of understanding the cGMP signaling pathway has grown, so has the awareness of the limited techniques with which to study the rapid intracellular cGMP kinetics. We have previously demonstrated the construction of cygnets, cGMP indicators using energy transfer comprised of cyan and yellow variants of green fluorescent protein flanked by conformationally sensitive cGMP receptor portion taken from the cGMP-dependent protein kinase *(7)*. Here, we report that cGMP binds to Cygnet-2.1, utilizing ECFP and Citrine, with an apparent equilibrium-binding constant of 600 nM causing a total fluorescence intensity ratio change of 45%. In contrast, cAMP could elicit a maximal 10% change in fluorescence resonance energy transfer (FRET) ratio, demonstrating an approx 500-fold selectivity for cGMP. When expressed in vascular smooth muscle cells, cygnets demonstrated even cytosolic distribution and nuclear exclusion. Cultured rat aortic smooth muscle cells, which exhibit a noncontractile, synthetic phenotype typically seen in response to atherosclerosis or vascular injury, responded to natriuretic peptide (BNP)-mediated activation of the particulate guanylyl cyclase. In conclusion, cygnets have facilitated the temporal resolution and evaluation of the contributions of cyclases and phosphodiesterases in determining overall cGMP accumulation, and the visualization of novel spatial dynamics that will contribute to more fully understanding the role of cGMP in the mediation of smooth muscle relaxation.

Key Words

Cygnet; cGMP; cGMP-indicator; FRET; smooth muscle.

From: *Methods in Molecular Biology, vol. 307: Phosphodiesterase Methods and Protocols*
Edited by: C. Lugnier © Humana Press Inc., Totowa, NJ

1. Introduction

With the development of wavelength mutations of green fluorescent protein (GFP), suitable for fluorescence resonance energy transfer (FRET), the design of noninvasive small-molecule indicators has become an experimental reality *(1,2)*. Examples of such indicators include cameleons for calcium *(3,4)* as well as detectors for cyclic adenosine monophosphate (cAMP) *(5,6)*. We have previously reported the use of such a FRET-based indicator system specifically tailored for another second-messenger molecule, the cyclic nucleotide cyclic guanosine 5′-monophosphate (cGMP) *(7,8)*. Our cGMP indicators, which we have named cygnets (cyclic GMP indicators using energy transfer), were designed utilizing protein kinase G (PKG) as the central cGMP sensor flanked by cyan fluorescent protein (CFP) and yellow fluorescent protein (YFP). We have chosen PKG because it binds cGMP with high affinity, undergoes a conformational change in response to cGMP, and is not restricted to membranes *(9–11)*. The validity of this intramolecular FRET approach was demonstrated by showing that cygnets are exclusively selective for cGMP, allow detection of intracellular cGMP in single living cells, are fully reversible to monitor fast spatial and temporal cGMP changes, and are minimally invasive when analyzing intracellular cGMP signaling events *(7,8)*. Here we provide a detailed methodological account of their construction, in vivo characterization, and application in cell culture.

2. Materials

1. Bac-to-Bac Baculovirus Expression System (Invitrogen, Carlsbad, CA).
2. *Escherichia coli* strain DH5α.
3. Oligonucleotide primers and enzymes for DNA manipulation.
4. DNA electrophoresis equipment.
5. Insect cell line SF9 (Invitrogen).
6. French Pressure cell.
7. Protease inhibitor cocktail (*see* **Subheading 3.2.3.**).
8. Buffer A: 50 m*M* potassium phosphate, pH 6.5, 10 m*M* dithiothreitol (DTT), 5 m*M* EDTA, 5 m*M* EGTA, 10 m*M* benzamidine at 4°C.
9. Buffer B: 50 m*M* potassium phosphate, pH 6.8, 1 m*M* EDTA, 2 m*M* benzamidine, 0.1 m*M* β-mercaptoethanol at 4°C.
10. cGMP, cAMP, and 8-AEA-cAMP-agarose column (Biolog, Bremen, Germany).
11. Sodium dodecyl sulfate-polyacrylamide gel electrophoresis (SDS-PAGE) equipment.
12. Buffer C: 250 m*M* 2(*N*-Morpholino) ethanesulfonic acid (MES), pH 6.9, 2 m*M* EGTA, 5 m*M* Mg-acetate, 50 m*M* NaCl, 10 µL of bovine serum albumin (BSA) (10 mg/mL), 10 µL of DTT (100 m*M*).
13. Buffer EDB: 5 m*M* MES, pH 7.0, 0.2 m*M* EDTA, 0.5 mg mL of BSA.
14. Phosphocellulose paper (Whatman).

15. Liquid scintillation analyzer (Packard BioScience, Downers Grove, IL).
16. Buffer D: 50 mM potassium phosphate, pH 7.0, 1 mM adenosine triphosphate (ATP), 2 mM MgCl$_2$.
17. Fluorescence spectrophotometer F-4500 (Hitachi, Tokyo, Japan).
18. RFL-6 (CCL-192) rat cell line (American Type Cell Collection [ATCC], Manassas, VA).
19. Ham's F12 medium (Cellgro by Mediatech, Herndon, VA).
20. Dulbecco's modified Eagle's medium (DMEM) (Cellgro by Mediatech).
21. Elastase (ICN, Aurora, OH).
22. Collagenase (Worthington, Lakewood, NJ).
23. FuGene 6 transfection reagent (Roche Molecular Biochemicals, Indianapolis, IN).
24. Glass-bottomed 35 × 10 mm cell culture dishes.
25. Buffer E: Hank's balanced salt solution (HBSS) with 20 mM HEPES, 2 g/L of glucose.
26. Reagents (s-Nitrosoglutathione [GSNO], natriuretic peptides, 3-isobutyl-1-methylxanthine) for manipulating intracellular cGMP (Calbiochem, San Diego, CA).
27. Dual emission fluorescence microscope.

3. Methods

The following methods outline the construction, insect cell and mammalian cell expression, purification, in vivo characterization, and single-cell imaging of Cygnet-1, Cygnet-2, and Cygnet-2.1 (**Fig. 1**).

3.1. Plasmids

The construction of the cygnet expression plasmids is described in **Subheadings 3.1.1.–3.1.4.** This includes the description of expression plasmids, the description of cGMP-dependent protein kinase cDNA, cloning, and site-directed mutagenesis for the kinase inactive Cygnet-2 and -2.1.

3.1.1. pFastBac

The Bac-to-Bac baculovirus expression system is a rapid and efficient method to generate recombinant baculoviruses *(12)*. pFastBac is a donor plasmid that contains site-specific transposition of an expression cassette into a baculovirus shuttle vector (bacmid). This vector allows the insertion of the gene of interest. Gene expression is controlled by an *Autographa californica* multiple nuclear polyhedrosis virus polyhedrin (PH) promotor. When this plasmid is transformed into *E. coli* strain DH10Bac, a recombinant bacmid is generated in this strain by transposition between the mini-Tn7 element on the pFastBac vector and the mini-*att*Tn7 target site on the bacmid *(13)*. After reaction of the transposition, the high molecular weight recombinant bacmid DNA is isolated by standard molecular biology techniques and transfected into insect cells to generate a recombinant baculovirus that is used for protein expression.

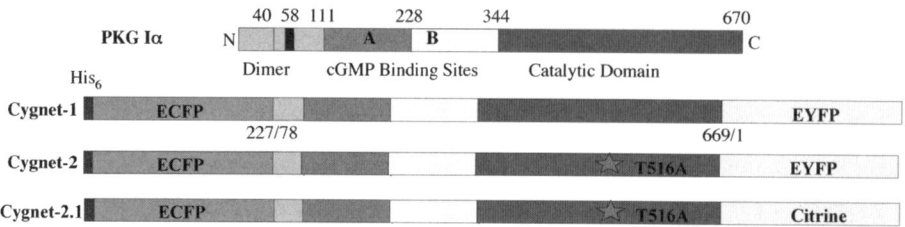

Fig. 1. Cygnet domain structures. N-terminal, Δ1-77 deletion mutants of PKG Iα *(14)* lacking the dimerization and autoinhibitory domain are sandwiched between enhanced CFP and YFP (Cygnet-1). Cygnet-2 represents the catalytically inactive mutant Δ1-77/Thr516Ala. In the successor Cygnet-2.1 EYFP is substituted with the pH-insensitive YFP version citrine *(15)*.

3.1.2. pcDNA3.1(–)

The pcDNA3.1(–) expression vector is designed for high-level stable and transient expression in a variety of mammalian hosts. It contains the human cytomegalovirus immediate-early promoter for high-level expression and an ampicillin resistance gene for cloning selection.

3.1.3. cDNA

A plasmid containing the cDNA for bovine cGMP-dependent protein kinase type Iα was kindly donated by Dr. F. Hofmann *(14)*. It contains the initiation methionine codon followed by 2010 nucleotides (nt) coding for the 670 amino acid PKG Iα protein along with a 3′ untranslated region. A plasmid containing yellow cameleon-2 and citrine were kindly donated by Dr. R. Y. Tsien *(3)*. Yellow cameleon-2 encloses the initiation methionine codon followed by 1479 nt coding for the 493 amino acid fusion protein with CFP, the Calmodulin receptor, and YFP. This cDNA is cut by *Hin*dIII and *Bam*HI and subcloned into pSP72 vector (Promega, Madison, WI) for construction of cygnets. Citrine is a variant of YFP, which is less sensitive to pH *(15)*. This cDNA is amplified by polymerase chain reaction (PCR) and substituted to the YFP portion of Cygnet-2 by digestion with *Sac*I and *Bam*HI by standard molecular biology techniques.

3.1.4. Cloning

An N-terminal truncated form of PKG type Iα (PKG/Δ1-77) cDNA was generated by PCR. Amplified DNA was subcloned into pSP72 plasmid using *Sph*I. The kinase inactive form of PKG/Δ1-77 was generated by PCR-based site-directed mutagenesis, substituting Thr 516 for Ala according to Feil et al. *(16)*. The PKG/Δ1-77 and PKG/Δ1-77/T516A genes were first fully digested with *Sph*I and subsequent partially digested with *Sac*I (*see* **Note 1**). After isolating the PKG/Δ1-77 and PKG/Δ1-77/T516A genes from agarose gels, they were subcloned into pSP72-yellow camelon-2 plasmid. Once the cygnet genes had been constructed in pSP72 plasmids, they were cloned into pFastBac and pcDNA3.1 plasmids for insect cell and mammalian expression, respectively. Because of the restraint of directional cloning, pFastBac/cygnet genes were first subcloned into pRSETA plasmids by digestion with *Xho*I and *Eco*RI. Susequently, the constructs were subcloned into pFastBac by digestion with *Bam*HI and *Eco*RI (**Fig. 2A**). Cygnet genes were directly cloned from pSP72/cygnet into pcDNA3.1(–) by digestion with *Xho*I and *Eco*RI (**Fig. 2B**). All cloning steps were verified by DNA sequencing.

3.2. Expression and Purification of Insect Cells

We employed the baculovirus-protein expression system to ensure functional cygnet expression. Cygnets, similar to wild-type I PKG (*17*), cannot be functionally expressed using bacterial expression systems. **Subheadings 3.2.1.–3.2.3.** describe the insect cell system, cygnet expression, and purification.

3.2.1. Sf9 Cells

Sf9 insect cells were grown in SF-900 II SFM medium containing 50 μg/mL of gentamicin, 2% pluronic, 2% fetal bovine serum (FBS) and maintained in suspension at a replication rate of 18–24 h (110 rpm at 28°C). Maximal viability and optimal protein expression performance were achieved between passage 5 and 12. Typically, 98% viability was confirmed using flow cytometry (Coulter Epics XL-MCL) and incubating cells in the presence of 2.5 μg/mL of propidium iodide. In preparation for baculovirus infection, cells were diluted to a density of 8×10^5 cells/mL and grown overnight to a density of $1.2–1.5 \times 10^6$ cells/mL in SF-900 II medium containing 5% FBS (*see* **Note 2**).

3.2.2. Expression of cGMP Indicators

Baculovirus stocks (Bac-to-Bac system, Invitrogen) carrying the Cygnet-2.1 gene with typical titers of approx 1×10^{10} plaque-forming units (PFU)/mL were obtained after three exhaustive 7-d amplifications (*see* **Note 3**). Cells were infected with virus stock to a final 1:50 ratio and incubated for 72 h. After centrifuging at 4000*g* for 10 min, the pellet was resuspended in 1 mL of cold buffer A plus protease inhibitors (50 μg/mL of Nα-p-tosyl-L-lysine chloromethyl ketone [TLCK], 100 μg/mL of *N*-p-tosyl-L-phenylalanyl chloromethyl ketone [TPCK], 100 μg/mL of soybean typsin inhibitor [SBTI], 50 μg/mL of antipain, 170 μg/mL phenylmethylsulfonyl fluoride [PMSF])/g wet cell pellet.

3.2.3. Purification

The cells were lysed using a French Pressure cell at 1200 psi, diluted with cold buffer A plus 50 μg/mL of TLCK, 100 μg/mL of TPCK, 100 μg/mL of SBTI, 50 μg/mL of antipain, and 170 μg/mL of PMSF to 2 mL/g wet pellet and

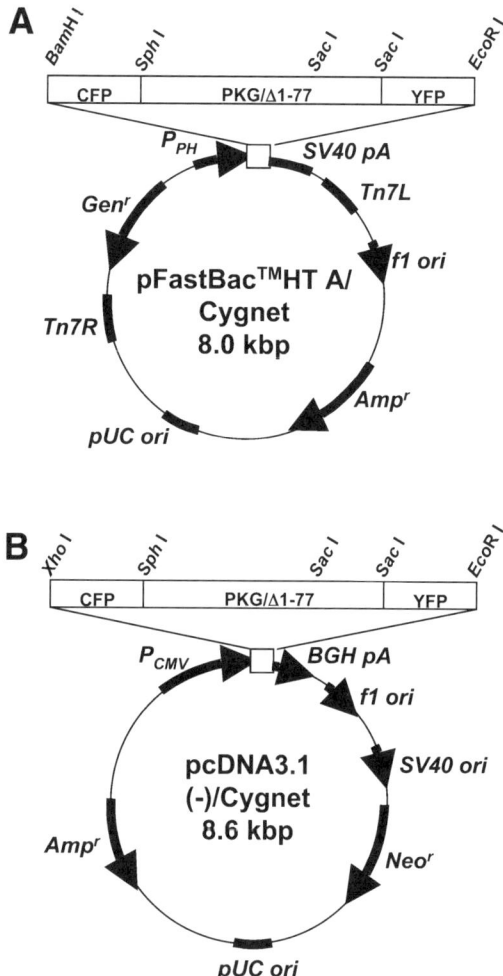

Fig. 2. Maps of cygnet expression plasmids: **(A)** Schematic diagram of donor plasmid pFastBacHTA-Cygnet for insect cell expression; **(B)** schematic diagram of expression plasmid pcDNA3.1(–)-Cygnet for mammalian expression.

centrifuged at 25,000 rpm at 4°C for 30 min. The clear supernatant was loaded onto a 2.5-mL 8-AEA-cAMP-agarose column at 4°C and washed with 20 vol of buffer A plus 50 µg/mL of TLCK, 100 µg/mL of TPCK, 100 µg/mL of SBTI, and 170 µg/mL of PMSF, followed by 20 vol of buffer A plus 500 mM NaCl and 500 vol of buffer A. Cygnet-2.1 was clearly visible as a greenish band in the column and the protein was recovered from the column using a discontinuous and isocratic elution profile (2.5 mL/fraction) at room temperature using buffer A plus 50 µg/mL of TLCK, 100 µg/mL of TPCK, 100 µg/mL of SBTI, 170 µg/mL

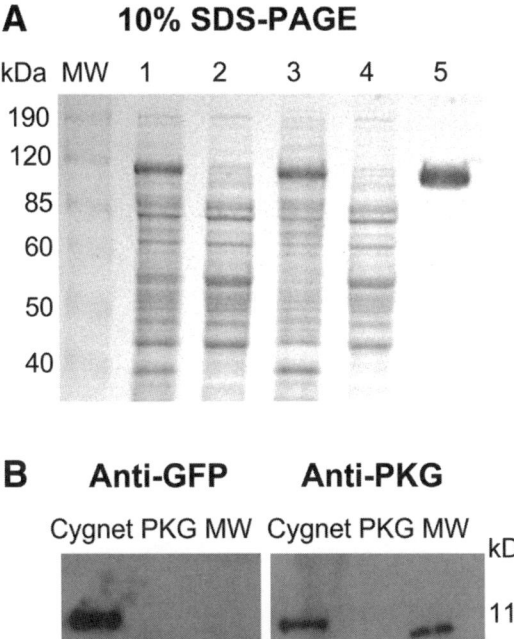

A   10% SDS-PAGE

B   Anti-GFP   Anti-PKG

Fig. 3. Sf9 expression and purification of Cygnet-2.1. (**A**) Ten percent SDS-PAGE showing 8-AEA-cAMP-agarose affinity-chromatography purification Cygnet-2.1 protein from Sf9 cells. Lane 1, 72-h Cygnet-2.1 baculovirus-infected crude Sf9 cell extract (20 µg); lane 2, soluble fraction (20 µg); lane 3, insoluble fraction (20 µg); lane 4, cAMP-agarose flowthrough (20 µg); lane 5, pooled 1 m*M* cAMP peak fractions showing the 123-kDa protein (5 µg). (**B**) Western blot analysis of purified Cygnet-2.1 using antibodies for PKG and GFP. The polyvinyl difluoride membrane was first probed with a type I-specific PKG antibody *(18)* (note the positive PKG control) and subsequently reprobed with a GFP antibody. MW, molecular weight.

of PMSF, and 1 m*M* cAMP (*see* **Note 4**). Peak fractions were pooled and dialyzed against a total volume of 10 L of buffer B, with changing of the buffer four to five times over a period of 48 h. Typical yields ranged from 1 to 5 mg/L of Sf9 cell suspension (**Fig. 3A**). Indicator integrity and purity were verified using immunoblotting with GFP and PKG antibodies (**Fig. 3B**).

3.3. In Vivo Characterization

In vivo characterization procedures are described in **Subheadings 3.3.1.–3.3.3.** These are kinase activity of cygnets, fluorescence spectral analysis, and titration of cyclic nucleotide.

3.3.1. Kinase Activity

Calculate with a 5- to 10-sample excess in **steps 2–4**.

1. Load a 96-well dish with 5 μL (1 μ*M*) of cGMP for each well in various concentrations.
2. Mix ^{32}P-ATP and ATP at a ratio of approx 1:100. Take a 10-μL sample and dilute it 1:10. Measure the activity of 10 μL of this stock on a P82 filter paper (300,000–400,000 cpm). The specific activity should be 300–400 cpm/pmol.
3. Prepare the following reaction mixture: 20 μL of buffer C, 5 μL of water, 5 μL of substrate (16 μ*M* peptide TQAKRKKSLAMA), and 5 μL of ^{32}P-ATP stock (1 m*M*) to a total of 35 μL. The mix can be prepared at the bench. Add the ATP (now mixed with ^{32}P-ATP) on the radiation bench.
4. Prepare the enzyme by diluting it in EDB and distribute it into a 96-well dish. Keep the dish on ice.
5. Place the cGMP-containing 96-well dish in a water bath (30°C). The dish should be half submerged so that it does not float.
6. Pipet 35 μL of the reaction mixture into each well of the 96-well dish.
7. Place the filtration dish on a vacuum device and turn on the vacuum. Adjust the strength of the vacuum to the lowest setting.
8. Start the reaction by pipetting 10 μL of enzyme solution into each well.
9. Stop the reaction by pipetting 25 μL from the 96-well dish onto the filtration dish. Wash immediately with 2 × 250 μL of 75 m*M* phosphoric acid.
10. Dry the filtration dish with hot air (using a hair dryer), punch out the filters, and count (**Fig. 4**).

3.3.2. Fluorescence Resonance Energy Transfer

We used 50 n*M* of cygnet protein to obtain typical fluorescence spectra.

1. Add 490 μL of ice-cold buffer D to a quartz cuvet, followed by 50 n*M* (final concentration) cygnet protein.
2. Add 2.5 μL of water (for control) or 60 μ*M* cGMP to the cuvet. The final volume should be 500 μL.
3. Set the cuvet in a fluorescence spetrophotometer. Excite each sample at 432 nm, and monitor emission intensities from 450 to 580 nm (**Fig. 5A**). The FRET ratio (475-nm intensity/525-nm intensity) should be about 0.7 without cGMP, and about 1.0 with cGMP. The change in FRET ratio should be 45–50% (*see* **Note 5**).

3.3.3. Titration of Cyclic Nucleotide

Each sample is prepared as described in **Subheading 3.3.2.** The ratios of the 475- to 525-nm emission intensities are plotted against the concentration

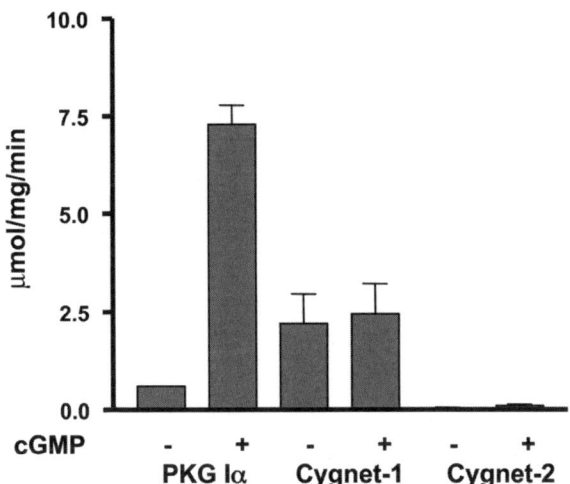

Fig. 4. Kinase activity of Cygnets. Phosphoryltransferase assay of PKG Iα and cygnets using peptide TQAKRKKSLAMA as substrate *(19)*. PKG shows strong cGMP dependency for kinase activity. Cygnet-1, reminiscent of Δ1-77 PKG *(20)*, is constitutively active as a kinase. However, similar to PKG *(16)*, the Thr516Ala catalytic domain mutants Cygnet-2 (data not shown) and Cygnet-2.1 show no kinase activity.

of cyclic nucleotides ranging from 1 n*M* to 250 μ*M* added to the sample to generate a titration curve (**Fig. 5B**). The maximum change in ratio for cGMP should be 45–50% at 15 μ*M* or higher concentration, and 8–11% for cAMP at 300 μ*M* or higher. The apparent K_D should be about 600 n*M* and 300 μ*M* for cGMP and cAMP, respectively.

3.4. Single-Cell Imaging

Rat fetal lung (RFL-6) fibroblast cells and rat aortic smooth muscle cells (RASMCs) were successfully used for single-cell imaging, which is described in **Subheadings 3.4.1.–3.4.4.** The procedures are mammalian cell cultivation, plasmid transfection and transient protein expression, and dual fluorescence imaging.

3.4.1. RFL-6 Fibroblast Cells

RFL-6 is an established cell line known to respond with high cGMP levels on stimulation *(21)*. Cells should be cultured according to the supplier (ATCC). RFL-6 cells are cultured in Ham's F12 medium supplemented with 20% FBS at 37°C, 5% CO_2. Cells should be subcultured every 5 to 6 d with 0.25% trypsin at a ratio of 1:4 and dished on either tissue culture dishes for maintenance or glass-bottomed dishes for imaging (*see* **Note 6**).

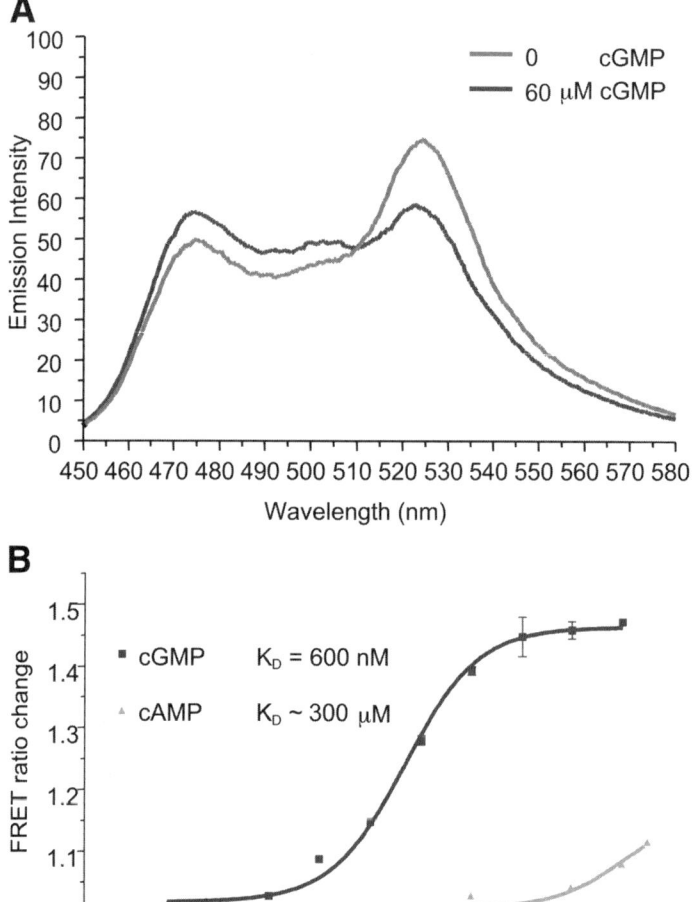

Fig. 5. In vivo characterization of Cygnet-2.1. (**A**) Fluorescence spectra (excited at 432 nm) of purified recombinant Cygnet-2.1 with zero and saturating (60 μ*M*) cGMP, respectively. Note the differential responses of enhanced cyan fluorescent protein (ECFP) emission at 475 nm and citrine emission at 525 nm *(7)*. (**B**) Titration curves for cGMP (maximal change in FRET ratio = 48%) and cAMP (maximal change in FRET ratio = 10%), when combining five independent measurements. The fitted curves correspond to the apparent dissociation constants for cGMP and cAMP of 600 n*M* and 300 μ*M*, respectively.

3.4.2. Rat Aortic Smooth Muscle Cells

To generate a rat aortic smooth muscle primary cell culture, two methods are available: explant and dissociation.

3.4.2.1. EXPLANT PROTOCOL (ADAPTED FROM REF. *22*)

1. Euthanize a mature Sprague-Dawley rat (250–350 g) by a lethal dose of pentobarbital sodium and exsanguination, and remove the thoracic aorta.
2. Remove fat and connective tissue from the aorta and slice into 1- to 2-mm rings with a sterile scalpel.
3. Create grid lines on a 60-mm cell culture dish with a scalpel, and embed each ring on the grid.
4. Add DMEM containing 10% FBS, 50 µg/mL of gentamicin, and 2.5 µg/mL of amphotericin B to the dish and incubate at 37°C in humidified 5% CO_2. Change the medium every 2 to 3 d. The smooth muscle cells will proliferate from the aortic explants in about 7–10 d.
5. When the cells reach confluency, remove the artery ring and subculture the cells onto the glass-bottomed dish using standard cell culture procedure. Once the cells have adhered to the dish, proceed to transfection (*see* **Subheading 3.4.3.**).

3.4.2.2. DISSOCIATION PROTOCOL (ADAPTED FROM REFS. *23* AND *24*)

1. After extracting the aorta, clean the aorta of all fat and connective tissue, and incubate in isolation medium supplemented with 1.25 U/mL of elastase and 175 U/mL of collagenase for 30–60 min in a shaking air incubator at 37°C.
2. Rinse the aorta with HBSS to remove endothelial and other nonadherent cells, and then remove the tunica adventitia.
3. Incubate the tunica media overnight in DMEM with 10% FBS and 100 µg/mL of penicillin/streptomycin.
4. The next morning, cut open the tunica media longitudinally and then cut laterally into 1-mm pieces. Digest the pieces in solution containing 175 U/mL of collagenase and 2.5 U/mL of elastase for 1 to 2 h.
5. Stop digestion with an equal volume of DMEM with 20% FBS, and pellet the cells by centrifuging at 900 rpm in a 4°C tabletop centrifuge. Resuspend the cells in DMEM with 20% FBS with 100 µg/mL of penicillin/streptomycin in a 60-mm dish. Incubate at 37°C in humidified 5% CO_2 until ready for transfection.

3.4.3. Transfection

1. Prior to transfection, plate RFL-6 cells or RASMCs on 35-mm glass-bottomed dishes with 50–60% confluency.
2. For each 35-mm dish, pipet 3 µL of FuGene 6 directly into 200 µL of serum-free DMEM in a plastic 1.5-mL microcentrifuge tube, while minimizing the contact of FuGene 6 with the wall of the tube.

3. After adding 1 mg of highly purified pcDNA3.1(–)-Cygnet-2.1 to the DMEM-FuGene and mixing by inversion, incubate at room temperature for 15 min to allow DNA/lipid complexes to form.

4. Without removing the medium from the 35-mm dishes, add the DNA/lipid mixture to the dishes dropwise using a pipet. Distribute the transfection mixture around the dishes by gently shaking the dishes side to side.

5. Incubate the cells at 37°C for 24–48 h before imaging (*see* **Notes 7** and **8**).

3.4.4. Imaging

The following dual imaging system was used for the experiments described: a microscope with a 40/1.30 oil Ph4DL objective (Nikon, Tokyo, Japan), an ORCA ER cooled charge-coupled device camera (Hamamatsu, Bridgewater, NJ), Lambda 10-2 Optical Filter Changers, an LS Xenon Arc Lamp and power supply (Sutter, Novato, CA), a Cameleons 2 filter set (Chroma, Rockingham, VT), and Metamorph and Metafluor 4.64 software (Universal Imaging, Media, PA). The Metafluor protocol was followed to obtain dual emission images (*see* **Note 9**).

1. Place a glass-bottomed dish with pcDNA3.1-Cygnet-2.1-transfected cells on a microscope table.

2. Locate the transfected cells and then focus the cells through the monitor.

3. In the Metafluor program, select "Acquire One" on the tool bar to get the image through both 475- and 535-nm emission channels (**Fig. 6A,B**). Pseudocolored images, which correlate color with a 475/535-emission ratio, can also be acquired during the data collection (**Fig. 6C,D**).

4. Define regions on the image, and then start the experiment by acquiring the images every 10–15 s. As the program acquires images, the intensity from each channel and ratio of (475 channel)/(535 channel) intensity values from the defined regions are plotted in graph 1 and graph 2, respectively (**Fig. 7A,B**).

5. Once a stable baseline is established, 1 μ*M* B-type natriuretic peptide (BNP) can be added to the dish. Continue monitoring the ratio and intensities for each channel (*see* **Note 10**). As cGMP is produced in the cells, a significant decrease in 535-nm intensity and a small increase in 475-nm intensity can be observed in graph 1 (**Fig. 7B**). As the result of these changes, an increase of 475/535-emission ratio will be seen in graph 2 (**Fig. 7A**). Typically, changes in ratio of 30% have been observed using BNP.

4. Notes

1. As shown in **Fig. 2**, the PKG Iα gene contains a *Sac*I digestion site. Since *Sac*I is required for the construction of cygnets, we performed partial digestions. The best results were achieved when 1 μg of pSP72-PKG/Δ1-77 and subsequent plasmids were incubated in a 20-μL reaction mixture with 10 U of *Sac*I for 7–10 min.

2. Maintenance of Sf9 suspension culture cells appears to be critical for functional protein expression. High doubling rate (≤24 h), low passage number (5–10), low

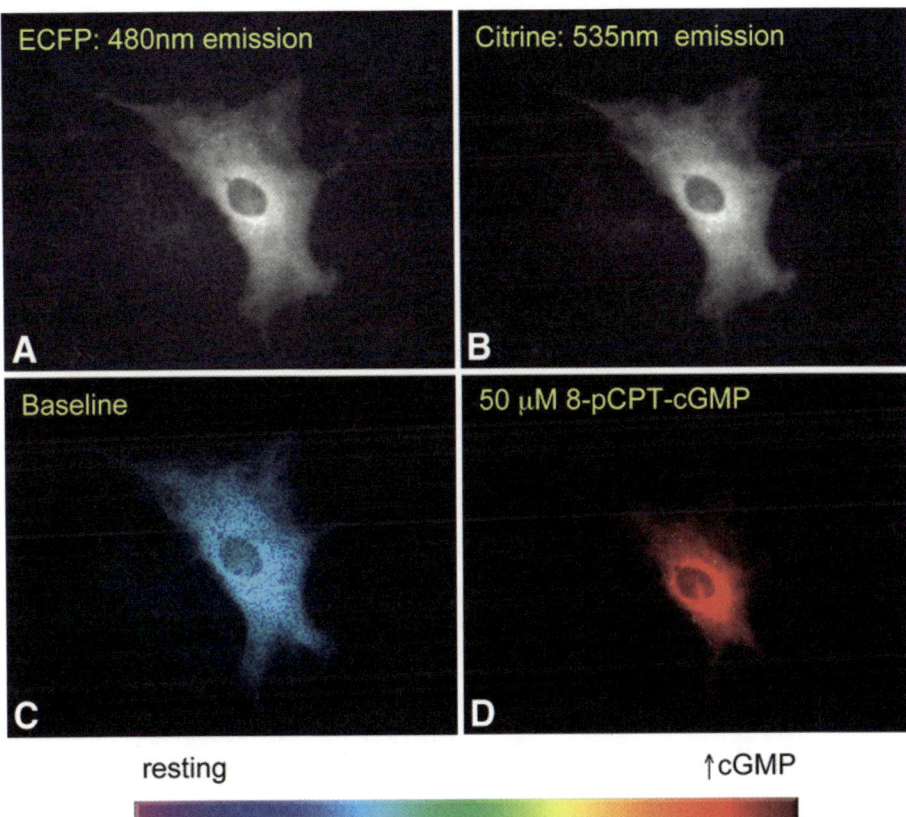

Fig. 6. Imaging of Cygnet-2.1 in vascular smooth muscle cells (VSMCs). Cygnet-2.1 demonstrates cytosolic localization and nuclear exclusion in cultured RASMCs (top) as shown by the fluorescence images of **(A)** ECFP (480-nm emission) and **(B)** enhanced yellow fluorescent protein (EYFP) = citrine (535-nm emission). Pseudocolor representations are shown of the 480- to 535-nm FRET ratio at **(C)** resting cGMP and **(D)** 50 µ*M* cell-permeable derivative 8-pCPT-cGMP. The change in color from blue to red indicates a 30% change in ratio.

shaking speed (110 rpm), moderate cell density (8×10^5 to 4×10^6), low FBS concentration (2%), and the presence of 2% pluronic to minimize shear stress are critical parameters for optimal Sf9 cell health. However, most important for optimal protein expression is our observation that prior to infection, cells need to recover overnight from dilution to the appropriate preparative volume. We routinely dilute cells to 8×10^5 in 2- to 4-L flasks (5% FBS) and continue shaking (overnight) until they reach a density of $1.2–1.5 \times 10^6$, at which point we infect.

3. We have found that three exhaustive amplifications (7 d each) yield a relatively constant virus titer of approx 1×10^{10} PFU. Similarly important, by using 1:50

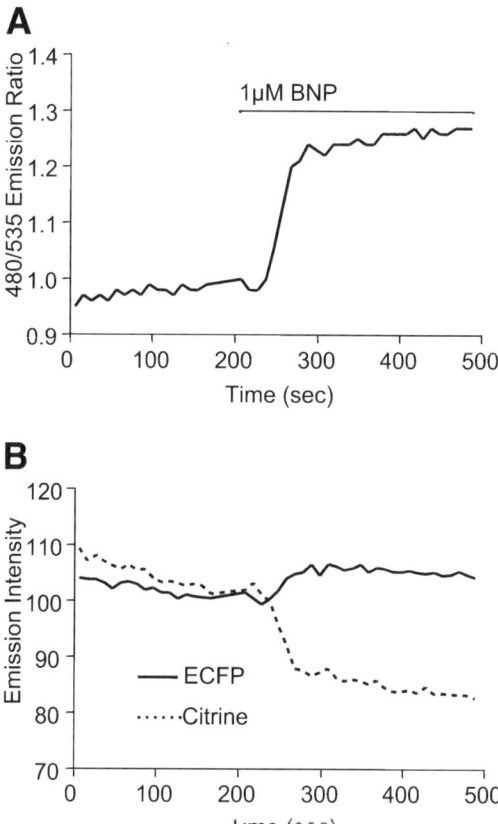

Fig. 7. Particulate guanylyl cyclase activity in a VSMC. (**A**) Application of 1 μ*M* natriuretic peptide BNP produces a FRET ratio change of approx 28% through stimulation of the particulate guanylyl cyclase (pGC) NPR-A. (**B**) The individual emission intensities of ECFP and citrine during pGC activation demonstrate an increase in ECFP and a decrease in citrine emissions.

virus dilutions for infection, our titer appears to be several orders of magnitude above the recommended dosage. However, it appears that FBS acts as a sink for the virus, and because we discovered that slightly elevated FBS levels (5%) are beneficial for protein expression, we abandoned the standard protocols.

4. Cygnet purification requires the use of 8-aminoethylamino-agarose to ensure elution with cAMP, rather then cGMP. Optimal protein recovery was obtained by employing a discontinuous isocratic elution with 1 m*M* cAMP at room temperature. Note that cygnets are less stable (aggregation, loss of FRET) at concentrations above 1 mg/mL. Purified cygnets should be stored at approx 0.5 mg/mL and 4°C and protected from light (1 mo) or frozen at –20°C in 50% glycerol (>12 mo).

5. To omit photobleaching, each sample should be analyzed only once.
6. Cells should not be used beyond passage number eight.
7. Transfection efficiency using pcDNA3/Cygnet-2.1 is often low, particularly in primary cells. In addition, we have found that in smooth muscle cells cygnet expression levels can vary substantially. Optimal expression may require alternative transfection reagents and optimization of transfection conditions, depending on the cell type.
8. It is important to compare the total cGMP binding capacity of untransfected and cygnet-transfected cells because intracellular fluctuations in cGMP can potentially be buffered by the expression of the indicator. We have used a type I–specific antibody of PKG *(18)* to determine whether the total PKG immunoreactivity in different passage cells and, therefore, cGMP binding capacity is elevated by the expression of cygnets. As a general rule, early passage cells are less likely to show relative cygnet overexpression, probably owing to high endogenous levels of PKG.
9. Fluorophore excitation should occur at the minimum intensity, frequency, and duration to minimize excessive illumination of cygnets, which may result in photobleaching of the GFP mutants. To establish that changes in FRET are not owing to abnormal behavior of YFP or CFP as the result of photobleaching or other phenomena such as pH-induced alterations, we routinely monitor the emission profiles of the individual fluorophores (**Fig. 7B**).
10. Prior to application, reagents should be premixed with imaging buffer. Adding the concentrated reagent directly onto the dish can result in uneven distribution.

Acknowledgments

We thank Dr. F. Hofmann for providing the bovine cGMP-dependent protein kinase type Iα cDNA, and Dr. R. Y. Tsien for providing the yellow cameleon-2 plasmid and citrine cDNA. This work was supported by National Science Foundation Grant MCB-9983097, the Lake Champlain Cancer Research Organization, the Totman Medical Research Trust, and American Heart Association Grant 9920260T.

References

1. Heim, R. and Tsien, R. Y. (1996) Engineering green fluorescent protein for improved brightness, longer wavelengths and fluorescence resonance energy transfer. *Curr. Biol.* **6,** 178–182.
2. Heim, R., Prasher, D. C., and Tsien, R. Y. (1994) Wavelength mutations and posttranslational autoxidation of green fluorescent protein. *Proc. Natl. Acad. Sci. USA* **91,** 12,501–12,504.
3. Miyawaki, A., Llopis, J., Heim, R., McCaffery, J. M., Adams, J. A., Ikura, M., and Tsien, R. Y. (1997) Fluorescent indicators for Ca^{2+} based on green fluorescent proteins and calmodulin. *Nature* **388,** 882–887.

4. Miyawaki, A., Griesbeck, O., Heim, R., and Tsien, R. Y. (1999) Dynamic and quantitative Ca2+ measurements using improved cameleons. *Proc. Natl. Acad. Sci. USA* **96,** 2135–2140.

5. Zaccolo, M. and Pozzan, T. (2002) Discrete microdomains with high concentration of cAMP in stimulated rat neonatal cardiac myocytes. *Science* **295,** 1711–1715.

6. Zaccolo, M., De Giorgi, F., Cho, C. Y., Feng, L., Knapp, T., Negulescu, P. A., Taylor, S. S., Tsien, R. Y., and Pozzan, T. (2000) A genetically encoded, fluorescent indicator for cyclic AMP in living cells. *Nat. Cell. Biol.* **2,** 25–29.

7. Honda, A., Adams, S. R., Sawyer, C. L., Lev-Ram, V., Tsien, R. Y., and Dostmann, W. R. G. (2001) Spatiotemporal dynamics of guanosine 3′,5′-cyclic monophosphate revealed by a genetically encoded, fluorescent indicator. *Proc. Natl. Acad. Sci. USA* **98,** 2437–2442.

8. Sawyer, C. L., Honda, A., and Dostmann, W. R. G. (2003) Cygnets: spatial and temporal analysis of intracellular cGMP. *Proc. West. Pharmacol. Soc.* **46,** 28–31.

9. Ruth, P., Landgraf, W., Keilbach, A., May, B., Egleme, C., and Hofmann, F. (1991) The activation of expressed cGMP-dependent protein kinase isozymes I alpha and I beta is determined by the different amino-termini. *Eur. J. Biochem.* **202,** 1339–1344.

10. Zhao, J., Trewhella, J., Corbin, J., Francis, S., Mitchell, R., Brushia, R., and Walsh, D. (1997) Progressive cyclic nucleotide–induced conformational changes in the cGMP-dependent protein kinase studied by small angle X-ray scattering in solution. *J. Biol. Chem.* **272,** 31,929–31,936.

11. Wall, M. E., Francis, S. H., Corbin, J. D., Grimes, K., Richie-Jannetta, R., Kotera, J., Macdonald, B. A., Gibson, R. R., and Trewhella, J. (2003) Mechanisms associated with cGMP binding and activation of cGMP-dependent protein kinase. *Proc. Natl. Acad. Sci. USA* **100,** 2380–2385.

12. Whitford, W. G. and Mertz, L. M. (1996) Multiplicity of bacoloviral infection and recombinant protein production in sf9 cells. *Focus* **18,** 3: 75–76.

13. Luckow, V. A., Lee, C. S., Barry, G. F., and Olins, P. O. (1993) Efficient generation of infectious recombinant baculoviruses by site-specific transposon-mediated insertion of foreign genes into a baculovirus genome propagated in *Escherichia coli. J. Virol.* **67,** 4566–4579.

14. Wernert, W., Flockerzi, V., and Hofmann, F. (1989), The cDNA of the two isoforms of bovine cGMP-dependent protein kinase. *FEBS Lett.* **251,** 191–196.

15. Griesbeck, O., Baird, G. S., Campbell, R. E., Zacharias, D. A., and Tsien, R. Y. (2001) Reducing the environmental sensitivity of yellow fluorescent protein: mechanism and applications. *J. Biol. Chem.* **276,** 29,188–29,194.

16. Feil, R., Kellermann, J., and Hofmann, F. (1995) Functional cGMP-dependent protein kinase is phosphorylated in its catalytic domain at threonine-516. *Biochemistry* **34,** 13,152–13,158.

17. Feil, R., Bigl, M., Ruth, P., and Hofmann, F. (1993) Expression of cGMP-dependent protein kinase in *Escherichia coli. Mol. Cell. Biochem.* **127–128,** 71–80.

18. Keilbach, A., Ruth, P., and Hofmann, F. (1992) Detection of cGMP dependent protein kinase isozymes by specific antibodies. *Eur. J. Biochem.* **208,** 467–473.

19. Dostmann, W. R. G., Nickl, C., Thiel, S., Tsigelny, I., Frank, R., and Tegge, W. (1999) Delineation of selective cyclic GMP–dependent protein kinase I substrate and inhibitor peptides based on combinatorial peptide libraries on paper. *Pharmacol. Ther.* **82,** 373–387.
20. Heil, W. G., Landgraf, W., and Hofmann, F. (1987) A catalytically active fragment of cGMP-dependent protein kinase: occupation of its cGMP-binding sites does not affect its phosphotransferase activity. *Eur. J. Biochem.* **168,** 117–121.
21. Ishii, K., Sheng, H., Warner, T. D., Forstermann, U., and Murad, F. (1991) A simple and sensitive bioassay method for detection of EDRF with RFL-6 rat lung fibroblasts. *Am. J. Physiol.* **261,** H598–H603.
22. Cornwell, T. L. and Lincoln, T. M. (1989) Regulation of intracellular Ca^{2+} levels in cultured vascular smooth muscle cells: reduction of Ca^{2+} by atriopeptin and 8-bromo-cyclic GMP is mediated by cyclic GMP–dependent protein kinase. *J. Biol. Chem.* **264,** 1146–1155.
23. Travo, P., Barret, G., and Burnstock, G. (1980) Differences in proliferation of primary cultures of vascular smooth muscle cells taken from male and female rats. *Blood Vessels* **17,** 110–116.
24. Korshunov, V. A. and Berk, B. C. (2003) Flow-induced vascular remodeling in the mouse: a model for carotid intima-media thickening. *Arterioscl. Thromb. Vasc. Biol.* **23,** 2185–2191.

4

High-Throughput Screening of Phosphodiesterase Activity in Living Cells

Thomas C. Rich and Jeffrey W. Karpen

Summary

Phosphodiesterases (PDEs) hydrolyze the second messengers cyclic adenosine monophosphate (cAMP) and cyclic guanosine 5′-monophosphate (cGMP) and play a crucial role in the termination and spatial segregation of cyclic nucleotide signals. Despite a wealth of molecular information, very little is known about how PDEs regulate cAMP and cGMP signals in living cells because conventional methods lack the necessary spatial and temporal resolution. We present here a sensitive optical method for monitoring cAMP levels and PDE activity near the membrane, using cyclic nucleotide-gated (CNG) ion channels as sensors. These channels are directly opened by the binding of cyclic nucleotides and allow cations to cross the membrane. The olfactory channel A subunit (CNGA2) has been genetically modified to improve its cAMP sensitivity and specificity. Channel activity is assessed by measuring Ca^{2+} influx using standard fluorometric techniques. In addition to studying PDEs in their native setting, the approach should be particularly useful in high-throughput screening assays to test for compounds that affect PDE activity, as well as the activities of the many G protein-coupled receptors that cause changes in intracellular cAMP.

Key Words

Phosphodiesterase; adenylyl cyclase; cyclic nucleotide-gated ion channels; G protein-coupled receptors; calcium influx; second messengers; adenosine monophosphate; single-cell fluorescence assays.

1. Introduction

Cyclic nucleotide phosphodiesterases (PDEs) are the crucial terminators of cyclic adenosine monophosphate (cAMP) and cyclic guanosine 5′-monophosphate (cGMP) signals. Since their original discovery *(1)*, PDEs have been classified into 11 families according to substrate specificity; regulation; pharmacology; and, more recently, amino acid homology *(2–4)*. Several studies have revealed

From: *Methods in Molecular Biology, vol. 307: Phosphodiesterase Methods and Protocols*
Edited by: C. Lugnier © Humana Press Inc., Totowa, NJ

differential regulation of PDE families by Ca^{2+}-calmodulin (Ca^{2+}-CaM), G proteins, phosphorylation, and cyclic nucleotides. The diversity of PDE families has led to the realization that PDE activity is a central element in the control of second-messenger signaling *(5,6)*. Yet, little is known about how PDE regulates cyclic nucleotide signals in vivo, or how these signals differentially regulate a wide variety of cellular targets. Here we describe a convenient optical assay for monitoring cAMP signals *in living cells*. The method uses genetically engineered cyclic nucleotide-gated (CNG) channels as cAMP sensors. On binding cAMP these ion channels open, allowing the flow of cations across the plasma membrane (*see* **Note 1**). CNG channel activity can be monitored by measuring Ca^{2+} influx in single cells or cell populations *(7–9)*. As such, this approach can be used with conventional fluorescence or standard high-throughput screening assays to evaluate PDE activity in living cells. Indeed, CNG channel-based biosensors will be particularly useful to screen for drugs that regulate PDEs, adenylyl cyclases (ACs), or G protein-coupled receptors (GPCRs) linked to the cAMP pathway.

2. Materials

1. Adenovirus constructs for expression of CNG channels (pertinent characteristics of four CNG channel constructs used for monitoring cAMP levels are described in **Subheading 3.2.**).
2. Primary or cultured cell lines in which measurements are to be conducted: The examples given here are in HEK-293 cells.
3. Minimal essential medium (MEM) supplemented with 26.2 mM NaHCO$_3$, 10% (v/v) fetal bovine serum (Gemini), penicillin (50 µg/mL), and streptomycin (50 µg/mL), pH 7.0 (*see* **Note 2**).
4. 100 mM Hydroxyurea stock solution: This is required only for cell lines in which adenovirus can readily replicate, including HEK-293 cells (*see* **Note 3**).
5. Buffer for evaluating apparent cAMP affinity of CNG channels: 130 mM NaCl; 2 mM HEPES; 0.02 mM EDTA; and 1 mM EGTA, pH 7.6 (*see* **Note 4**).
6. Electrophysiology setup for excised, inside-out patch experiments (*see* **Note 4**).
7. Assay buffer for Ca^{2+}-imaging experiments: 145 mM NaCl; 11 mM D-glucose; 10 mM HEPES; 4 mM KCl; 1 mM CaCl$_2$; 1 mM MgCl$_2$; and 1 mg/mL of bovine serum albumin (BSA), pH 7.4.
8. Loading buffer: Ham's F-10 medium supplemented with 1 mg/mL of BSA fraction V and 20 mM HEPES, pH 7.4.
9. Fura-2/AM and pluronic F-127 (Molecular Probes, Eugene, OR).
10. Ca^{2+}-imaging setup for either single cells or cell populations (e.g., the LS-50B spectrofluorimeter from Perkin Elmer, Shelton, CT; or the FLIPR^2 fluorometric imaging plate reader from Molecular Devices, Sunnyvale, CA).
11. Activators of AC and inhibitors of PDE: Several compounds, including forskolin, 3-isobutyl-1-methylxanthine (IBMX), and 4-(3-butoxy-4-methoxybenzyl)-2-imidazolidinone (RO-20-1724), are added to control solutions from concentrated

dimethyl sulfoxide (DMSO) stocks, with final concentrations as indicated (final DMSO concentrations were <0.5%).
12. Data analysis software. Appropriate software should be available from the manufacturer of the Ca^{2+}-imaging setup. We typically use the MATLAB software package (Mathworks, Natick, MA) to design analysis software in-house.

3. Methods

The following methods outline an approach to detect changes in cAMP levels in living cells using modified olfactory CNG channels (*see* **Note 1**). The approach is applicable to the study of either cell populations or single cells. As such, this assay is well suited for high-throughput screening of agents that regulate G protein-signaling pathways.

3.1. Cell Culture and CNG Channel Expression

It is important to note that CNG channel expression in different cell types requires different multiplicity of infection (MOI) and incubation times. **Table 1** summarizes the conditions for optimal expression of CNG channels in several cell types.

1. Maintain HEK-293 cells in supplemented MEM (*see* **Subheading 2.**, **item 3**) at 37°C in a humidified atmosphere of 95% air and 5% CO_2.
2. Plate the cells at approx 60% confluence in 100-mm culture dishes 24 h prior to infection with the CNG channel-encoding adenovirus constructs (MOI = 10 plaque-forming units [PFU]/cell).
3. Two hours postinfection, add hydroxyurea from a stock solution to the medium at a final concentration of 2 mM to partially inhibit viral replication (*see* **Note 3**).
4. Twenty-four hours postinfection, detach the cells with phosphate-buffered saline containing 0.03% EDTA (PBS-EDTA), resuspend in serum-containing medium, allow to recover for 1 h, and assay within 12 h. We use PBS-EDTA rather than trypsin-containing solutions so as not to modify extracellular regions of the CNG channels.

3.2. Characterization of Sensitivity of CNG Channels Using Excised Patch Technique

For high-throughput screening of PDE activity, it is not necessary to evaluate the sensitivity of the CNG channels because they have been studied in detail in HEK-293 cells (*7,10*). However, when a more detailed analysis of the underlying cAMP signal is required, it may be necessary to evaluate the sensitivity of channel constructs in the cell type of interest. Thus, we outline the following method (*see* **Note 4**).

To assess the cyclic nucleotide sensitivity of CNG channel constructs, excised, inside-out patch recordings are made using an Axopatch-200A patch-clamp

Table 1
Conditions for Expression of CNG Channels Using Adenovirus
Constructs in Various Cell Types

	Source	MOI (PFU/cell)	Incubation time (h)
Cultured cell lines			
HEK-293[a]	Human embryonic kidney	10	24
C6-2B	Rat glioma	100	48–72
GH3 and GH4C1	Rat pituitary tumor	50	48
PC12	Rat neuroendocrine	100	48
A7r5	Rat smooth muscle	100	24–48
Primary cultures			
Vascular smooth muscle	Rat, mouse	100–200	48–72
Neonatal ventricular myocytes	Rat	100–200	24–48
Adult ventricular myocytes	Rat	100–200	24–48

[a]Expression of CNG channels in HEK-293 cells requires the addition of 1 to 2 mM hydroxyurea 2 h postinfection to inhibit viral replication (*see* **Note 3**).

amplifier (Axon). These experiments are typically conducted at room temperature (20–22°C). Pipets are pulled from acid-washed, borosilicate glass tubes using a micropipet puller (P-87; Sutter, Novato, CA) and heat polished (MF830 microforge; Narishige, Tokyo, Japan). They are then backfilled with the buffer described in **Subheading 2., item 5** and lowered onto a cell, and a GΩ seal is formed. Patches are excised by shearing cells from the pipet with a jet of liquid. Ionic currents through the CNG channels are typically elicited by 250-ms pulses to membrane potentials of +50 and –50 mV from a holding potential of 0 mV. Current records are sampled at five times the filter setting and stored on an IBM-compatible computer. They are then corrected for errors owing to series resistance using the standard relation $V = IR$, in which V is the membrane potential, I is the current, and R is the series resistance. Typically, the series resistance is between 2 and 4 MΩ. Cyclic nucleotide–induced currents are obtained from the difference between currents in the presence and absence of cyclic nucleotides. Dose-response curves for cAMP and cGMP are obtained by examining the effects of several (typically >10) concentrations of cyclic nucleotide at both +50 and –50 mV on the same patch. Dose-response relations are then analyzed using the Hill equation, $I/I_{max} = [\text{cNMP}]^N/([\text{cNMP}]^N + K_{1/2}^N)$, in which I/I_{max} is the fraction of maximal current; cNMP represents cyclic nucleotide; $K_{1/2}$ is the concentration that gives a half-maximal current; and N is the Hill coefficient, an index of cooperativity.

We have used this approach to assess the cAMP-sensing ability of several CNG channel constructs *(7,11)*. The design of these constructs was guided by the rich history of structure-function studies on this and homologous ion channels *(12–16)*. Several channel properties, as well as the range of cAMP levels that they can readily detect in intact cells, are summarized in **Table 2**. Other properties of these constructs are described elsewhere *(7,11)*. Note that the C460W/E583M and the Δ61-90/C460W/E583M constructs are particularly useful for measuring cAMP levels.

3.3. Detection of cAMP in Cell Populations and Single Cells

Homomultimeric CNG channels comprising rat olfactory channel α (CNGA2) subunits are permeable to Ca^{2+} *(17)*. We have used this property to detect changes in local cAMP concentration in living cells *(7–11)*. In this assay, an increase in local cAMP concentration causes activation of CNG channels and a subsequent increase in Ca^{2+} entry. Changes in the rate of Ca^{2+} influx in response to stimuli reflect changes in cAMP levels. Ca^{2+} influx can be readily monitored using calcium dyes such as fura-2 *(18)* *(see* **Note 5**). Care must be taken either to minimize the effects of other sources of Ca^{2+} entry into the cell or to separate the Ca^{2+} influx through CNG channels *(see* **Note 6**). With this approach, cAMP levels and, thus, PDE activity can be monitored in a variety of cell lines.

3.3.1. Fura-2 Loading of HEK-293 Cells

In this assay Ca^{2+} dyes are used to detect the activation of CNG channels. These channels are rapidly activated (<0.1 s) following increases in cAMP levels. Here we describe appropriate conditions for loading HEK-293 cells with fura-2 *(see* **Note 5**).

It is best to shield fura-2 from the light. We typically load cells in a 15-mL conical tube wrapped in aluminum foil in a dimly lit room. Cells are loaded with 4 μ*M* fura-2/AM (a membrane-permeant form) and 0.02% pluronic F-127 at room temperature for 30–40 min. We use approx 10^7 cells/mL of loading buffer *(see* **Subheading 2., item 8**). Under these conditions, about 35 μ*M* fura-2 accumulates in the cells. After loading fura-2, the cells are washed twice and resuspended in assay buffer (3 to 4 × 10^6 cells/3 mL).

We have compiled conditions for the loading of fura-2/AM in several cell types (**Table 3**).

3.3.2. Evaluation of PDE Activity in Populations of HEK-293 Cells

CNG channels provide a continuous readout of cAMP levels in subcellular compartments near the plasma membrane *(7,10,11)*. This approach can be used to either (1) determine which PDE type(s) regulate cAMP in this pool, or (2) screen novel PDE inhibitors as follows:

Table 2
Characteristics of Rat Olfactory CNGA2 Channels Expressed in HEK-293 Cells[a]

CNG channel construct	$K_{1/2}^{cAMP}$ (μM)	N^{cAMP}	$K_{1/2}^{cGMP}$ (μM)	N^{cGMP}	$I_{max}^{cGMP}/I_{max}^{cAMP}$	Readily detected cAMP levels (μM)
WT	36 ± 5	2.2 ± 0.1	1.3 ± 0.4	2.5 ± 0.4	1.0	4–200
E583M	9 ± 2	2.2 ± 0.1	32 ± 4	1.9 ± 0.2	0.35 ± 0.16	0.9–45
C460W/E583M	0.89 ± 0.23	2.2 ± 0.3	6.2 ± 1	2.7 ± 0.1	0.50 ± 0.10	0.09–4.5
Δ61-90/C460W/E583M	10.5 ± 0.1	2.2 ± 0.1	16 ± 1	2.6 ± 0.2	0.09 ± 0.06	1–50

[a]The high-cAMP-affinity C460W/E583M channel allows cAMP to be measured in the low end of the physiological range. Experiments using this construct should be conducted in either Ca^{2+}-free solutions or in conditions when intracellular Ca^{2+} is highly buffered (e.g., with high intracellular concentrations of fura-2) (7,11), because its cAMP sensitivity is lowered by the binding of Ca^{2+}-CaM to a short region of the N-terminus, amino acids 61–90 (16). To remove Ca^{2+}-CaM regulation we have deleted this region. The resulting construct, the Δ61-90/C460W/E583M channel, allows the measurement of cAMP at the high end of the physiological range and is not modulated by Ca^{2+}-CaM (7). Data were fit with the Hill equation as described in **Subheading 3.2.** and are presented as the mean ± SD of at least three experiments. $I_{max}^{cGMP}/I_{max}^{cAMP}$ is the current induced by saturating cGMP divided by the current induced by saturating cAMP.

Table 3
Conditions for Loading Fura-2/AM in Various Cell Types[a]

	[Fura-2/AM] (μM)	[Pluronic] (μM)	Loading time (min)	References
Cultured cell lines				
HEK-293	4	0–8	30–40	*7,10*
C6-2B	2	4–8	20–30	*9,27*
GH4C1	4–5	0–8	60–70	*7,28*
PC12	2–4	0	60	*29*
A7r5	2	0–10	90–120	*30*
Primary cultures				
Vascular smooth muscle	0.1–2	0–0.8	20–60	*31,32*
Neonatal ventricular myocytes	2–5	0–8	60–120	*33*
Adult ventricular myocytes	4–10	0–1	20–40	*34,35*

[a]The ranges of loading conditions represent our experience as well as the prior experience of others (*see* **Note 5**). Pluronic is a detergent that enhances the loading of fura-2/AM in some cell types. For example, in our hands, pluronic had little effect on fura loading in HEK-293 cells but greatly enhanced loading in C6-2B cells. In all cases, loading was done at room temperature. The sources of the cells are the same as those in **Table 1.**

C460W/E583M or Δ61-90/C460W/E583M channels are used to monitor forskolin-stimulated cAMP accumulation in the presence and absence of a series of PDE inhibitors. Cells are loaded with fura-2 as described in **Subheading 3.3.1.** Initially, PDE inhibitor concentrations should be about fivefold higher than the IC_{50}. A table of published IC_{50}s of PDE inhibitors can be found in **ref. 7**. If warranted, more detailed dose–response relations may be used to estimate the K_I of the inhibitor in this assay (*see* **Subheading 3.4.**). Either vehicle or PDE inhibitors are added at 0 s and forskolin is added at 180 s (*see* **Notes 7** and **8**). The forskolin concentration should be high enough to trigger small responses (without saturating the channels).

A typical set of experiments (done serially in about 2 h) is shown in **Fig. 1**. This experiment could easily be conducted in a multiwell dish. In this example, little or no forskolin-induced Ca^{2+} influx was observed in the absence of PDE inhibitors, whereas in the presence of IBMX (10 or 100 μM) or the PDE type IV-specific inhibitor RO-20-1724 (10 μM) significant forskolin-induced Ca^{2+} influx was observed. Inhibitors specific to other PDE families did not affect forskolin-induced Ca^{2+} influx (*see* **Note 9**).

These techniques are readily applicable to standard high-throughput screening platforms such as the FLIPR[2] fluorometric imaging plate reader from Molecular Devices (www.moleculardevices.com). This system is designed for measurement of intracellular Ca^{2+} in 384-well dishes. Indeed, combining high throughput Ca^{2+}-screening systems with CNG channel-based cAMP sensors is well suited for screening drugs that regulate PDE activity as well as the activities of AC or GPCRs linked to the cAMP pathway.

3.3.3. Detection of cAMP in Single Cells

We use a similar approach to detect cAMP levels in single cells. In this assay cells are plated on laminin-coated cover slips.

1. After 24 h transfect the cells using the protocol described in **Subheading 3.1.** At 24–72 h postinfection the cells are ready to assay (*see* **Table 1**).
2. Add a fura-2/AM stock solution to 0.5 mL of loading buffer (final concentration of 4 μM) and briefly sonicate to disperse the fura-2/AM.
3. Attach cover slips to the perfusion chamber using Dow Corning high-vacuum grease. This forms a tight, solution-resistant seal between the chamber and the cover slips.
4. Directly pipet 0.5 mL of the loading buffer containing 4 μM fura-2/AM into the chamber, and place the cells in a darkened room to load (**Table 3**). After loading, cAMP activity in single cells can be monitored using a variety of commercially available fluorescent imaging systems.

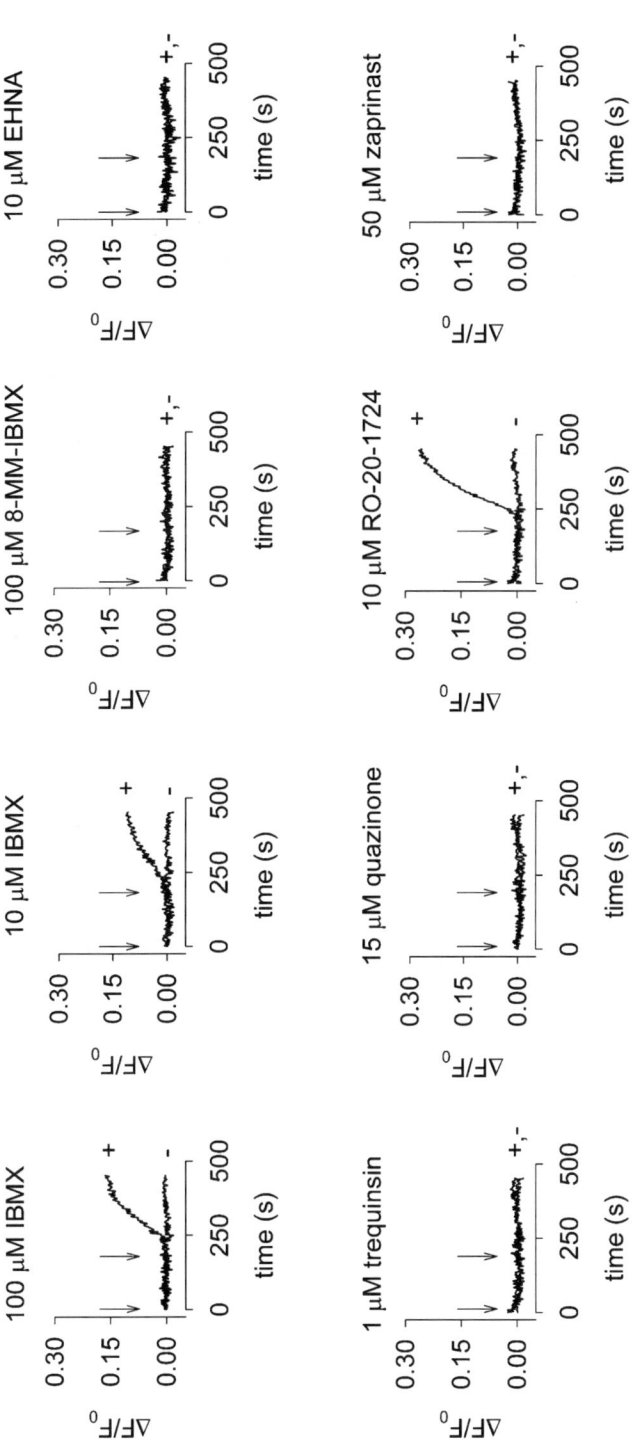

Fig. 1. Effects of different PDE inhibitors on forskolin-induced Ca^{2+} influx in HEK-293 cells. C460W/E583M channels were used to monitor cAMP accumulation in the presence and absence of PDE inhibitors. PDE inhibitors were added at 0 s and 1 μM forskolin at 180 s (indicated by arrows). Only the nonspecific PDE inhibitor IBMX and the PDE type IV-specific inhibitor RO-20-1724 influenced the time course of forskolin-induced Ca^{2+} influx. In general, there was little variability in responses within a single batch of cells. (Reproduced from **ref. 7** by copyright permission of Rockefeller University Press.)

3.3.4. Assay of PDE Type IV Activity in Vascular Smooth Muscle Cells

Vascular smooth muscle cells (VSMCs) were isolated from adult rats and transfected with the C460W/E583M CNG channel construct as described in **Subheading 3.1.** Forty-eight hours postinfection, the VSMCs were loaded with 4 μ*M* fura-2/AM for 1 h. The chamber was attached to a continuous perfusion system (~15-s switch time), and intracellular Ca^{2+} levels were monitored using a single-cell fluorescent imaging system. The preparation was perfused with 10 μ*M* forskolin (0.5 min), and intracellular Ca^{2+} levels rose to a steady plateau, indicating a marked increase in local cAMP concentration (**Fig. 2**). Subsequent perfusion with a solution containing 10 μ*M* forskolin and 5 μ*M* rolipram (5 min) triggered a sustained increase in intracellular Ca^{2+}, indicating a substantial reduction in PDE type IV activity. One micromolar nimodipine was present throughout the experiment to inhibit Ca^{2+} influx through voltage-gated Ca^{2+} channels. Little or no response was observed in control cells (cells not expressing CNG channels).

This example demonstrates that CNG channel-based sensors offer a convenient approach to monitoring the effects of PDE activity on cAMP signals near the surface membrane.

3.4. Estimation of K_I of PDE Inhibitors in Living Cells

In many cases, it is useful to examine the dose response of PDE inhibitors in order to estimate the K_I. To do this in living cells, a quantitative framework to assess the relationship among cAMP synthesis, hydrolysis, and redistribution throughout the cell must be developed. Thus, we adopted the following formula:

$$\frac{d[\text{cAMP}]}{dt} = C - \frac{V_{\max} \cdot [\text{cAMP}]}{K_m \left(1 + \dfrac{[I]}{K_I}\right) + [\text{cAMP}]} - k_f \cdot [\text{cAMP}] \tag{1}$$

in which C is the steady-state rate of cAMP synthesis by AC, $V_{\max}$ is the maximal rate of cAMP hydrolysis, K_m is the Michaelis constant for PDE, and k_f is the rate constant of cAMP flux out of the microdomain.

To estimate K_I for PDE inhibitors we made two assumptions. First, we assumed that at low levels of AC stimulation and PDE inhibition, the concentration of local cAMP is low and diffusion of cAMP out of the microdomains where it is produced is negligible. Second, we assumed that cAMP levels reach steady state shortly after AC stimulation (i.e., equal rates of synthesis and hydrolysis). With these assumptions, **Eq. 1** can be simplified to:

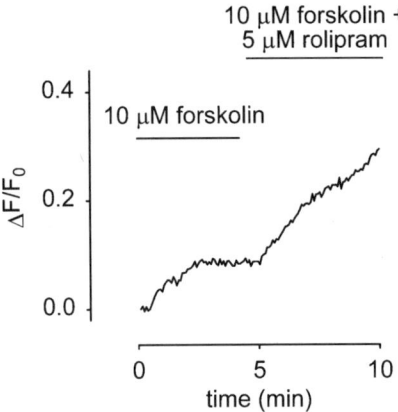

Fig. 2. Effect of specific PDE inhibitor on forskolin-induced cAMP accumulation in single VSMC from rat aorta. VSMCs were transfected with the C460W/E583M CNG channel construct and loaded with fura-2 as described in the text. Perfusion of this cell with 10 μ*M* forskolin (0.5 min) triggered a cAMP-induced increase in Ca^{2+} influx through CNG channels. Subsequent perfusion with 10 μ*M* forskolin and 5 μ*M* rolipram (5 min) triggered a sustained increase in Ca^{2+} influx through CNG channels, indicating a substantial reduction in PDE type IV activity.

$$[\text{cAMP}] = \frac{C \cdot K_m}{V_{\max} - C} \cdot \left(1 + \frac{[I]}{K_I}\right) \qquad (2)$$

Interestingly, this equation reveals that when the inhibitor concentration is equal to K_I the cAMP concentration is twice that in the absence of inhibitor.

At low cAMP concentrations ($<K_{1/2}$ for the channel), the cAMP concentration is proportional to the square root of the Ca^{2+} influx rate (the Hill coefficient for channel activation is approx 2). Thus, K_I can be estimated using a linear fit to the square root of the slopes of the fluorescence traces as a function of inhibitor concentration.

High-cAMP-affinity C460W/E583M channels were used to detect changes in cAMP concentration following pretreatment with PDE inhibitors (added at 0 s) and modest forskolin stimulation (0.5 μ*M* added at 180 s). IBMX, RO-20-1724, and rolipram were completely equilibrated across the plasma membrane of HEK-293 cells in less than 180 s (*see* **Notes 7** and **8**).

Dose-response relations for the three PDE inhibitors are shown in **Fig 3**. Based on fits to the data with **Eq. 2** (see insets in **Fig. 3A–C**), the K_I values were estimated to be 11 ± 2 μ*M* (IBMX), 0.13 ± 0.02 μ*M* (RO-20-1724), and 0.07 ± 0.02 μ*M* (rolipram), *n* = 4, which are consistent with published IC_{50}

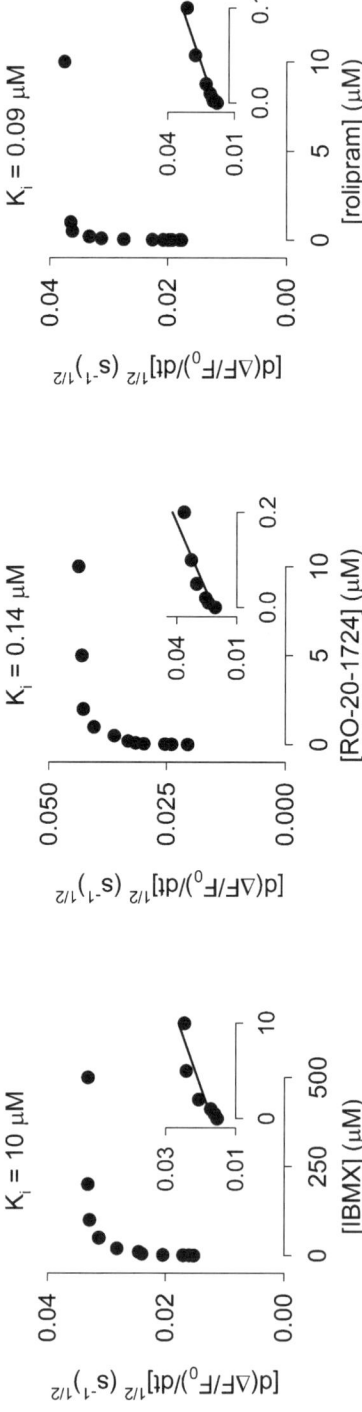

Fig. 3. Estimates of K_I for three different PDE inhibitors reveal that PDE type IV regulates local cAMP levels in HEK-293 cells. These data were evaluated from experiments similar to those depicted in **Fig. 1**. The linear fits used to estimate K_I values are shown in the insets (*see* **Subheading 3.4.** for explanation). The K_Is of the PDE inhibitors were 10 μM (IBMX), 0.14 μM (RO-20-1724), and 0.09 μM (rolipram) for the experiments shown. (Reproduced from **ref. 7** by copyright permission of Rockefeller University Press.)

values from in vitro experiments. These data demonstrate that with this approach the K_I of PDE inhibitors can be estimated in intact cells (*see* **Note 10**).

4. Notes

1. This approach was inspired by the field of retinal phototransduction, the best-studied second-messenger signaling system, in which elegant biochemical studies have been complemented by real-time measurements of cGMP signals using endogenous CNG channels *(19–21)*. CNG channels are directly opened by the binding of cyclic nucleotides. They were discovered in retinal photoreceptor cells and olfactory receptor neurons, where they generate the electrical response to light and odorants. The native retinal channel is cGMP specific, whereas the native olfactory channel is equally sensitive to cAMP and cGMP. Native CNG channels consist of A- and B-subunits, both of which bind cyclic nucleotides, although most A-subunits form functional channels on their own. We have modified an olfactory channel A-subunit (CNGA2) to improve its sensitivity and selectivity for cAMP.

2. In this chapter, we describe the conditions for expression of CNG channels using adenovirus in HEK-293 cells. We have successfully used CNG channels to monitor cAMP signals in several other cell types. The infection conditions are provided in **Table 1**, and the extracellular medium was modified as appropriate; for examples, *see* **refs. *7*** and ***10***.

3. Higher hydroxyurea concentrations (≥10 m*M*) are required to completely inhibit viral replication. However, such high concentrations affect cellular adenosine triphosphate (ATP) pools and can significantly alter second-messenger signaling. We have determined that 1 to 2 m*M* hydroxyurea sufficiently inhibits viral replication without significantly altering ATP pools in HEK-293 cells. As a further control for potential effects of hydroxyurea, the experiments described here can be repeated with CNG channels expressed using other transfection techniques such as $Ca_3^{2+}(PO_4)_2$ precipitation *(22)*. We have done these controls and found that the lower hydroxyurea concentrations have little or no effect on cAMP signals *(7,10,11)*. Note that we use adenovirus to transfect cells because it allows relatively uniform CNG channel expression in a majority (≥70%) of cells.

4. For the screening of PDE activity, it is not necessary to determine the affinity of the different CNG channel-based sensors in a particular cell type. One can simply rely on previously published affinities to obtain an estimate for the range of the cAMP signals being monitored. However, excised membrane patches are also useful for determining whether novel compounds influence CNG channel activity *(7,10,11)*. Note also that the description of electrophysiological techniques used to assess the ligand sensitivity of CNG channels is sufficient for investigators familiar with electrophysiological techniques. Those not familiar with these techniques or the equipment may wish to refer to **refs. *23–25***.

5. Other Ca^{2+} indicators such as fluo-3 or indo-1 are equally appropriate for these measurements. Available equipment and accompanying filter sets dictate which of these probes is most appropriate. In our experience, the AM esters of these dyes load with similar efficacy in a given cell type.

6. In this assay it is necessary to separate Ca^{2+} influx through CNG channels from other mechanisms that increase cellular Ca^{2+} levels (e.g., L-type Ca^{2+} channels or Ca^{2+}-induced Ca^{2+} release). A reasonable estimate of cAMP-induced activation of CNG channels can be obtained by comparing the responses of CNG channel-expressing and control cells to PDE inhibitors and forskolin (or other AC agonists). This should suffice if a quantitative interpretation of the time course of Ca^{2+} influx through CNG channels is not required (e.g., for high-throughput screening). If quantitative interpretation of the time course is required, it may be necessary to inhibit other mechanisms that increase cellular Ca^{2+} levels. Nimodipine (1–10 μ*M*) can be used to block voltage-gated Ca^{2+} channels without blocking CNG channels *(7)*. Agonist-induced release from Ca^{2+} stores can be inhibited by the addition of compounds such as thapsigargin or ryanodine prior to the assay. Ca^{2+} influx through CNGA2 channels can be distinguished from other mechanisms that raise intracellular Ca^{2+} by blocking CNGA2 channels with pseudechetoxin *(26)*.

7. The rate at which PDE inhibitors cross the plasma membrane can vary considerably, depending on both the inhibitor used and the cell type. We found that in this assay 180 s is enough time to equilibrate IBMX, rolipram, and RO-20-1724 across the surface membrane of HEK-293 cells. This can be readily tested by the addition of a low forskolin concentration (e.g., 1 μ*M*) at various times following the addition of a subsaturating concentration of a PDE inhibitor (e.g., 10 μ*M* IBMX).

8. To extract quantitative information from these experiments, it is necessary that the mixing or perfusion time be considerably faster than the process being measured. The perfusion time in our single-cell imaging system is approx 15 s. The mixing time in our cell population measurements (stirred cuvet) is on the order of 5 s. Note that little or no kinetic information can be obtained if additions are made into unstirred cuvets or bath chambers that are not well mixed (e.g., at best one can assess only if a response occurred).

9. Two known PDE types (VIII and IX) are insensitive to IBMX as well as most other PDE inhibitors. Unfortunately, dipyridamole, the inhibitor to which the cAMP-specific PDE type VIII is most sensitive, cannot be used in this assay because it fluoresces.

10. In these studies, we are concerned with relative Ca^{2+} influx rates through CNG channels, which report changes in cAMP levels. This requires that fura-2 detect a fixed proportion of the entering Ca^{2+} in a given experiment. For this to be true, the concentration of unbound fura-2 should not change appreciably on Ca^{2+} binding, and cellular Ca^{2+} buffers that are not overwhelmed by fura-2 should also not be significantly depleted by Ca^{2+} binding. For experiments in which quantitative information was extracted (e.g., the in vivo estimate of K_I for PDE inhibitors), it is necessary to work at low levels of Ca^{2+} influx *(7)*. For example, the fluorescence changes used to estimate K_I (**Fig. 3**) were generally less than 10% of the saturated fura-2 response, indicating that unbound fura-2 was predominant.

References

1. Drummond, G. I. and Perrot-Yee, S. (1961) Enzymatic hydrolysis of adenosine 3′,5′-phosphoric acid. *J. Biol. Chem.* **236**, 1126–1129.

2. Beavo, J. A. (1988) Multiple isozymes of cyclic nucleotide phosphodiesterase. *Adv. Second Messenger Phosphoprotein Res.* **22,** 1–38.
3. Beavo, J. A. (1995) Cyclic nucleotide phosphodiesterases: functional implications of multiple isoforms. *Physiol. Rev.* **75,** 725–748.
4. Conti, M. and Jin, S. L. (1999) The molecular biology of cyclic nucleotide phosphodiesterases. *Prog. Nucleic Acid Res. Mol. Biol.* **63,** 1–38.
5. Mehats, C., Andersen, C. B., Filopanti, M., Jin, S. L., and Conti, M. (2002) Cyclic nucleotide phosphodiesterases and their role in endocrine cell signaling. *Trends Endocrinol. Metab.* **13,** 29–35.
6. Houslay, M. D., Sullivan, M., and Bolger, G. B. (1998) The multienzyme PDE4 cyclic adenosine monophosphate–specific phosphodiesterase family: intracellular targeting, regulation, and selective inhibition by compounds exerting anti-inflammatory and antidepressant actions. *Adv. Pharmacol.* **44,** 225–342.
7. Rich, T. C., Tse, T. E., Rohan, J. G., Schaack, J., and Karpen, J. W. (2001) In vivo assessment of local phosphodiesterase activity using tailored cyclic nucleotide–gated channels as cAMP sensors. *J. Gen. Physiol.* **118,** 63–77.
8. Rich, T. C. and Karpen, J. W. (2002) Cyclic AMP sensors in living cells: what signals can they actually measure? *Ann. Biomed. Eng.* **30,** 1088–1099.
9. Fagan, K. A., Schaack, J., Zweifach, A., and Cooper, D. M. F. (2001) Adenovirus encoded cyclic nucleotide–gated channels: a new methodology for monitoring cAMP in living cells. *FEBS Lett.* **500,** 85–90.
10. Rich, T. C., Fagan, K. A., Nakata, H., Schaack, J., Cooper, D. M. F., and Karpen, J. W. (2000) Cyclic nucleotide–gated channels colocalize with adenylyl cyclase in regions of restricted cAMP diffusion. *J. Gen. Physiol.* **116,** 147–161.
11. Rich, T. C., Fagan, K. A., Tse, T. E., Schaack, J., Cooper, D. M. F., and Karpen, J. W. (2001) A uniform extracellular stimulus triggers distinct cAMP signals in different compartments of a simple cell. *Proc. Natl. Acad. Sci. USA* **98,** 13,049–13,054.
12. Varnum, M. D., Black, K. D., and Zagotta, W. N. (1995) Molecular mechanism for ligand discrimination of cyclic nucleotide–gated channels. *Neuron* **15,** 619–625.
13. Brown, R. L., Snow, S. D., and Haley, T. L. (1998) Movement of gating machinery during the activation of rod cyclic nucleotide–gated channels. *Biophys. J.* **75,** 825–833.
14. Gordon, S. E., Varnum, M. D., and Zagotta, W. N. (1997) Direct interaction between amino- and carboxyl-terminal domains of cyclic nucleotide–gated channels. *Neuron* **19,** 431–441.
15. Zong, X., Zucker, H., Hofmann, F., and Biel, M. (1998) Three amino acids in the C-linker are major determinants of gating in cyclic nucleotide–gated channels. *EMBO J.* **17,** 353–362.
16. Liu, M., Chen, T. Y., Ahamed, B., Li, J., and Yau, K.-W. (1994) Calcium-calmodulin modulation of the olfactory cyclic nucleotide–gated cation channel. *Science* **266,** 1348–1354.
17. Frings, S., Seifert, R., Godde, M., and Kaupp, U. B. (1995) Profoundly different calcium permeation and blockage determine the specific function of distinct cyclic nucleotide–gated channels. *Neuron* **15,** 169–179.

18. Grynkiewicz, G., Poenie, M., and Tsien, R. Y. (1985) A new generation of Ca^{2+} indicators with greatly improved fluorescence properties. *J. Biol. Chem.* **260,** 3440–3450.
19. Stryer, L. (1991) Visual excitation and recovery. *J. Biol. Chem.* **26,** 10,711–10,714.
20. Yau, K.-W. (1994) Phototransduction mechanism in retinal rods and cones: The Friedenwald Lecture. *Invest. Ophthalmol. Vis. Sci.* **35,** 9–32.
21. Molday, R. S. (1998) Photoreceptor membrane proteins, phototransduction, and retinal degenerative diseases: The Friedenwald Lecture. Invest. *Ophthalmol. Vis. Sci.* **39,** 2493–2513.
22. Jordan, M., Schallhorn, A., and Wurm, F. M. (1996) Transfecting mammalian cells: optimization of critical parameters affecting calcium-phosphate precipitate formation. *Nucleic Acids Res.* **24,** 596–601.
23. Sakmann, B. and Neher, E. (1995) *Single Channel Recording*, Plenum, New York.
24. Gordon, S. E. (2000) Using site-directed mutagenesis and modification of cGMP-gated ion channels expressed in Xenopus oocytes to study the structural basis of their functional properties. *Methods Enzymol.* **316,** 772–785.
25. Hille, B. (2001) *Ionic Channels of Excitable Membranes*, Sinauer, Sunderland, MA.
26. Brown, R. L., Haley, T. L., West, K. A., and Crabb, J. W. (1999) Pseudechetoxin: a peptide blocker of cyclic nucleotide–gated channels. *Proc. Natl. Acad. Sci. USA* **96,** 754–759.
27. Fagan, K. A., Mahey, R., and Cooper, D. M. F. (1996) Functional co-localization of transfected Ca^{2+}-stimulated adenylyl cyclases with capacitative Ca^{2+} entry sites. *J. Biol. Chem.* **271,** 12,438–12,444.
28. Villalobos, C. and Garcia-Sancho, J. (1996) Caffeine-induced oscillations of cytosolic Ca^{2+} in GH3 pituitary cells are not due to Ca^{2+} release from intracellular stores but to enhanced Ca^{2+} influx through voltage-gated Ca^{2+} channels. *Pflügers Arch. Eur. J. Physiol.* **431,** 371–378.
29. Taylor, S. C., Green, K. N., Carpenter, E., and Peers, C. (2000) Protein kinase C evokes quantal catecholamine release from PC12 cells via activation of L-type Ca^{2+} channels. *J. Biol. Chem.* **275,** 26,786–26,791.
30. Byron, K. L. and Taylor, C. W. (1993) Spontaneous Ca^{2+} spiking in a vascular smooth muscle cell line is independent of the release of intracellular Ca^{2+} stores. *J. Biol. Chem.* **268,** 6945–6952.
31. Purdy, K. E. and Arendshorst, W. J. (1999) Prostaglandins buffer ANG II-mediated increases in cytosolic calcium in preglomerular VSMC. *Am. J. Physiol.* **277,** F850–F858.
32. Olschewski, A., Hong, Z., Nelson, D. P., and Weir, E. K. (2002) Graded response of K+ current, membrane potential, and $[Ca^{2+}]_i$ to hypoxia in pulmonary arterial smooth muscle. *Am. J. Physiol.* **283,** L1143–L1150.
33. Eble, D. M., Qi, M., Waldschmidt, S., Lucchesi, P. A., Byron, K. L., and Samarel, A. M. (1998) Contractile activity is required for sarcomeric assembly in phenylephrine-induced cardiac myocyte hypertrophy. *Am. J. Physiol.* **274,** C1226–C1237.

34. Wang, H. X., Lau, S. Y., Huang, S. J., Kwan, C. Y., and Wong, T. M. (1997) Cobra venom cardiotoxin induces perturbations of cytosolic calcium homeostasis and hypercontracture in adult rat ventricular myocytes. *J. Mol. Cell. Cardiol.* **29,** 2759–2770.

35. Yoshida, H., Tanonaka, K., Miyamoto, Y., Abe, T., Takahashi, M., Anand-Srivastava, M. B., and Takeo, S. (2001) Characterization of cardiac myocyte and tissue beta-adrenergic signal transduction in rats with heart failure. *Cardiovasc. Res.* **50,** 34–45.

5

Assessment of Phosphodiesterase Isozyme Contribution in Cell and Tissue Extracts

Thérèse Keravis, Rima Thaseldar-Roumié, and Claire Lugnier

Summary

Cyclic nucleotide phosphodiesterases (PDEs), which are ubiquitously distributed in mammalian tissues, play a major role in cell signaling by hydrolyzing cyclic adenosine $3',5'$-monophosphate and cyclic guanosine $3',5'$-monophosphate. Owing to their diversity, which allows specific distribution at the cellular and subcellular level, PDEs can selectively regulate various cellular functions. We present here a convenient and sensitive radioenzymatic assay for characterizing and determining the contribution of the various PDE families in cell and tissue extracts. This assay is based on the knowledge and use of chosen PDE family-specific inhibitors in order to determine the distinct PDE isozyme contribution in the overall cyclic nucleotide hydrolyzing activity. It can be used to characterize total, cytosolic, and membrane-associated PDE activities, as well as PDEs associated with purified subcellular structures. This approach is useful for comparing data of control and treated extracts and is therefore quite valuable for viewing the PDE status in different physiopathological conditions.

Key Words

Cyclic nucleotide phosphodiesterase activity; tissues; inhibitor; resin; zaprinast; nimodipine.

1. Introduction

Cyclic nucleotide phosphodiesterases (PDEs), which inactivate cyclic adenosine $3',5'$-monophosphate (cAMP) and cyclic guanosine $3',5'$-monophosphate (cGMP), play a major role in the control of cell signaling. They have been classified into 11 families (PDE1 to PDE11) according to their gene, substrate specificity, regulation, and pharmacology (for review, *see* refs. *1–3*). PDE1, PDE2, PDE3, PDE4, PDE7, PDE8, PDE10, and PDE11 hydrolyze cAMP. PDE1, PDE2, PDE3, PDE5, PDE6, PDE9, PDE10, and PDE11 hydrolyze cGMP. Their diversity and their specific distribution at cellular and subcellular levels allow a

From: *Methods in Molecular Biology, vol. 307: Phosphodiesterase Methods and Protocols*
Edited by: C. Lugnier © Humana Press Inc., Totowa, NJ

fine regulation of different cellular functions. Among these families, PDE1 to PDE5 are the most representative and the best-characterized isozymes present in cells and tissues. As shown in **Table 1**, PDE1 is activated by Ca^{2+}-calmodulin $(Ca^{2+}$-CaM$)$. PDE2 is allosterically activated by cGMP. PDE3, which mainly hydrolyzes cAMP, is competitively inhibited by cGMP, because it hydrolyzes cGMP at a lower rate than cAMP. PDE4 specifically hydrolyzes cAMP and is insensitive to cGMP. PDE5 specifically hyrolyzes cGMP *(4–6)*. Knowledge of the participation of these various PDE isozymes in cyclic nucleotide PDE activity will help in understanding the regulation of cAMP and cGMP levels in tissues or cells in response to hormonal stimulation or in physiopathological states. Herein, we describe a convenient radioenzymatic approach that determines the nature and proportion of a given PDE isozyme activity by using its specific inhibitor. This approach can be used to evaluate PDE activity in cytosolic and membrane fractions as well as purified subcellular structures of normal and pathological tissues, thus contributing to the ascertainment of the functional role of PDEs in various physiopathological states.

2. Materials

1. [2,8-^{3}H] adenosine 3′,5′-cyclic phosphate, ammonium salt, 30–50 Ci/mmol (no. TRK498, Amersham).
2. [8-^{3}H] guanosine 3′,5′-cyclic phosphate, ammonium salt, 5–25 Ci/mmol (no. TRK392, Amersham).
3. Thin-layer chromatography (TLC) plates (20 × 20 cm), Silicagel 60 F_{254} (no. 1.05715, Merck).
4. TLC chromatography solvent: 70 isopropanol/15 NH_4OH/15 H_2O (v/v/v).
5. QAE-Sephadex™ A-25 (no. 17-0190-02, Amersham).
6. Econo-column, polypropylene (0.8 × 4 cm) (no. 7311550, Bio-Rad, Hercules, CA).
7. Buffer A: 20 mM Tris-HCl, pH 7.5; 2.0 mM Mg acetate; 5.0 mM EGTA, 1.0 mM dithiothreitol, 10 μg/mL of leupeptin, 10 μg/mL of soya trypsin inhibitor, 2000 U/mL of aprotinin, and 0.33 mM Pefabloc.
8. Formate solution: 0.4 M $HCOONH_4$ to be diluted 20-fold and adjusted to pH 7.4 just before use.
9. Tris-BSA: 20 mM Tris-HCl, pH 7.5, 2.5 mg of bovine serum albumin (BSA)/mL.
10. Tris-Mg-Ca: 400 mM Tris-HCl, pH 7.5, 20 mM Mg acetate, 0.1 mM $CaCl_2$.
11. EGTA-Mg: 10 mM EGTA, 10 mM Mg acetate, to be adjusted to pH 7.5.
12. Mg-CaM: 10 mM Mg acetate, 200 nM purified CaM *(7)*.
13. SV: 5 mg of Crotalus Atrox snake venom/mL (no. V700-1G, Sigma-Aldrich, St. Louis, MO).
14. PDE-stop: 100 mM Tris-HCl, pH 7.5, 50 mM EDTA, 30 mM theophylline, 5 mM cAMP, 5 mM cGMP.
15. SV-stop: 15 mM EDTA, 0.1 mM adenosine, 0.1 mM guanosine, to be adjusted to pH 6.5.

Table 1
Classification of PDE Family

PDE family	Substrate	Property	Specific inhibitors
PDE1	cAMP, cGMP	Ca-CaM-activated	Nimodipine (13,14)
PDE2	cAMP, cGMP	cGMP activated	EHNA (15)
PDE3	cAMP, cGMP	cGMP inhibited	Cilostamide (16,17), milrinone (17,18)
PDE4	cAMP	cGMP insensitive	Rolipram (16,19), RO-20-1724 (16,19)
PDE5	cGMP	PKA/PKG phosphorylated[a]	Zaprinast (16,19), DMPPO (20), sildenafil (21)
PDE6	cGMP	Transducin activated	Zaprinast (6), DMPPO (20), sildenafil (21)
PDE7	cAMP	Rolipram insensitive	BMS-586353 (22)
PDE8	cAMP	IBMX insensitive	n/a
PDE9	cGMP	IBMX insensitive	n/a
PDE10	cAMP, cGMP	n/a	Papaverine (23)
PDE11	cAMP, cGMP	n/a	n/a

[a]PKA/PKG, protein kinase A/protein kinase G; n/a, not available.

16. PDE inhibitors: cilostamide, nimodipine, erythro-9-(2-hydroxy-3-nonyl) adenine) (EHNA), rolipram, zaprinast, and 3-isobutyl-1-methylxanthine (IBMX) from Sigma-Aldrich.
17. Ultra-Turrax homogenizer.
18. Mortar and pestle.
19. Glass-glass potter homogenizer.
20. Well-regulated 30°C water bath.
21. β Liquid scintillation counter.
22. Scintillation fluid: Ultima Gold™ XR (no. 6013119, Perkin Elmer®).
23. Millipore water, 18 MΩ.

3. Methods

3.1. Tissue Homogenization

1. Hand powder fresh or frozen (stored at −80°C) weighted tissues in a mortar with a pestle under liquid nitrogen. Collect the resulting powder onto aluminum paper.
2. Homogenize the powdered tissue in buffer A (5 v/w; *see* **Note 1**) first with an ultra-turrax (six times for 10 s at 2000 rpm) and thereafter with a glass–glass potter homogenizer.
3. Aliquot (100–500 µL) the homogenate in Eppendorf tubes, and store at −80°C for protein determination according to Lowry et al. *(8)* and for PDE assay (*see* **Subheading 3.2.**).

3.2. PDE Assay

PDE activities are measured by a two-step radioenzymatic assay as previously described *(9–11)* at a substrate concentration of 1 µM cAMP or cGMP in the presence of 15,000 cpm [³H]-cAMP or [³H]-cGMP as a tracer, respectively (**Fig. 1**). During the first incubation (30°C for 30 min) performed in a final volume of 250 µL, cyclic nucleotides are hydrolyzed into their respective nucleotide monophosphates. This incubation is stopped by the addition of an excess of cAMP and cGMP in the presence of EDTA and a nonselective PDE inhibitor (theophylline) that inactivates PDE activity. The second incubation performed in the presence of an excess of 5′-nucleotidase (snake venom) results in the formation of adenosine or guanosine. This second incubation is stopped by the addition of an excess of adenosine and guanosine in the presence of EDTA. The enzymatic reaction products are separated by anion-exchange column chromatography on QAE-Sephadex A-25 formate *(12)*. The amount of resulting [³H]-nucleoside is determined by liquid scintillation.

3.2.1. Purification of Radiolabeled cAMP or cGMP

[³H]-cAMP and [³H]-cGMP are purified by TLC on Silicagel 60 F_{254} TLC plates.

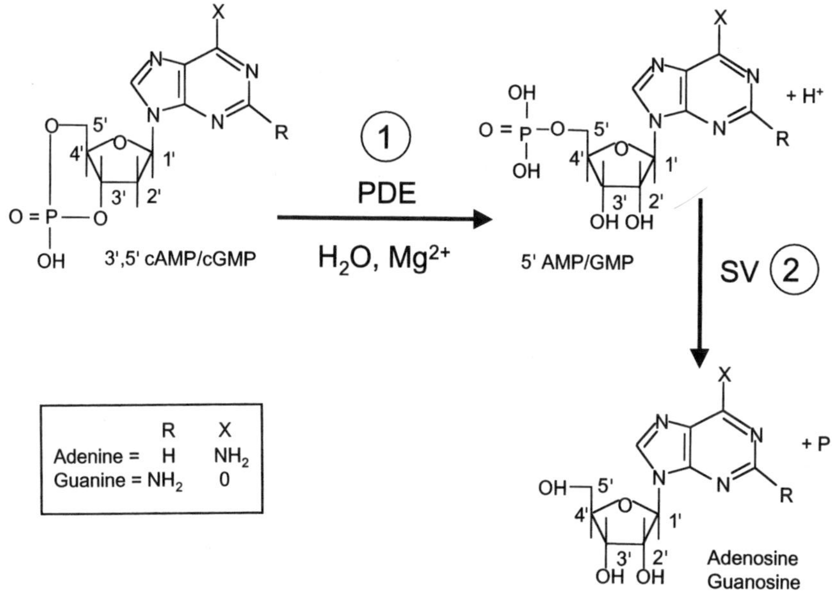

Fig. 1. Principle of two-step PDE assay. In the first step, cyclic nucleotides are converted into 5′ nucleotides by PDE. In the second step, an excess of 5′ nucleotidase (snake venom) produces the corresponding nucleosides that are recovered by QAE-Sephadex A-25 column chromatography.

1. Along a horizontal line drawn 2 cm from the bottom of a TLC plate, apply 2 μL of 10 m*M* unlabeled cyclic nucleotides as migration markers.
2. Separately spot 80 μL of [³H]-cAMP and [³H]-cGMP by successive additions of 10 μL under drying air.
3. Place the TLC plate in a vertical developing tank containing TLC chromatography solvent and develop for approx 4 h, until the solvent front is 4 cm beneath the top of the plate. Then dry the plate.
4. Under an ultraviolet light (254 nm), locate the positions of the cyclic nucleotide standards, as well as those of the labeled cyclic nucleotides, and surround with pencil. Using a razor blade, scrape off the corresponding region for each labeled cyclic nucleotide onto aluminum paper, and transfer the silicagel powder into a 1.5-mL Eppendorf tube.
5. To extract the cyclic nucleotide, add 1 mL of 50% ethanol to the tube, thoroughly mix, and centrifuge to settle down the silicagel; repeat this extraction a second time with 0.4 mL. Then combine the supernatants.
6. Count a sample (10 μL) of the purified labeled cyclic nucleotide. Adjust the supernatant with 50% ethanol to give 15,000 cpm/μL and store at −20°C (*see* **Note 2**).

3.2.2. QAE-Sephadex A-25 Formate Anion-Exchange Resin (12)

The use of anion-exchange resin to recover [^{3}H]-adenosine or [^{3}H]-guanosine is critical for the accuracy of the assay. This method, contrary to that described by Thompson et al. *(9)*, allows recovering and regenerating the resin, which can then be used over a long period of time. Because QAE-Sephadex A-25 is expensive, the ability to recycle is quite valuable.

3.2.2.1. PREPARATION AND EQUILIBRATION OF RESIN

1. Allow 250 g of QAE-Sephadex A-25 (in the chloride form) to swell in 5 L of Millipore water overnight at 4°C.
2. Convert the resin from the OH⁻ form to the formate form using **steps 3–8** in **Subheading 3.2.2.2.**
3. Store the resin at +4°C as a 50% slurry.

3.2.2.2. REGENERATION OF RESIN

Used resin collected from the assay columns is stored at +4°C and regenerated in a large glass column (35 × 8 cm) with a sintered-glass bottom by successive treatments:

1. Treat with 0.1 *N* HCl until the pH of the effluent is 1.0 to 2.0.
2. Treat with Millipore water until the pH of the effluent is 5.0 to 6.0.
3. Treat with 0.2 *N* NaOH until the effluent is Cl⁻ free, as checked with AgNO$_3$ under acidic conditions.
4. Treat with Millipore water until the pH of the effluent is 6.0 to 7.0.
5. Treat with 0.2 *N* formic acid until the pH of the effluent is about 2.5.
6. Treat with 2 *M* ammonium formate until the pH of the effluent is 6.4 to 6.7.
7. Treat with Millipore water until the pH of the effluent is about 4.0.
8. Suspend the resin in 100 m*M* ammonium formate and store at 4°C.

3.2.3. Two-Step PDE Assay

3.2.3.1. ENZYMATIC REACTIONS

Adding reactive constituents sequentially by 25- or 50-µL fractions constitutes the 250-µL final assay medium, allowing maximum flexibility and convenience for several assay purposes. During this procedure test tubes are kept in an ice-water bath.

1. Add 25 µL of Millipore water, solvent, or drug (made up soluble with solvent such as dimethyl sulfoxide [DMSO] or ethanol) at 10X concentration to be tested.
2. Add 25 µL of Millipore water or additional drug.
3. Add 25 µL of Tris-Mg-Ca.
4. Add 25 µL of Mg-EGTA or Mg-CaM.
5. Add 50 µL of Tris-BSA.

6. 50 µL of tissue homogenate (PDE extract) diluted with Tris-BSA to give about 15% of the substrate hydrolyzed in control conditions (*see* **Note 3**).
7. Vortex the test tubes.
8. Initiate the enzymatic reaction by adding 50 µL of substrate solution (25 µL of 10 µ*M* cAMP or cGMP + 24 µL of Millipore water + 1 µL of purified [³H]-cAMP or [³H]-cGMP, 15,000 cpm), gently vortex, and incubate at 30°C.
9. Stop the reaction by adding 25 µL of PDE-stop, and after vortexing the test tubes, return them to the ice-water bath.
10. Initiate the nucleotidase reaction by adding 20 µL of SV; vortex the tubes and incubate for 20 min at 30°C.
11. Stop the reaction by adding 300 µL of SV-stop.

3.2.3.2. QAE-SEPHADEX FORMATE ANION-EXCHANGE COLUMN CHROMATOGRAPHY

This procedure allows separation of the dephosphorylated products (adenosine or guanosine) from the nonhydrolyzed cyclic nucleotide *(12)*.

1. For each test tube, fill a chromatography column with 2 mL of 50% resin slurry (*see* **Subheading 3.2.2.**) in order to get 1 mL of settled resin.
2. Equilibrate the columns by adding 3 mL of formate solution.
3. By using a homemade rack (**Fig. 2**), place each column above a scintillation vial.
4. Pour the test tubes onto their corresponding columns.
5. Rinse the test tubes with 1.5 mL of formate solution and pour again onto the columns.
6. Add 7 mL of scintillation fluid to each vial, cap the vials, vigorously vortex, and then place in a β liquid scintillation counter.
7. Recover the resin from the columns and store at +4°C until regeneration; wash the columns exhaustively with hot tap water.

3.2.4. Expression of Data

To determine PDE activity, four test tubes without PDE extract (50 µL of Tris-BSA) are run in each assay procedure until **step 11** in **Subheading 3.2.3.1.** Then the two first tubes, used to determine the total radioactivity (R_t), are directly poured into scintillation vials and 1.5 mL of formate is added; the two other tubes, used to determine blank assay (R_o), are treated similarly to the other test tubes.

The percentage of cyclic nucleotide phosphate (CNP) hydrolyzed is calculated as follows:

$$\% \text{ CNP hydrolyzed} = R_x - R_o/R_T - R_o$$

in which R_x is the cpm of the sample, R_o is the cpm of the blank assay, and R_T is the cpm of the total radioactivity.

The PDE activity of the sample is calculated as follows:

$$\text{PDE activity (pmol/[min•mg])} = \frac{(\% \text{ CNP})(0.25 \times 10^{-3})(1 \times 10^{-6}M)}{t \times (0.050C)}$$

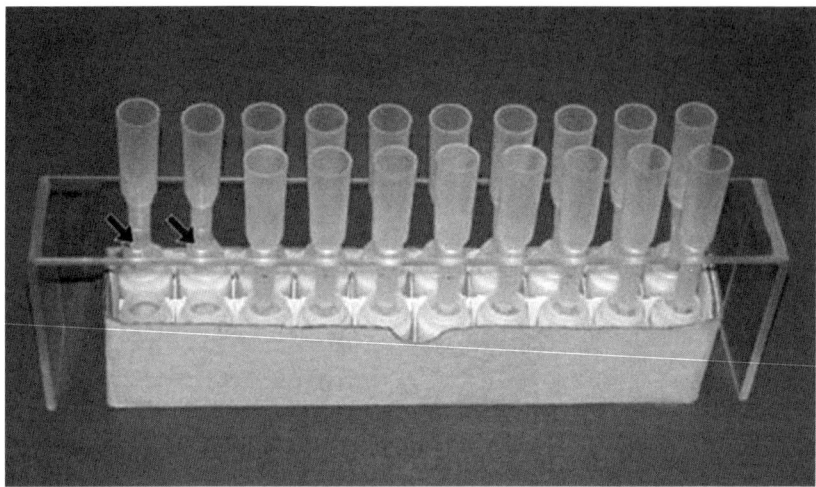

Fig. 2. Column chromatography with direct elution of each tube's assay medium in scintillation vials. Note the absence of columns in the two first holes of the first row of the column rack; the corresponding scintillation vials are used for determination of total radioactivity.

in which *t* is the incubation time of the first enzymatic reaction (min), and *C* is the protein concentration (mg/mL).

3.3. Characterization of PDE Family Contribution

To determine the contribution of one PDE isozyme in total cAMP or cGMP hydrolytic activity in tissue, both total PDE hydrolytic activity and residual PDE activities in the presence of specific inhibitors are determined in the same assay. A PDE isozyme family activity is obtained by determining the activity that is sensitive to its corresponding specific inhibitor (*see* **Table 1**).

cAMP-hydrolyzing PDE isozymes are determined with 1 µ*M* cAMP as substrate in the presence of 1 m*M* EGTA by using 10 µ*M* rolipram for PDE4 and 1 µ*M* cilostamide for PDE3 (*see* **Note 4**).

The proportion of cGMP-hydrolyzing PDE isozymes is determined with 1 µ*M* cGMP as substrate in the presence of 1 m*M* EGTA by using 20 µ*M* EHNA for PDE2, 10 µ*M* nimodipine for PDE1, 1 µ*M* cilostamide for PDE3, and 10 µ*M* zaprinast (or 0.1 µ*M* 1,3-dimethyl-6-(2-propoxy-5-methanesulfonylamido(phenyl)pyrazol[3,4-d]pyrimidin-4-(5H)-one [DMPPO] or sildenafil) for PDE5 (*see* **Notes 5** and **6**).

In the same assay, the solvent (for total PDE activity) and the various inhibitors are added in the assay as previously indicated (*see* **step 1** in **Subheading 3.2.3.1.**). The whole assay procedure must be run as stated in **Subheading 3.**

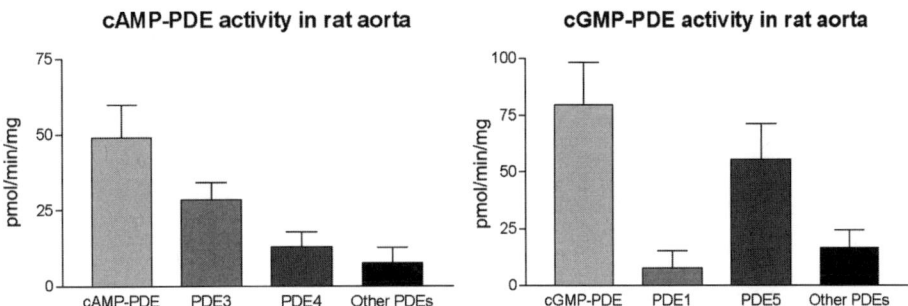

Fig. 3. Characterization of PDE isozyme activities in presence of EGTA in whole rat aorta (unpublished data). PDE1 and PDE5 activities were determined as cGMP-PDE activities sensitive to 10 μ*M* nimodipine and 0.1 μ*M* DMPPO, respectively. PDE3 and PDE4 activities were determined as cAMP-PDE activities sensitive to 1 μ*M* cilostamide and 10 μ*M* rolipram, respectively. The remaining cAMP-PDE activity (Other PDEs) is mainly owing to PDE1 activity, which hydrolyzes similarly cAMP and cGMP in its basal state *(13)*. The remaining cGMP-PDE activity (Other PDEs) is mainly owing to PDE3.

The PDE activity of one isozyme family (PDEx) is expressed as follows (*see* **Note 6**):

$$PDEx \ (pmol/[min \bullet mg]) = PDE_T - PDEn$$

in which PDE_T is the total PDE activity, and PDEn is the activity assessed in the presence of specific PDEx inhibitor. The characterization of PDE isozyme activity in whole rat aorta (**Fig. 3**) is given as an example.

4. Notes

1. To study subcellular distribution of PDE isozymes (cytosolic and membrane-bound PDEs or associated to isolated subcellular structures), tissue powder must be homogenized in an isotonic buffer, i.e., buffer A with 0.25 *M* sucrose. To study PDE isozymes in cells, fresh or frozen (stored at –80°C), cell pellets are homogenized in a minimal volume of 500 μL and sonicated six times for 10 s at +4°C.
2. Purified tritium-labeled cAMP and cGMP can be stored for 2 mo at –20°C. When blank values represent more than 1% of total radioactivity, labeled cyclic nucleotides must be purified again.
3. Product accumulation should be determined to be linear with time and with enzyme dilution under all conditions studied.
4. For technical convenience, in this protocol we have chosen to determine PDE2 participation by evaluating the effect of EHNA toward its cGMP activity. This may directly reflect PDE2 activity in its stimulated state. It should also be possible to determine the participation of PDE2 in total cAMP hydrolytic activity. In this

case, since EHNA is effective only on cGMP-stimulated PDE2, it is necessary to perform the assay in the presence of cilostamide (1 μM) + cGMP (5 μM) + EHNA (20 μM). This must be done under conditions in which the final concentration of DMSO will not exceed 1%. The difference between the hydrolytic activity in the presence of 1 μM cilostamide and the activity obtained in the presence of the PDE inhibitor cocktail will give the cAMP-PDE activity for PDE2 in its activated state. Usually, PDE3 and PDE4 activities represent more than 80% of the total cAMP-PDE activity. Residual cAMP-PDE activity (R) could be assigned to PDE1, PDE2, PDE7, PDE8, PDE10, and PDE11. One could evaluate the participation of PDE1 activity in cAMP hydrolysis by using 10 μM nimodipine as previously reported *(24)*, keeping in mind that 10 μM nimodipine is able to inhibit by 25% the purified PDE4 activity. Characterization and participation of PDE8 can be evaluated by using 100 μM IBMX, because PDE1, PDE2, PDE7, PDE10, and PDE11, but not PDE8, are sensitive to IBMX. In the case where the cAMP-PDE activity measured in presence of 100 μM IBMX would be equal to the residual cAMP-PDE activity (R), this latter one could then be ascribed to PDE8.

5. Nimodipine (10 μM) is a dihydropyridine compound, well known as a calcium antagonist by its inhibitory effect on L-type calcium channel. This compound can be used on tissue and cell homogenates to determine PDE1 participation with no information relative to the basal or CaM-activated states, because it acts on both states with the same potency *(5,13)*. Vinpocetine, referred to as a specific PDE1 inhibitor, is unable to inhibit bovine vascular and rat cardiac PDE1 *(5)*. Its inhibitory effect depends on the PDE1 subtype splice variant studied (IC_{50} from 10 to 100 μM; *[25]*). Vinpocetine is also able to inhibit PDE7B with an IC_{50} of 60 μM *(26)*. In addition, it was shown to be a direct activator of BK(Ca) channels *(27)*. Another compound used as PDE1 inhibitor, 8 methoxymethyl IBMX, is a poor selective PDE1 inhibitor, because it inhibits in the same range of concentrations PDE1 and PDE5 (IC_{50} of 8 and 10 μM, respectively; *[28]*).

6. As we have previously reported, zaprinast (M&B 22,948), which was first described as a CaM-PDE inhibitor *(5)*, is in fact more potent as a PDE5 inhibitor (cGMP-PDE; *[17,19]*). Zaprinast, which is commercially available, should be used at 10 μM to minimize its effect on PDE1. More specific and potent inhibitors such as DMPPO *(20)* and sildenafil *(21)* can now be obtained from pharmaceutical companies. They are more convenient and should be used at 0.1 μM to evaluate PDE5. Nevertheless, it must be noted that zaprinast, DMPPO, and sildenafil potently inhibit PDE6, which is known to be mainly present in retina *(1–6)* and pineal gland *(29)*. They also inhibit PDE9, PDE10, and PDE11 to a lesser degree *(6)*. The development of specific inhibitors for PDE6 to PDE11 will allow further characterization, thus enhancing the knowledge of their functional role, as recently shown for a PDE7 inhibitor conceived by Bristol-Meyer-Squibb *(22)*.

7. For additional characterization of PDE isozymes, cAMP-PDE and cGMP-PDE activities could be determined in the presence of calcium/CaM to evaluate the contribution of PDE isozymes to CaM-activated PDE activities. In that case, 25 μL of Mg-CaM is added instead of Mg-EGTA in **step 4** in **Subheading 3.2.3.1.**

Furthermore, comparison of the whole PDE activity with and without CaM gives an index of PDE1 contribution.

Acknowledgment

We are very grateful to Cara Farley for proofreading the manuscript.

References

1. Beavo, J. A. (1995) Cyclic nucleotide phosphodiesterases: functional implications of multiple isoforms. *Physiol. Rev.* **75,** 725–748.
2. Conti, M. and Jin, S. L. (2000) The molecular biology of cyclic nucleotide phosphodiesterases. *Prog. Nucleic Acid Res. Mol. Biol.* **63,** 1–35.
3. Francis, S. H., Turko, I. V., and Corbin, J. D. (2001) Cyclic nucleotide phosphodiesterases: relating structure function. *Prog. Nucleic Acid Res. Mol. Biol.* **65,** 1–52.
4. Dousa, T. P. (1999) Cyclic-3′,5′-nucleotide phosphodiesterase isozymes in cell biology and pathophysiology of the kidney. *Kidney Int.* **55,** 29–62.
5. Stoclet, J. C., Keravis, T., Komas, N., and Lugnier, C. (1995) Cyclic nucleotide phosphodiesterases as therapeutic targets in cardiovascular diseases. *Exp. Opin. Investig. Drugs* **4,** 1081–1100.
6. Essayan, D. M. (2001) Cyclic nucleotide phosphodiesterases. *J. Allergy Clin. Immunol.* **108,** 671–680.
7. Follenius, A. and Gerard, D. (1984) Fluorescence investigations of calmodulin hydrophobic sites. *Biophys. Res. Commun.* **119,** 1154–1160.
8. Lowry, O. H., Rosebrough, N. J., Farr, A. L., and Randall, R. J. (1951) Protein measurement with the folin phenol reagent. *J. Biol. Chem.* **193,** 265–275.
9. Thompson, W. J., Brooker, G., and Appleman, M. M. (1974) Assay of cyclic nucleotide phosphodiesterases with radioactive substrates. *Methods Enzymol.* **38,** 205–212.
10. Keravis, T. M., Wells, J. N., and Hardman, J. G. (1980) Cyclic nucleotide phosphodiesterase activities from pig coronary arteries: lack of interconvertibility of major forms. *Biochim. Biophys. Acta* **613,** 116–129.
11. Kincaid, R. L. and Manganiello, V. C. (1988) Assay of cyclic nucleotide phosphodiesterase using radiolabeled and fluorescent substrates *Methods Enzymol.* **159,** 457–470.
12. Schultz, G., Böhme, E., and Hardman, J. G. (1974) Separation and purification of cyclic nucleotides by ion exchange resin column chromatography. *Methods Enzymol.* **38,** 13–16.
13. Lugnier, C., Follenius, A., Gerard, D., and Stoclet, J. C. (1984) Bepridil and flunarizine as calmodulin inhibitors. *Eur. J. Pharmacol.* **98,** 157, 158.
14. Lugnier, C. and Komas, N. (1993) Modulation of vascular cyclic nucleotide phosphodiesterases by cyclic GMP: role in vasodilatation. *Eur. Heart J.* **14(Suppl. I),** 141–148.
15. Podzuweit, T., Nennstiel, P., and Muller, A. (1995) Isozyme selective inhibition of cGMP-stimulated cyclic nucleotide phosphodiesterases by erythro-9-(2-hydroxy-3-nonyl) adenine. *Cell Signal* **7,** 733–738.

16. Lugnier, C., Stierle, A., Beretz, A., Schoeffter, P., Le Bec, A., Wermuth, C. G., Cazenave, J.-P., and Stoclet, J.-C. (1983) Tissue and substrate specificity of inhibition by alkoxy-aryl-lactams of platelet and arterial smooth muscle cyclic nucleotide phosphodiesterase. *Biochem. Biophys. Res. Commun.* **113**, 954–959.

17. Komas, N., Lugnier, C., and Stoclet, J. C. (1991) Endothelium-dependent and independent relaxation of the rat aorta by cyclic nucleotide phosphodiesterase inhibitors. *Br. J. Pharmacol.* **104**, 495–503.

18. Bristol, J. A., Sircar, I., Moos, W. H., Evans, D. B., and Weishaar, R. E. (1984) Cardiotonic agents. 1. 4,5-Dihydro-6-[4-(1H-imidazol-1-yl)phenyl]-3 (2H)-pyridazinones: novel positive inotropic agents for the treatment of congestive heart failure. *J. Med. Chem.* **27**, 1099–1101.

19. Lugnier, C., Schoeffter, P., Le Bec, A., Strouthou, E., and Stoclet, J. C. (1986) Selective inhibition of cyclic nucleotide phosphodiesterases of human, bovine and rat aorta. *Biochem. Pharmacol.* **35**, 1743–1751.

20. Coste, H. and Grondin, P. (1995) Characterization of a novel potent and specific inhibitor of type V phosphodiesterase. *Biochem. Pharmacol.* **50**, 1577–1585.

21. Boolell, M., Allen, M. J., Ballard, S. A., Gepi-Attee, S., Muirhead, G. J., Naylor, A. M., Osterloh, I. H., and Gingell, C. (1996) Sildenafil: an orally active type 5 cyclic GMP-specific phosphodiesterase inhibitor for the treatment of penile erectile dysfunction. *Int. J. Impot. Res.* **8**, 47–52.

22. Yang, G., McIntyre, K. W., Townsed, R. M., Shen, H. H., Pitts, W. J., Dodd, J. H., Nadler, S. G., McKinnon, M., and Watson, A. J. (2003) Phosphodiesterase 7A-deficient mice have functional T cells. *J. Immunol.* **171**, 6414–6420.

23. Lebel, L. A., Menniti, F. S., and Schmidt, C. J. (2003) Therapeutic use of selective PDE10 inhibitors. Int. Pat. Appl. WO O3/093499 A2.

24. Georget, M., Mateo, P., Vandecasteele, G., Lipskaia, L., Defer, N., Hanoune, J., Hoerter, J., Lugnier, C., and Fischmeister, R. (2003) Cyclic AMP compartmentation due to increased cAMP-phosphodiesterase activity in transgenic mice with a cardiac-directed expression of the human adenylyl cyclase type 8 (AC8). *FASEB J.* **17**, 1380–1391.

25. Yan, C., Zhao, A. Z., Bentley, J. K., and Beavo, J. A. (1996) The calmodulin-dependent phosphodiesterase gene PDE1C encodes several functionally different splice variants in a tissue-specific manner. *J. Biol. Chem.* **271**, 25,699–25,706.

26. Sasaki, T., Kotera, J., Yuasa, K., and Omori, K. (2000) Identification of human PDE7B, a cAMP-specific phosphodiesterase. *Biochem. Biophys. Res. Commun.* **271**, 575–583.

27. Wu, S. N., Li, H. F., and Chiang, H. T. (2001) Vinpocetine-induced stimulation of calcium-activated potassium currents in rat pituitary GH3 cells. *Biochem. Pharmacol.* **61**, 877–892.

28. Ahn, H. S., Bercovici, A., Boykow, G., et al. (1997) Potent tetracyclic guanine inhibitors of PDE1 and PDE5 cyclic guanosine monophosphate phosphodiesterases with oral antihypertensive activity. *J. Med. Chem.* **40**, 2196–2210.

29. Morin, F., Lugnier, C., Kameni, J., and Voisin, P. (2001) Expression and role of phosphodiesterase 6 in the chicken pineal gland. *J. Neurochem.* **78**, 88–99.

6

Localization of Cyclic Guanosine 3′,5′-Monophosphate-Hydrolyzing Phosphodiesterase Type 9 in Rat Brain by Nonradioactive *In Situ* Hybridization

Wilma C. G. van Staveren and Marjanne Markerink-van Ittersum

Summary

In this chapter, we describe a protocol for the localization of the cyclic guanosine 3′,5′-monophosphate-specific phosphodiesterase type 9 (PDE9) mRNA in the adult rat brain that uses digoxigenin-labeled riboprobes in a nonradioactive *in situ* hybridization (ISH). The three different riboprobes used all showed similar PDE9 mRNA expression patterns, detecting PDE9 in cell bodies throughout the whole brain. By using immunocytochemical double labeling of the ISH sections with the neuronal marker NeuN or the glial cell marker glial fibrillary acidic protein, the cells expressing PDE9 mRNA were further characterized. Double-labeling experiments revealed that PDE9 was predominantly expressed in neurons.

Key Words

In situ hybridization; protocol; phosphodiesterase type 9; riboprobes; rat; brain; NeuN; glial fibrillary acidic protein.

1. Introduction

The second messenger cyclic guanosine 3′,5′-monophosphate (cGMP) plays an important role in neurotransmission in the central nervous system. cGMP exerts its action by influencing the activity of a number of intracellular targets, among them 3′,5′-cyclic nucleotide phosphodiesterases (PDEs). To date, 11 different PDE families (PDE1–PDE11) have been identified based on their specific characteristics *(1)*. Because differential expression of PDE families can create fundamental diversity among cell types, knowledge about the cellular localization of PDEs is important to obtain more insight into the role of PDEs in cell functioning in complex tissues such as the brain. The localization of a

From: *Methods in Molecular Biology, vol. 307: Phosphodiesterase Methods and Protocols*
Edited by: C. Lugnier © Humana Press Inc., Totowa, NJ

number of PDE families has been studied using radioactive *in situ* hybridization (ISH) *(2–10)*. Here, we describe a protocol for the localization of the cGMP-hydrolyzing PDE family type 9 (PDE9) in the rat brain using nonradioactive ISH. Nonradioactive ISH has the following advantages above radioactive ISH: working with radioisotopes can be avoided, a high resolution of the signal can be obtained, and no long developmental procedures are required. We also describe double staining of ISH sections with the neuronal and glial cell markers NeuN and glial fibrillary acidic protein (GFAP), respectively, to characterize the cell types expressing PDE9 mRNA.

2. Materials

1. Diethylpyrocarbonate (DEPC)-treated water (1000 mL): Add 1 mL of DEPC (Sigma Chemical Corporation, St. Louis, MO) to 1000 mL of distilled deionized water (mQ) and shake thoroughly. Dissolve DEPC overnight in an incubator at 37°C. Then autoclave and store at room temperature.
2. 4% Paraformaldehyde/1X phosphate-buffered saline (PBS), pH 7.4 (200 mL): Weigh 8 g of paraformaldehyde and add 100 mL of DEPC-treated water. Add three droplets of RNase-free 1 M NaOH (Merck), and then heat up to 60°C while stirring until the solution is clear. Filter the solution and add 80 mL of DEPC-treated water and 20 mL of 10X PBS.
3. 10X PBS (1000 mL): Weigh 80 g of NaCl (Merck), 2 g of KCl (Merck), 11.5 g of Na_2HPO_4 (Merck), and 2.4 g of KH_2PO_4 (Merck) and dissolve in 800 mL of mQ. Adjust to 1000 mL with mQ. Add 1 mL of DEPC, shake thoroughly, and incubate overnight in an incubator at 37°C. Autoclave and store at room temperature.
4. 0.1 M Triethanolamine (500 mL): To 400 mL of mQ, add 6.9 mL of undiluted triethanolamine (7.21 M) (Merck). Adjust the pH to 8.0 and adjust the volume to 500 mL with mQ. Add 0.5 mL of DEPC and incubate overnight at 37°C. Autoclave and store at room temperature.
5. 20X Saline sodium citrate (SSC) (1000 mL): Weigh 175.3 g of NaCl and 88.2 g of trisodiumcitrate dihydrate (Merck), and dissolve in 800 mL of mQ. Add 1 mL of DEPC and incubate overnight at 37°C. Autoclave and store at room temperature.
6. Deionized formamide (1000 mL): Add 50 g of AG 501-X8 resin (Bio-Rad, Hercules, CA) to 1000 mL of formamide (Merck). Stir overnight at room temperature. Filter the solution, aliquot in 50-mL tubes, and store at –20°C.
7. 50% Dextran sulfate (20 mL): Weigh 10 g of dextran sulfate sodium salt (Fluka) and add 10 mL of DEPC-treated water. Vortex to dissolve overnight at room temperature. Adjust the volume to 20 mL with DEPC-treated water and store aliquots at –20°C.
8. 50X Denhardt's solution (100 mL): Weigh 1 g of albumin fraction V (from bovine serum) (Merck), 1 g of Ficoll 400 (Merck), and 1 g of polyvinylpyrrolidone (Sigma Chemical Corporation). Dissolve in 80 mL of DEPC-treated mQ. Adjust the volume to 100 mL with DEPC-treated water, and store aliquots at –20°C.

9. Buffer I (1000 mL) (0.1 M maleic acid, 0.15 M NaCl, pH 7.5): Weigh 11.61 g of maleic acid (Fluka) and dissolve in 800 mL of mQ. Add 6 g of NaOH, dissolve, and adjust the pH to 7.5 with 5 M NaOH. Add gently (to prevent formation of precipitates) while stirring 8.76 g of NaCl and dissolve. Adjust the volume to 1000 mL with mQ and store at room temperature.

10. Buffer II (50 mL) (0.1 M maleic acid, 0.15 M NaCl, 1% blocking reagent, pH 7.5): To 50 mL of buffer I, add 0.5 g of blocking reagent (final concentration of 1% [w/v]), (cat. no. 1096176; Roche Molecular Biochemicals, Indianapolis, IN). Stir for approx 2 h at room temperature. Store at 4°C; it can be kept for about 3 wk.

11. Buffer III (1000 mL) (0.1 M Tris, 0.1 M NaCl, 0.05 M MgCl$_2$, pH 9.5): Weigh 12.11 g of Tris (Merck), dissolve in 800 mL of mQ, and adjust the pH to 9.5. Add 5.84 g of NaCl and 10.17 g of MgCl$_2$ (Calbiochem) and dissolve. Adjust the volume to 1000 mL with mQ and store at room temperature.

12. Buffer IV (20 ml) (50 mM Tris, 100 mM NaCl, 50 mM MgCl$_2$, 1 mM levamisole): To 17.5 mL of mQ, add 1 mL of 1 M Tris, pH 9.5. Add 0.4 mL of 5 M NaCl. Add shortly before use 1 mL of 1 M MgCl$_2$ and 200 µL of 100 mM levamisole (Sigma Chemical Corporation). Be sure that levamisole is prepared freshly and dissolved in mQ.

13. Buffer V (1000 mL) (10 mM Tris, 1 mM EDTA, pH 8.0): Weigh 1.21 g of Tris and 0.37 g of EDTA disodium salt dihydrate (Calbiochem) and dissolve in 800 mL of mQ. Adjust the pH to 8.0, adjust the volume to 1000 mL with mQ, and store at room temperature.

14. 10X TBS (1000 mL) (Tris-buffered saline, pH 7.6): Weigh 60.6 g of Tris and 88 g of NaCl and dissolve in 800 mL of mQ. Adjust the pH to 7.6, adjust the volume to 1000 mL with mQ, and store at 4°C.

15. 0.1 M EDTA, pH 8.0 (500 mL): Weigh 18.6 g of EDTA and dissolve in 400 mL of mQ. Adjust the pH to 8.0 and adjust the volume to 500 mL with mQ. Add 0.5 mL of DEPC and incubate overnight at 37°C. Autoclave and store at room temperature.

16. Hybridization mix (1 mL): To 0.5 mL of deionized formamide (final concentration of 50% [v/v]), add 100 µL of tRNA from *Escherichia coli* MRE 600 (Roche Molecular Biochemicals) (final concentration of 1 mg/mL). Then add 200 µL of 50% dextran sulfate (final concentration of 10% [v/v]) and 100 µL of 20X SSC (final concentration of 2X), followed by the addition of 20 µL of 50X Denhardt's solution (final concentration of 1X). Next, add 25 µL of salmon sperm DNA (10 mg/mL) (Invitrogen, Carlsbad, CA) (final concentration of 250 µg). Before adding DNA, denature the strands by placing for 5 min at 96°C and put on ice until use. Add 55 µL of probe (final concentration of 200 ng/mL) to the mixture.

3. Methods

For the localization of mRNAs in tissues, it is important to work with materials and solutions that are prepared RNase free. Furthermore, gloves must be worn throughout the procedure and they must be changed frequently to prevent contamination with RNases of the skin. Glass materials needed for the preparation of solutions, such as bottles, beakers, and cylinders, are baked overnight at 180°C, before use. Materials such as knives, tweezers, tubes, funnels, and pipet tips are autoclaved before use. If necessary, materials can also be cleaned using 10% sodium dodecyl sulfate (Fluka) followed by rinsing in DEPC-treated water and thereafter in 70% ethanol. RNase-free solutions can be prepared by adding DEPC to the solution or by preparing solutions with DEPC-treated water. DEPC is necessary to inactivate present RNases; however, DEPC in turn needs to be inactivated by autoclaving.

3.1. Preparation of Probes

For development of probes, probe lengths varying from 200 to 1000 bp can be used. A detailed description of nonradioactive ISH has been described previously (11). It is important when developing probes to avoid cross hybridization with other mRNAs. Therefore, before constructing the probes, sequences were analyzed on homology with other protein sequences with the program BLAST (http://www.ncbi.nlm.nih.gov/BLAST). For detection of PDE9 mRNA, three different probes with various lengths were used. The parts used for the localization of PDE9 derived from the mouse PDE9A1 (AF031147) were nucleotides 10–357, nucleotides 10–828, and nucleotides 879–1437. A more detailed description of the probes and constructs used were described recently (12). These tested probes gave identical results. Briefly, probes were developed by a PCR reaction on purified mouse PDE9A1 with primers containing two different restriction sites. After digestion of the PCR product, each part was ligated into a vector containing RNA polymerase sites. Before the start of production of the digoxigenin (DIG)-labeled riboprobes, each construct was linearized using restriction enzymes for the formation of both sense and antisense probes.

3.1.1. DIG Labeling of Probes

1. Add the following to 5 µg of linearized template DNA in 10 µL of DEPC-treated water:
   a. 2 µL of DIG-RNA labeling mix (10X) (Roche Molecular Biochemicals).
   b. 2 µL of 10X transcription buffer (Roche Molecular Biochemicals).
   c. 0.4 µL of Rnasin (40 U/µL) (Promega, Madison, NJ).
   d. 2 µL of RNA polymerase (T7 RNA polymerase [20 U/µL]) (Roche Molecular Biochemicals) or Sp6 RNA polymerase (20 U/µL) (Roche Molecular Biochemicals).
   e. 3.6 µL of DEPC-treated water.
2. Mix gently and incubate for 2 h at 37°C.
3. Add 0.5 µL of DNase I (10 U/µL) (Roche Molecular Biochemicals) and incubate for 30 min at 37°C to remove the template.

4. Add 4 µL of 0.1 *M* EDTA (pH 8.0).
5. Add 0.5 µL of 10 mg/mL tRNA from *E. coli* MRE 600.
6. Add 130 µL of DEPC-treated water.
7. Add to the total volume (155 µL), 387.50 µL of 100% ethanol (Fluka) and 15.50 µL of 3 *M* sodium acetate (Merck).
8. Mix gently and store at –80°C.

3.1.2. Determination of Riboprobe Concentration

1. Take tubes containing antisense and sense probes from –80°C and centrifuge at 16,755*g* at 4°C for 15 min to pellet the probes.
2. Remove the supernatant and place the tubes at 50°C for about 5 min to remove excess fluid.
3. Dissolve the probe in 100 µL of DEPC-treated mQ by placing at 50°C for approx 15 min.
4. For determination of probe concentration, dilutions of the probe were made and spotted onto positively charged nylon membranes (Roche Molecular Biochemicals) and compared to a DIG-labeled control RNA (5 µg/50 µL) (Roche Molecular Biochemicals) as described previously *(11)*. Take 1 µL for determination of the amount of produced probe and keep the rest on ice until determination is ready.
5. After estimation of the probe content, aliquot the probes and store at –80°C after the addition of 2.5X 100% ethanol (v/v) and 0.1X sodium acetate.

3.2. Preparation of Tissue Sections

Animals were decapitated and their brains dissected as quickly as possible and thereafter frozen using a CO_2 bottle containing a pressure tube (Hoek Loos, Schiedam, The Netherlands). Brains were wrapped in tin foil and stored at –80°C until sectioning. Brains were frozen quickly on a tissue holder using Tissue Tek O.C.T. compound (Sakura, Torrance, CA) and CO_2. Brains were placed in a cryostat and adjusted to its temperature for 30 to 60 min. Before the start of sectioning, parts of the cryostat that were in contact with the tissue were cleaned using 70% ethanol. Serial frozen sagittal or coronal sections (14 µm) were cut at –20°C using the cryostat. Sections were cut onto Bregma using Paxinos *(13)*. Sections were then transferred onto SuperFrost Plus slides (Menzel-Glaser, Braunschweig, Germany) by gently thaw mounting them on the slides. Slides containing the sections were placed into an RNase-free container (Emergo 048770) and thereafter stored at –80°C until use.

3.3. Pretreatment of Tissue Sections

1. Place sections in a slide holder (Klinipath number 10-30) and quickly thaw them by placing them for 5 to 10 min in an incubator at 50°C. Sections need to be quickly thawed to prevent RNases from being active.
2. Postfix the sections for 20 min at room temperature by placing the slides in 4% paraformaldehyde (Merck)/1X PBS while shaking.

3. Wash the sections three times shortly in 1X PBS at room temperature.
4. Transfer the sections to 0.1 *M* triethanolamine and add shortly before incubation 375 µL (in 100 mL of triethanolamine) of acetic anhydride (Merck), and incubate for 5 min at room temperature while shaking, to reduce aspecific background staining.
5. Add again 375 µL of acetic anhydride and incubate for another 5 min while shaking.
6. Wash the slides two times for 5 min with 2X SSC at room temperature while shaking.
7. Wash the slides at 37°C with 2X SSC containing 50% (v/v) formamide. The slides can be left in this buffer until the start of hybridization.

3.4. Hybridization

1. Prepare the incubation chamber needed for hybridization by placing tissues moistened with water into the chamber.
2. Prepare hybridization mix on ice as described in **Subheading 2., item 16**. For the addition of probe, take tubes containing antisense and sense probes from –80°C, and centrifuge and dissolve the probes in DEPC-treated water as described in **Subheading 3.1.2.** Mix the hybridization solution by gently tumbling the tubes and place at 68°C for 15 min or longer when required. Mix again immediately before use.
3. Take the slides from 2X SSC/50% formamide and discard most of the fluid by shaking the slides a few times, and dry the excess fluid around the section with a tissue.
4. Place the slides horizontally in a humidified container and add 100–200 µL of hybridization mix, depending on the size of the section. Place a cover slip or Parafilm on the section to prevent drying of the section.
5. Cover the container with a lid, and place the humidified chamber overnight in an incubator at 55°C.

3.5. Posthybridization and Development

1. Before the start of the washing procedure, prepare two water baths, one placed at 55°C and the other at 37°C. All washing solutions should be prepared and placed

---

Fig. 1. *(opposite page)* Localization of PDE9 mRNA in sagittal section from rat brain stem (**A–C**) and cerebellum (**D,E**). Sections were hybridized with PDE9 antisense probes (A, B and D, E) and double stained with the neuronal marker NeuN (B) or the glial cell marker GFAP (E). The section shown in (C) was hybridized with sense probe. G, granule cell layer; M, molecular layer; PC, Purkinje cell layer. Arrows indicate neurons. Bar = 100 µm for all pictures. The pictures were taken using an Olympus AX70 microscope equipped with a cooled charge-coupled device Olympus Digital video camera F-view. Pictures showing PDE9 antisense (A) and PDE9 sense (C) were taken using identical settings. All images were stored digitally using the computer program Analysis (Soft Imaging System, Münster, Germany) and arranged using the program Adobe Photoshop 5.5.

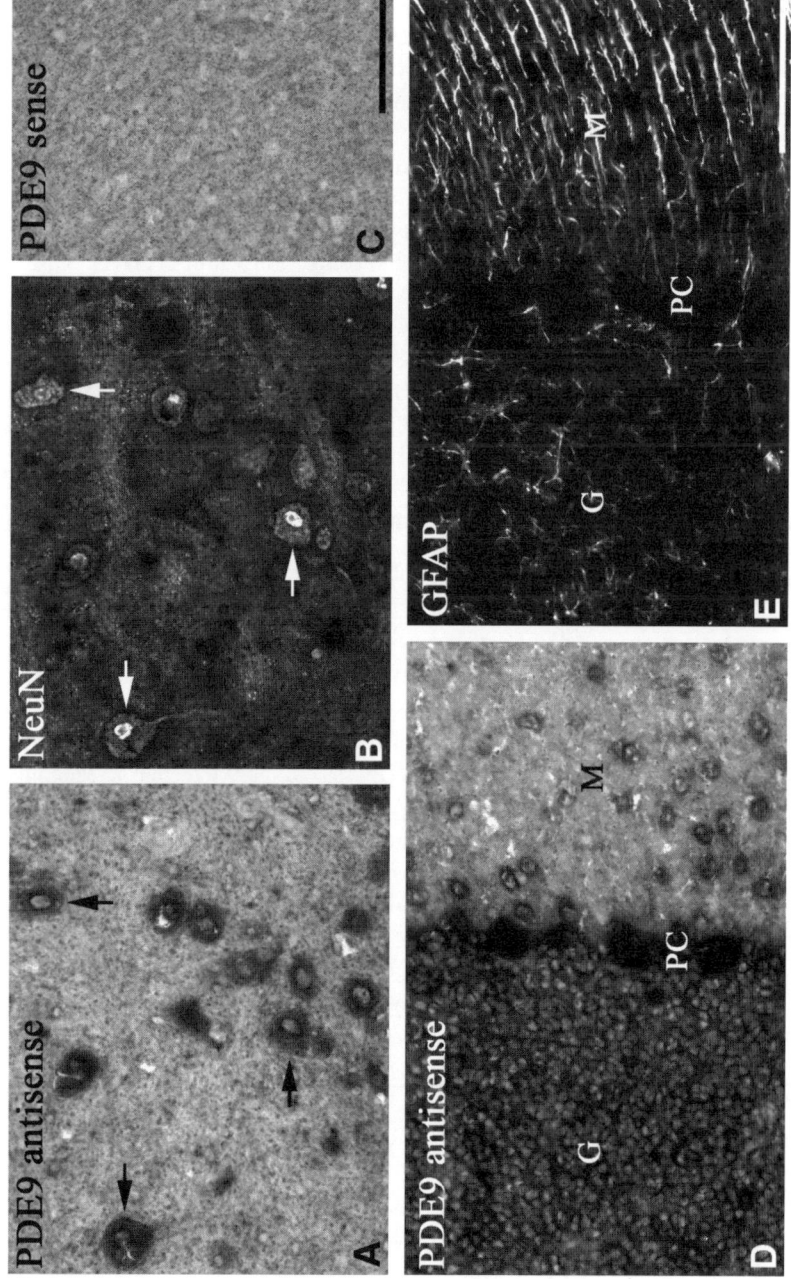

in the water bath some time before starting the washing of the slides, to adjust the buffers' temperature (*see* **Note 1**).

2. Take the slides from the humid chamber and gently remove the cover slip and the excess hybridization mix, and place the slides in 2X SSC/50% deionized formamide at 55°C for 20 min while shaking.
3. Wash the slides in 1X SSC/50% deionized formamide at 55°C for 20 min while shaking.
4. Wash the slides in 0.1X SSC/50% deionized formamide at 55°C for 20 min while shaking.
5. Place the slides in 2X SSC containing 1 mM EDTA and 2 U/mL of RNase T1 (100 U/µL) (Roche Molecular Biochemicals) for 15 min at 37°C. This step is necessary to eliminate single-stranded (unhybridized) probe (*see* **Note 2**).
6. Wash the sections for 20 min with 1X SSC at 45°C while shaking.
7. Wash the sections for 10 min with 2X SSC at room temperature while shaking.
8. Incubate the slides for 5 min with buffer I at room temperature while shaking.
9. Place the slides for 2 to 3 h at room temperature with buffer II containing 5% (v/v) sheep serum (Sigma Chemical Corporation).
10. Incubate the sections overnight at 4°C with a 1:2000 dilution of anti-DIG-alkaline phosphatase (Roche Molecular Biochemicals) in buffer II containing 1% sheep serum.
11. Wash the slides three times with buffer I.
12. Wash for 10 min with 1X TBS containing 0.025% (v/v) Tween-20 (Merck).
13. Wash three times for 5 min with 1X TBS.
14. Wash two times for 5 min in buffer III.
15. Stain the sections with freshly prepared NBT-BCIP solution by adding 4.5 µL of nitroblue tetrazolium salt (NBT) (Roche Molecular Biochemicals) and 3.5 µL of 5-bromo-4-chloro-3-indolyl-phosphate (BCIP) (Roche Molecular Biochemicals) to 1 mL of buffer IV. Development of the staining can vary from 1 to 2 d and should occur in the dark at room temperature (*see* **Note 3**). Replace the buffer with freshly prepared buffer after the first day. Evaluate microscopically during the development procedure the intensity of the staining (*see* **Note 4**).
16. Stop the color reaction by placing the sections in 10 mM Tris containing 1 mM EDTA (buffer V). After washing in TBS, sections can be mounted with 80% glycerol/1X TBS or used for double labeling with cellular makers as described in **Subheading 3.6.**

3.6. Double Immunostaining With Neuronal and Glial Cell Markers

1. Place the slides in 1X TBS and wash three times for 10 min.
2. Incubate the sections overnight at 4°C in a humidified chamber with the neuronal marker mouse antineuronal nuclei (NeuN) (Chemicon, Temecula, CA) diluted 1:50 with 1X TBS containing 0.3% Triton (TBS-T). Stain astrocytes using a monoclonal antibody against anti-GFAP (Clone G-A-5; Sigma Chemical Corporation) diluted 1:1000 in TBS-T (*see* **Note 5**).

3. Wash the sections while shaking with TBS-T, TBS, and TBS-T, each step for 10 min at room temperature.
4. Incubate the sections for 1 to 2 h in the dark at room temperature in a humidified chamber with 1:800 Cy3-conjugated affinipure donkey anti–mouse IgG (Jackson, West Grove, PA), diluted in TBS-T.
5. Wash the sections while shaking in TBS-T and two times with 1X TBS, each step for 10 min at room temperature.
6. Mount the sections with 80% glycerol (Merck)/1X TBS, and evaluate the staining microscopically (**Fig. 1**).

4. Notes

1. It is important that during the procedure, sections treated with antisense probes and sense probes are kept separate and washed in different containers to avoid contamination.
2. It is only necessary to work with RNase-free solutions and materials until **step 5**.
3. In some protocols polyvinyl alcohol (PVA) is used during development of the staining, leading to an enhancement of the signal. In our experience, signal intensity did not differ significantly whether incubation was done with or without PVA. However, we observed that during development of the staining in the presence of PVA, tissue sections derived from embryonic or early postnatal stages showed a strong nonspecific background. This background was not present when these sections of young animals were developed without PVA.
4. When developing the staining, it is important to evaluate the staining microscopically and to note that longer developments can also lead to an increased background staining.
5. To increase the intensity of the antibody staining, sections can also be incubated with the primary antibodies for 2 to 3 d. Furthermore, note that we tested a number of different antibodies, such as parvalbumin, the acetylcholine transporter, and the β1-subunit, on ISH sections. However, these antibodies did not give a positive signal on this material. Therefore, we concluded that the tissue treatment for ISH is apparently not suitable for preserving the integrity of all epitopes.

References

1. Beavo, J. A. (1995) Cyclic nucleotide phosphodiesterases: functional implications of multiple isoforms. *Physiol. Rev.* **75,** 725–748.
2. Yan, C., Bentley, J. K., Sonnenburg, W. K., and Beavo, J. A. (1994) Differential expression of the 61 kDa and 63 kDa calmodulin-dependent phosphodiesterases in the mouse brain. *J. Neurosci.* **14,** 973–984.
3. Furuyama, T., Iwahashi, Y., Tano, Y., Takagi, H., and Inagaki, S. (1994) Localization of 63-kDa calmodulin-stimulated phosphodiesterase mRNA in the rat brain by in situ hybridization histochemistry. *Brain Res. Mol. Brain Res.* **26,** 331–336.

4. Polli, J. W. and Kincaid, R. L. (1994) Expression of a calmodulin-dependent phosphodiesterase isoform (PDE1B1) correlates with brain regions having extensive dopaminergic innervation. *J. Neurosci.* **14,** 1251–1261.
 5. Repaske, D. R., Corbin, J. G., Conti, M., and Goy, M. F. (1993) A cyclic GMP-stimulated cyclic nucleotide phosphodiesterase gene is highly expressed in the limbic system of the rat brain. *Neuroscience* **56,** 673–686.
 6. Reinhardt, R. R. and Bondy, C. A. (1996) Differential cellular pattern of gene expression for two distinct cGMP-inhibited cyclic nucleotide phosphodiesterases in developing and mature rat brain. *Neuroscience* **72,** 567–578.
 7. Iwahashi, Y., Furuyama, T., Tano, Y., Ishimoto, I., Shimomura, Y., and Inagaki, S. (1996) Differential distribution of mRNA encoding cAMP-specific phosphodiesterase isoforms in the rat brain. *Brain Res. Mol. Brain Res.* **38,** 14–24.
 8. Kotera, J., Yanaka, N., Fujishige, K., Imai, Y., Akatsuka, H., Ishizuka, T., Kawashima, K., and Omori, K. (1997). Expression of rat cGMP-binding cGMP-specific phosphodiesterase mRNA in Purkinje cell layers during postnatal neuronal development. *Eur. J. Biochem.* **249,** 434–442.
 9. Miro, X., Perez-Torres, S., Palacios, J. M., Puigdomenech, P., and Mengod, G. (2001) Differential distribution of cAMP-specific phosphodiesterase 7A mRNA in rat brain and peripheral organs. *Synapse* **40,** 201–214.
10. Andreeva, S. G., Dikkes, P., Epstein, P. M., and Rosenberg, P. A. (2001) Expression of cGMP-specific phosphodiesterase 9A mRNA in the rat brain. *J. Neurosci.* **21,** 9068–9076.
11. Roche Molecular Biochemicals Applied Science. (2002) *DIG Application Manual for Nonradioactive* In Situ *Hybridization,* 3rd ed. Roche Molecular Biochemicals Diagnostics GmbH, Penzberg, Germany.
12. Van Staveren, W. C. G., Glick, J., Markerink-van Ittersum, M., Shimizu, M., Beavo, J. A., Steinbusch, H. W. M., and De Vente, J. (2002) Cloning and localization of the cGMP-specific phosphodiesterase type 9 in the rat brain. *J. Neurocytol.* **31,** 729–741.
13. Paxinos, G. and Watson, C. (1986) *The Rat Brain in Stereotaxic Coordinates,* 2nd ed., Academic, Sydney, Australia.

7

Determination of Ca²⁺/Calmodulin-Stimulated Phosphodiesterase Activity in Intact Cells

Chen Yan

Summary

Ca^{2+}/calmodulin (CaM)-stimulated phosphodiesterases (PDEs) constitute a large family (PDE1 family) of enzymes. All members of the PDE1 family can be stimulated by Ca^{2+} in the presence of CaM in vitro. It has been shown that the Ca^{2+}/CaM-stimulated PDE activity present in the vessel wall or vascular smooth muscle cells can be stimulated in vivo by contracting reagents that increase intracellular Ca^{2+} concentrations. We describe in detail a technique used to estimate the extent of PDE1 activation in vivo by measuring in vitro the PDE activity that represents the extent of association between Ca^{2+}-CaM and PDE1 in vivo. The technique involves the extraction and rapid assay of enzyme activity at a low temperature and in the presence of trifluoperazine to minimize the changes in association between Ca^{2+}-CaM and the PDE1 family member during cell lysis and assaying activity. This technique can be used to measure Ca^{2+}/CaM-stimulated PDE activity in cultured cells or tissues.

Key Words

Calcium; calmodulin; phosphodiesterase; trifluoperazine; intact cell.

1. Introduction

Ca^{2+}/calmodulin (CaM)-stimulated phosphodiesterases (PDEs) constitute a large family of enzymes (PDE1 family) encoded by three distinct genes: *PDE1A*, *PDE1B*, and *PDE1C*. Multiple N- or C-terminal splice variants have also been identified for each gene. Currently, at least 14 PDE1A, 2 PDE1B, and 5 PDE1C transcripts have been described. In vitro, the activity of all members of the PDE1 family can be stimulated by Ca^{2+} in the presence of CaM and compared to the basal state in which Ca^{2+} is removed by the chelator EGTA. The relative sensitivity to Ca^{2+} appears distinct among members of PDE1 (1) and the fold of activation may also vary from 2- to 10-fold, depending on the source or purity of the enzyme preparation. For example, bovine PDE1A1 and

From: *Methods in Molecular Biology, vol. 307: Phosphodiesterase Methods and Protocols*
Edited by: C. Lugnier © Humana Press Inc., Totowa, NJ

PDE1A2 have an identical protein sequence except for the first 18 amino acids at the N-terminus, but the half-maximal activation of PDE1A1 by CaM is 0.1 nM whereas for PDE1A2 it is 10 times higher *(2)*. All Ca^{2+}/CaM-stimulated PDEs contain Ca^{2+}/CaM-binding domains, and binding of Ca^{2+}-CaM to these domains is required for full activation of these PDEs. The results from studies of the structural organization of the Ca^{2+}/CaM-binding domains of bovine PDE1A2 suggest that the N-terminus of PDE1A2 contains two CaM-binding sites separated by a putative inhibitory domain located between these two CaM-binding sites *(2)*. However, the molecular details for activation of Ca^{2+}/CaM-stimulated PDEs by Ca^{2+} and CaM are still not clear.

Ca^{2+}/CaM-stimulated PDEs play an important role in Ca^{2+}-mediated regulation of intracellular cyclic adenosine 5′-monophosphate and cyclic guanosine 5′-monophosphate (cGMP) levels *(3)*. The association of Ca^{2+}/CaM with PDE1 is believed to be a reversible process that is dependent on intracellular Ca^{2+} concentration. This provides a mechanism to regulate dynamically PDE1 activity. Therefore, it is important to determine the extent of PDE1 activation in vivo in response to a stimulus that increases intracellular Ca^{2+} concentration. A technique has been developed and used to estimate the fraction of PDE1 in a Ca^{2+}-CaM-activated state in an intact cell. Using this technique, it has been shown that Ca^{2+}-CaM-stimulated PDE activity in arterial strips or vascular smooth muscle cells (VSMCs) was stimulated by contracting reagents such as KCl, histamine, or angiotensin II (Ang II) *(3–5)*.

The technique involves the rapid extraction and assay of enzyme activity at a low temperature and in the presence of trifluoperazine (TFP) to minimize the changes in association between Ca^{2+}-CaM and the PDE1 family member *(5,6)*. At a low temperature, the ternary Ca^{2+}-CaM-PDE complex dissociates very slowly *(5)*. TFP has been shown to bind to Ca^{2+}-CaM complexes that are not associated with an enzyme but does not bind to Ca^{2+}-CaM-enzyme complexes *(6)*, therefore preventing the reassociation of Ca^{2+}-CaM with an enzyme during homogenization and activity assay. Thus, using this technique one is able to estimate the extent of activation of the enzyme in cells just prior to lysis. From this technique, three PDE activity measurements from the same sample are carried out under different conditions: (1) the activity in the presence of TFP, (2) the activity with CaM added for full activation, and (3) the activity with EGTA added to determine Ca^{2+}-independent PDE activity. The percentage of maximal Ca^{2+}-CaM-stimulated PDE activity can be calculated from these three PDE activity measurements.

2. Materials

1. Homogenization buffer: 10 mM Tris-HCl, pH 7.5, 3 mM magnesium acetate, 125 μM TFP.
2. Dounce homogenizer (VWR).

3. Mortar and pestle (VWR).
4. 125 μM TFP (Sigma Chemical Corporation, St. Louis, MO), freshly prepared each time and protected from light.
5. 100 μM CaM (Sigma Chemical Corporation).
6. 4 mM EGTA.
7. 10X PDE assay buffer: 480 mM Tris-HCl, pH 7.5, 20 mM magnesium acetate, 100 μM CaCl$_2$, 1 mg/mL of bovine serum albumin.
8. Substrate mixture (50 μL for each assay): 1X PDE assay buffer, 100,000 cpm of ^{3}H-cGMP (NEN), 1 μM cGMP (Sigma Chemical Corporation).
9. Glass tube (12 × 75 mm).
10. DEAE-Sephadex A-25 resin (Sigma Chemical Corporation).
11. 5-in. P.P. Chromatography columns (Evergreen).
12. Water baths at 30 and 100°C.
13. Snake venom (2.5 mg/mL) (Sigma Chemical Corporation) dissolved in H$_2$O.
14. High-salt buffer: 20 mM Tris-HCl, pH 7.5, 500 mM NaCl.
15. Low-salt buffer: 20 mM Tris-HCl, pH 7.5.
16. EcoLite scintillation cocktail.

3. Methods

The following methods describe preparation of cell lysate and tissue homogenate, assay of PDE activity, and calculation of the extent of PDE activation. These methods are based on previously described methods with minor modifications *(4,5)*.

3.1. Cell Lysis and Tissue Homogenization

3.1.1. Preparation of Cell Lysate

1. Culture cells such as VSMCs from rat aorta, and grow in culture dishes (100-mm diameter) with Dulbecco's modified Eagle's medium containing 10% fetal bovine serum as previously described *(7)*.
2. Serum starve the cells for 48 h and treat with stimuli such as Ang II.
3. Rapidly remove the culture medium and rapidly wash the cells twice with 10 mL of ice-cold phosphate-buffered saline.
4. Add 1 mL of ice-cold homogenization buffer into the dish, and rapidly freeze the cells on the dish on a dry ice–ethanol bath for 30 s (*see* **Note 1**).
5. Scrape the frozen cell slurry off the dish, transfer into an ice-cold dounce homogenizer, and homogenize with approx 25 strokes in an ice bath.
6. Immediately assay aliquots (175 μL) of cell lysates for PDE activity as described in **Subheading 3.2.** before preparing the next sample.

3.1.2. Preparation of Tissue Homogenate

1. Quickly freeze tissues (such as aortas) in liquid nitrogen and store at −80°C.
2. Place the frozen aortas (80–100 mg) in a mortar cooled in liquid nitrogen.
3. Powder the tissues with a mortar and pestle, and then transfer to a tube containing 25 vol of ice-cold homogenization buffer.

4. Immediately homogenize the tissue suspension for 5 s at speed setting 5 two times with an ice-cold Polytron homogenizer.
5. Immediately assay aliquots (175 µL) of homogenates for PDE activity as described in **Subheading 3.2.** before preparing the next sample.

3.2. Assay of Ca²⁺/CaM-Stimulated PDE Activity

The PDE assay includes four main steps. The first step is to convert the substrate ³H-cGMP into ³H-5'GMP by PDE. The second step is to convert ³H-5'GMP into ³H-guanosine by 5'-nucleotidase present in snake venom. The third step is to separate modified product from the remaining substrate and measure the amount of product. The fourth step is to calculate PDE activity. The overall major procedures are shown in **Fig. 1.**

In the first step, aliquots (175 µL) of each cell lysate or homogenate are added to 12 × 75 mm glass tubes containing 25 µL of either (1) 125 µ*M* TFP, (2) 100 µ*M* CaM, or (3) 4 m*M* EGTA (*see* **Note 2**). Duplicates are made for each sample and each assay condition. The reaction in the first set of tubes (125 µ*M* TFP) is assayed immediately after homogenization. The reactions in the second and third set of tubes, containing CaM and EGTA, respectively, are preincubated for 5 min at 30°C and then cooled to 0°C before beginning the assay. This is done to achieve the maximum activation by excess Ca²⁺/CaM and inactivation by EGTA. All reactions are started by adding 50 µL of the substrate mixture and incubating for 5 min at 0°C and terminated by boiling for 1 min and cooled down at room temperature.

In the second step, after allowing the tubes to cool, 10 µL of 2.5 mg/mL of snake venom is added to the reaction mixtures, and the tubes are then incubated for 10 min at 30°C. Each reaction is then diluted by adding 250 µL of low-salt buffer and kept at room temperature.

In the third step, the DEAE-Sephadex A-25 columns are prepared by adding 1 mL of preswollen A-25 resin slurry to each column to make a packed-bed volume of approx 0.6–0.7 mL. The A-25 resin slurry is prepared by mixing 20 g of A-25 resin with 200 mL of deionized water at least 2–4 h before use. The A-25 resin slurry can be stored at 4°C. The columns are then washed with 8 mL of low-salt buffer and are ready to use after the low-salt buffer drains out completely. Columns are reusable. To regenerate columns, columns are washed with 8 mL of high-salt buffer followed by 8 mL of low-salt buffer. Columns should be discarded if the background reading becomes high, repeats are not consistent, or the flow through the column slows down significantly.

Each diluted reaction mixture from the second step is applied onto individual A-25 columns that empty into a scintillation vial. The columns then are washed with 0.5 mL of low-salt buffer three times, resulting in approx 2 mL

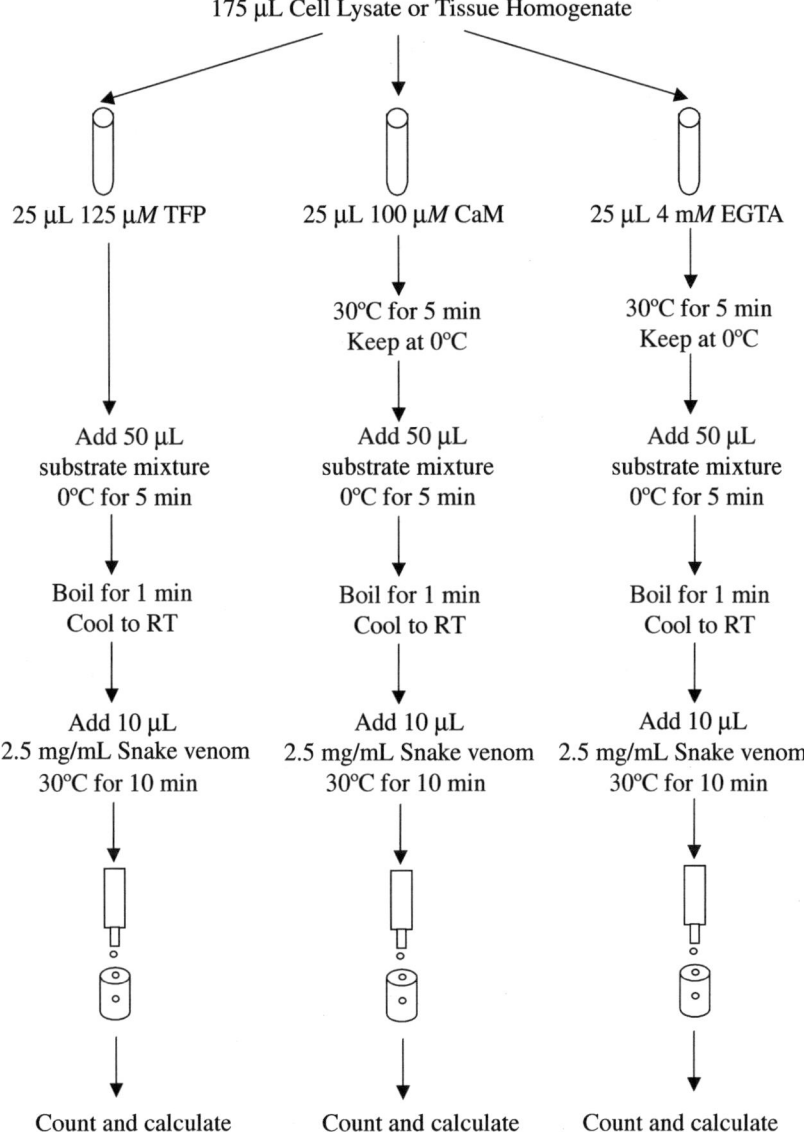

Fig. 1. Summarization of major steps of method used for measurement of Ca²⁺/CaM-stimulated PDE activity in intact cells. RT, room temperature.

collected in the scintillation vial. To each vial 4.5 mL of liquid scintillation cocktail is added and counted for 1 min in a scintillation counter.

In addition to the assay tubes containing experimental samples, four blank controls are required, in which 175 mL of homogenization buffer is used to replace cell lysates or tissue homogenates. Two of these tubes are not applied to columns but are added directly to scintillation vials, which serve as the "total radioactivity controls." The other two tubes are treated exactly the same as the other assay tubes and serve as "background controls."

In the fourth step, radioactivity for the sample tubes and the total radioactivity controls are corrected by subtraction of the mean value obtained from the "background controls." The PDE activity of samples is calculated by converting counts per minute into picomoles of cGMP hydrolyzed using the mean radioactivity value of the total radioactivity controls, which represents 250 pmol of substrate/vial.

3.3. Quantification of Ca²⁺/CaM-Stimulated PDE Activity

At this point three measurements of PDE activities are obtained: (1) the PDE activity with TFP added, (2) the PDE activity with CaM added in an amount sufficient to stimulate the enzyme fully even in the presence of TFP, and (3) the PDE activity with EGTA added for estimation of Ca^{2+}/CaM-independent PDE activity. The percentage of maximal Ca^{2+}/CaM-stimulated PDE activity is then computed as follows:

$$\frac{\text{Activity without CaM (1)} - \text{Activity with EGTA (3)}}{\text{Activity with CaM (2)} - \text{Activity with EGTA (3)}} \times 100$$

As an example, **Fig. 2** shows the results of an assay for the stimulation of Ca^{2+}/CaM-stimulated PDE activity in intact VSMCs on treatment with Ang II for 1 min.

4. Notes

1. It is critical that the lysis of cells or homogenization of tissue is performed very rapidly in the presence of TFP and at a cold temperature. If not done correctly, the PDE activity of the sample prepared with and without TFP or assayed by adding TFP or CaM will be indistinguishable. The Ca^{2+}/CaM-stimulated PDE isoform present in arteries and rat aortic VSMCs is PDE1A1, which has a high affinity to Ca^{2+} or CaM. The EC_{50} for Ca^{2+} and CaM is about 0.2 μM and 0.1 nM, respectively. Therefore, PDE1A1 has a high tendency to be activated at low levels of Ca^{2+} and CaM, and, thus, experiments measuring PDE1A1 require the presence of high levels of TFP and more careful processing of cell lysates or tissue homogenization.
2. The amount of cell lysates or tissue homogenates used for PDE assays may need to be diluted because it is necessary to ensure that the enzyme gives 5–30% hydrolysis of 1 μM cGMP.
3. We have always measured about 50% activation of Ca^{2+}/CaM-stimulated PDE under unstimulated conditions (**Fig. 2**) using this method. It is still not clear

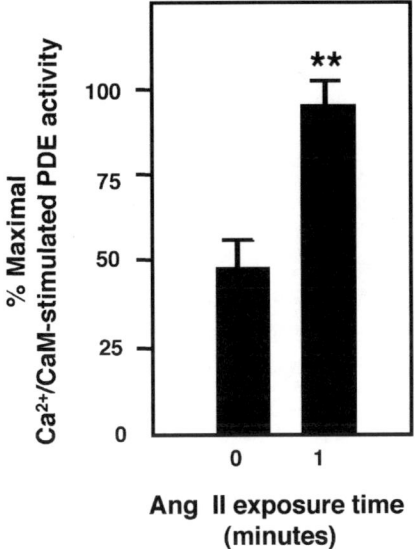

Fig. 2. Effect of Ang II on activation of Ca^{2+}/CaM-stimulated PDE activity in rat aortic VSMCs (*see* **Note 3**). Growth-arrested VSMCs were treated with or without 400 n*M* Ang II for 1 min. The extent of Ca^{2+}/CaM-stimulated PDE activity was measured using the method described in this chapter. Values are the mean ± SEM (*n* = 3). **p <0.01 vs control.

whether this high Ca^{2+}/CaM-stimulated PDE activity under basal conditions is a reflection of the extent of activation of the enzyme in the intact cells, or a reflection of the extent of activation that occurs during homogenization.

Acknowledgments

I thank James Surapisitchat for helpful comments on the manuscript. This work was supported in part by American Association Research Grant 0030302T.

References

1. Yan, C., Zhao, A. Z., Bentley, J. K., and Beavo, J. A. (1996) The calmodulin-dependent phosphodiesterase gene PDE1C encodes several functionally different splice variants in a tissue-specific manner. *J. Biol. Chem.* **271,** 25,699–25,706.
2. Sonnenburg, W. K., Seger, D., Kwak, K. S., Huang, J., Charbonneau, H., and Beavo, J. A. (1995) Identification of inhibitory and calmodulin-binding domains of the PDE1A1 and PDE1A2 calmodulin-stimulated cyclic nucleotide phosphodiesterases. *J. Biol. Chem.* **270,** 30,989–31,000.
3. Kim, D., Rybalkin, S. D., Pi, X., Wang, Y., Zhang, C., Munzel, T., Beavo, J. A., Berk, B. C., and Yan, C. (2001) Upregulation of phosphodiesterase 1A1 expression

is associated with the development of nitrate tolerance. *Circulation* **104,** 2338–2343.

4. Miller, J. R. and Wells, J. N. (1987) Effects of isoproterenol on active force and Ca^{2+} X calmodulin-sensitive phosphodiesterase activity in porcine coronary artery. *Biochem. Pharmacol.* **36,** 1819–1824.

5. Saitoh, Y., Hardman, J. G., and Wells, J. N. (1985) Differences in the association of calmodulin with cyclic nucleotide phosphodiesterase in relaxed and contracted arterial strips. *Biochemistry* **24,** 1613–1618.

6. LaPorte, D. C., Wierman, B. M., and Storm, D. R. (1980) Calcium-induced exposure of a hydrophobic surface on calmodulin. *Biochemistry* **19,** 3814–3819.

7. Travo, P., Barrett, G., and Burnstock, G. (1980) Differences in proliferation of primary cultures of vascular smooth muscle cells taken from male and female rats. *Blood Vessels* **17,** 110–116.

8

Adenovirus-Mediated Overexpression of Murine Cyclic Nucleotide Phosphodiesterase 3B

Faiyaz Ahmad, Linda Härndahl, Yan Tang,
Lena Stenson Holst, and Vincent C. Manganiello

Summary

To construct the recombinant adenovirus vector containing the cDNA for recombinant mouse cyclic nucleotide phosphodiesterase 3B (mPDE3B), the cDNA for mPDE3B was subcloned into pACCMV.pLpA. Subsequently, this recombinant plasmid, pACCMV.mPDE3B, was cotransfected with pJM17 plasmid containing the adenoviral genome into 293 human embryonic kidney cells, and the replication-deficient adenovirus AdCMV.mPDE3B was generated via homologous recombination. Large-scale preparation of adenovirus yielded 10^{11}–10^{13} viral particles/mL and could be quantitated by real-time polymerase chain reaction using iCycler (Bio-Rad). Efficiency of gene transfer was assessed by infecting FDCP2 or H4IIE cells with a recombinant adenovirus expressing β-galactosidase (β-gal); greater than 75% of cells were infected. Expression of mPDE3B in H4IIE hepatoma cells, FDCP2 hematopoietic cells, and β-cells from isolated pancreatic islets was detected by Western blot analysis. In lysates from FDCP2 cells and H4IIE hepatoma cells infected with recombinant adenoviral mPDE3B constructs, mPDE3B activity was increased 10- to 30-fold compared with the activity in lysates from cells infected with β-gal adenovirus. Stimulation of FDCP2 cells infected with mPDE3B adenovirus with insulin (100 nM, 10 min) resulted in an approx 1.7-fold increase in endogenous PDE3B and recombinant wild-type PDE3B activities. Infection of rat pancreatic islets resulted in a 5- to 10-fold increase in PDE3B expression and activity and subsequent blunting of insulin secretion. Thus, adenovirus-mediated gene transfer is effective for studying expression and regulation of recombinant PDE3 in insulin-responsive cells as well as insulin-secreting cells.

Key Words

PDE3B; phosphorylation; adenovirus purification; real-time polymerase chain reaction; HEK 293 cells; insulin release.

1. Introduction

Cyclic nucleotide phosphodiesterases (PDEs) are a heterogeneous group of structurally related enzymes that hydrolyze cyclic adenosine monophosphate

From: *Methods in Molecular Biology, vol. 307: Phosphodiesterase Methods and Protocols*
Edited by: C. Lugnier © Humana Press Inc., Totowa, NJ

(cAMP) and cyclic guanosine 5′-monophosphate (cGMP). Eleven PDE gene families (PDEs 1–11) have been identified *(1–5)*. They differ in primary sequences, substrate specificity, responses to effectors and specific inhibitors, and mechanisms of regulation *(1–5)*. Representatives of several PDE families are usually found in the same cells but differ in amounts, proportions, and subcellular locations *(5–7)*. Isoforms of PDE3 (PDE3A-B), encoded by two genes, are characterized by high affinities for both cAMP and cGMP (K_m values of ~0.1–0.8); sensitivity to specific inhibitors such as cilostamide, enoximone, and lixazinone; and phosphorylation and activation in response to agents that increase cAMP and to insulin *(8,9)*.

To pursue structure/function studies and to further understand mechanisms for regulation of membrane-associated PDE3B in intact cells, we are utilizing adenovirus-mediated gene transfer of recombinant PDE3B in cultured cells. The methods described herein outline cloning of the mouse cyclic nucleotide PDE3B (mPDE3B) cDNA into the adenovirus transfer plasmid pACCMV; cotransfection of mPDE3B adenovirus transfer plasmid together with adenovirus plasmid pJM17 for recombination in HEK 293 cells; large-scale preparation and CsCl purification of recombinant mPDE3B adenovirus; quantification of mPDE3B, mPDE3B604, and β-galactosidase (β-gal) adenovirus; and expression of mPDE3B and mPDE3B604 in cultured cells and isolated pancreatic islets.

2. Materials

1. pACCMV, pJM17, mPDE3B, and mPDE3B604 cDNA plasmids.
2. *Escherichia coli*, XL-1 Blue (Stratagene, La Jolla, CA).
3. Restriction endonucleases (Roche Molecular Biochemicals, Indianapolis, IN): *Xho*I, *Sal*I, *Xba*I, *Kpn*I, and *Hin*dIII.
4. HEK 293 (for viral amplification), and suitable cell lines such as FDCP2 myeloid cells and H4IIE hepatoma cells that can be easily infected by adenovirus.
5. Supplies for growing *E. coli*: Luria-Bertani (LB) broth, agar, Petri dishes, etc.
6. CsCl (Sigma Chemical Corporation, St. Louis, MO), Tris buffer (Digene, Columbia, MD).
7. Centrifuges and rotors: Beckman (Fullerton, CA) SW 41, Sorvall RC 5C plus (NEN, Boston, MA), and SLA 1500 rotors.
8. Agarose gel electrophoresis using the RunOne electrophoresis cell (Embitec, San Diego, CA).
9. PolyFect transfection reagent (cat. no. 301307; Qiagen, Valenica, CA).
10. Sodium dodecyl sulfate-polyacrylamide gel electrophoresis (SDS-PAGE) equipment, gels, and Western blot equipment from Invitrogen (Carlsbad, CA).
11. PDE3B antibodies (C-terminal, or Flag-PDE3B antibodies).
12. Dialysis cassettes (Pierce, Rockford, IL).

13. LB Agar Amp: *E. coli* FastMedia agar dishes were made according to the Fermentas protocol.
14. LB broth: 500 mL (Quality Biologicals, Gaithersburg, MD).
15. CsCl (1.25 g/mL): 54 g of CsCl. Bring weight to 200 g using 1X TN buffer (25 mM Tris; 140 mM NaCl; 5 mM KCl; 0.7 mM Na$_2$HPO$_4$, pH 7.5).
16. CsCl (1.33 g/mL): 66.66 g of CsCl. Bring weight to 200 g using 1X TN buffer.
17. CsCl (1.4 g/mL): 77.66 g of CsCl. Bring weight to 200 g using 1X TN buffer. Check the density of solutions by weighing 1 mL. Adjust the density by adding 1X TN buffer or CsCl powder as necessary.
18. Dialysis buffer: 10% glycerol in buffer containing 10 mM Tris; 1 mM MgCl$_2$, pH 7.4. Filter sterilize and store refrigerated at 4°C.
19. Dulbecco's modified Eagle's medium (DMEM) (Biosource International, Camarillo, CA): Store refrigerated at 4°C. Add 50 mL of fetal bovine serum (FBS) and 5 mL of penicillin-streptomycin (pen/strep)/500 mL when used.
20. Minimal essential medium (MEM) and RPMI-1640 cell culture medium (Life Technologies, Grand Island, NY).
21. Buffer A: 20 mM HEPES; 250 mM sucrose; 1 mM EDTA; 10 mM pyrophosphate; 5 mM NaF; 1 mM phenylmethylsulfonyl fluoride; 1 mM Na$_3$VO$_4$; aprotinin; 10 µg/mL each of leupeptin and pepstatin, pH 7.5 (Final).
22. Buffer B: 50 mM N-Tris(Hydroxymethyl)methyl-2-aminoethane-sulfonic acid (TES), pH 7.4; 250 mM sucrose; 1 mM EDTA; 0.1 mM EGTA; 10 µg/mL each of leupeptin and antipain, and 1 µg/mL of pepstatin.
23. Modified Krebs Ringer Bicarbon buffer (KRBB): 10 mM HEPES, pH 7.5; 120 mM NaCl; 5 mM NaHCO$_3$; 5 mM KCl; 1.2 mM KH$_2$PO$_4$; 2.5 mM CaCl$_2$; 1.2 mM MgSO$_4$; 0.2% bovine serum albumin (BSA).
24. WEHI-3B cells (American Type Culture Collection, Rockville, MD).

3. Methods

In the following sections, we describe the generation and purification of recombinant mPDE3B adenovirus, and expression of mPDE3B in cultured cells.

3.1. Subcloning of mPDE3B cDNA Into Adenovirus Transfer Plasmid pACCMV

pACCMV.pLpA, a vector that generates high levels of expression, was transformed into *E. coli*, amplified, purified, and used as parent vector for subsequent constructions. An *Xho*I fragment containing the cDNA for wild-type (WT) mPDE3B and its truncated recombinant mPDE3B (mPDE3B604) were ligated in sense orientation into the compatible *Sal*I site of pAcCMV.pLpA, and adenovirus vectors containing these cDNAs were generated. Sense orientation was verified by restriction with *Xba*I, yielding fragments of 2.5 and 9.6 kb for mPDE3B and of 0.7 and 9.6 kb for mPDE3B604.

3.2. Cotransfection of mPDE3B Adenovirus Transfer Plasmid Together With Adenovirus Plasmid pJM17 for Recombination in HEK 293 Cells

This protocol is conducted according to instructions from Qiagen.

1. Seed HEK 293 cells (8×10^5/60-mm dish) in 5 mL of DMEM medium 1 d before transfection. The cells should be 70–80% confluent on the day of transfection.
2. Mix together 3 µg of pACCMV.pLpA DNA containing the cloned cDNA fragment (mPDE3B or mPDE3B604) and 3 µg of pJM17 DNA in TE buffer (10 mM Tris; 1 mM EDTA, pH 8.0) (minimum DNA concentration: 0.1 µg/µL), and dilute to a total volume of 150 µL with cell growth medium (containing no serum, proteins, or antibiotics). Serum, proteins, and antibiotics present during this step may interfere with complex formation and will significantly decrease transfection efficiency. Centrifuge the solution for 30 s to bring down drops from the top of the tube.
3. Add 25 µL of PolyFect (Qiagen) transfection reagent to the DNA solution from **step 2**. Mix the DNA and PolyFect solutions by pipetting up and down three to four times.
4. After mixing the DNA solution and PolyFect, incubate samples for 5–10 min at room temperature (20–25°C) to allow complex formation.
5. After complex formation, remove the growth medium from the HEK 293 cell mono-layers in 60-mm dishes, and wash the cells once with 4 mL of phosphate-buffered saline (PBS). Add fresh growth medium containing serum and antibiotics (3 mL).
6. Add 1 mL of cell growth medium (containing serum and antibiotics) to the reaction tube containing the transfection complexes (pACCMV containing cDNA, pJM17, and PolyFect solution mixture). Transfer the mixture to the cells in the 60-mm dishes. After complex formation, serum and antibiotics no longer interfere with trans-fection and will significantly enhance the transfection efficiency of PolyFect reagent.
7. Incubate the cells with the complexes at 37°C and 5% CO_2 to allow for homolo-gous recombination. When the recombination event occurs in a cotransfected HEK 293 cell, the recombinant virus completes its life cycle, resulting in cell lysis and the formation of plaques of dead cells (may occur in 2–4 wk), with the first plaque appearing as a visible "hole" in the monolayer. This initial plaque will enlarge as viral infection progresses, so by 3–5 d after the appearance of the plaque, the monolayer should be completely lysed.
8. After the monolayer lyses, collect the medium and lyse any remaining intact cells by several freeze/thaw cycles (using dry ice and a 37°C water bath, respectively). Pellet the cell debris by centrifugation, and store the crude viral lysate (CVL) supernatant at –20°C in growth medium (**Fig. 1**).

3.3. Large-Scale Virus Preparation and CsCl Purification of Recombinant mPDE3B Adenovirus

3.3.1. Infection of HEK 293 Cells

1. Start with 25 dishes (15 cm) of 293 cells grown to confluence in DMEM supple-mented with 10% FBS and pen/strep.
2. Thaw virus stock at 37°C and place on ice.

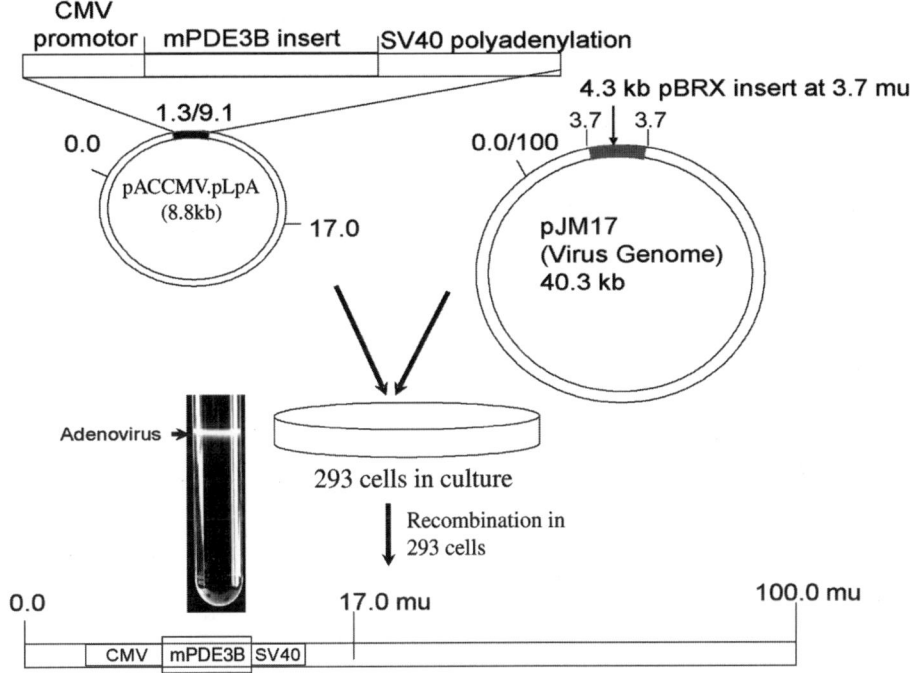

Fig. 1. Recombination in HEK 293 cells for generating mPDE3B adenovirus. mPDE3B cDNA was cloned into the pACCMV.pLpA vector. This recombinant pAC was cotransfected with pJM17 vector into HEK 293 cells. The resulting recombinant adenovirus transcription unit contains cytomegalovirus (CMV) promoter/enhancer, mPDE3B cDNA, and a polyadenylation cassette (1 mU equals 360 bp). The replication-deficient adenovirus AdCMV.mPDE3B was amplified in HEK 293 cells and purified via centrifugation in CsCl as described in **Subheading 3.3.3.**

3. Dilute an appropriate volume of crude virus (based on known approximate titer as determined by real-time polymerase chain reaction (PCR) using HEK 293 CVL as control) to achieve a final volume of 30 mL.
4. Remove medium from the 15-cm dishes, and add 11.5 mL of fresh serum-free medium with 1 mL of crude virus to each dish. Incubate the dish for 2 h in an incubator at 37°C/5% CO_2. Add back 12.5 mL of medium containing 10% serum to each dish (5% final serum concentration).
5. Incubate until all the cells have lifted from the surface of the dish, which takes 5–10 d after infection with approx 1×10^2 virus particles/cell (*see* **Note 1**).

3.3.2. Harvesting of Cells

HEK 293 cells can be harvested by scraping with medium still on the dish.

1. Remove the medium and cells from each dish with a 25-mL pipet, and transfer into sterile 50-mL tubes.

2. After all the dishes have been harvested, centrifuge in an SLA 1500 rotor at 4000 rpm (2400*g*) for 10 min at 4°C to collect the cell pellets, and discard the supernatant medium.
3. Transfer the virus-containing cell pellet into sterile 50-mL tubes, and release viral particles by five cycles of freezing and thawing (using dry ice and a 37°C water bath, respectively) in TN buffer.
4. Centrifuge the CVL at 8000 rpm (10,000*g* at 4°C) for 10 min in an SLA-1500 rotor. Collect the virus-containing supernatant (CVL) for CsCl purification and quantitation by real-time PCR or plaque assay.

3.3.3. First CsCl Gradient

1. In each of six 12-mL ultraclear ultracentrifuge tubes (cat. no. 344059; Beckman), aliquot 2.5 mL of low-density (1.25 g/mL) CsCl. Underlay the low-density CsCl with 2.5 mL of high-density (1.4 g/mL) CsCl. Carefully add the CVL onto the top of the low-density CsCl layer. Add no more than 7 mL of CVL in each tube.
2. If necessary, balance the tubes with 1X TN buffer, and centrifuge at 150,000*g* for 1 h at 20°C in a Beckman SW 41 rotor.
3. Carefully collect the virus band (in a sterile 15-mL tube) by piercing the side of the tube with a sterile 20-gage needle.
4. Quantify the collected volume of virus and slowly add an equal volume of 1X TN buffer, mixing while adding.

3.3.4. Second CsCl Gradient

1. In each of two 12-mL tubes, aliquot 8 mL of CsCl (1.33 g/mL), and overlay the CsCl with purified adenovirus (~4 mL) recovered after the first spin. Use the second tube as a balance.
2. Centrifuge overnight (~16 h) at 150,000*g* and 20°C in a Beckman SW 41 rotor.

3.3.5. Dialysis

1. Stop the centrifuge the next morning without using the brake. Collect the virus band (**Fig. 1**). Using a 3-mL syringe and 1.5-in. 20-gage needle, withdraw the virus by side puncture of the tube, and carefully inject into a dialysis cassette (Pierce).
2. Attach a styrofoam floater to the top of the cassette where the virus suspension was injected and place in a sterile 1000-mL beaker (with magnetic stir bar). Fill the beaker with dialysis buffer and place it on a stir dish at 4°C in a cold room. After 30 min, change the dialysis buffer. Continue to change the dialysis buffer four more times at 1-h intervals. Dialysis may be completed in 4–6 h.
3. Remove the cassette from the beaker. Using another 3-mL syringe with a needle, remove the virus from the cassette, and transfer the purified virus into a sterile 5-mL tube on ice. Aliquot the purified virus into 1.5-mL Eppendorf tubes in appropriate sizes.

3.4. Quantification of Adenovirus

3.4.1. Viral Particle Number

The DNA concentration of CsCl-purified virus can be determined by monitoring the absorbance on any standard ultraviolet spectrophotometer using $A_{260/280}$ readings. The relationship between virus particle number and absorbance is 1 A_{260} unit equals 1.1×10^{12} viral particles.

3.4.2. Real-Time PCR

Real-time PCR has made it possible to quantitate DNA copies or RNA transcripts (isolated from tissues or cell extracts). During the real-time PCR, a fluorescent reporter is used to monitor the PCR as it progress. With the progress of the PCR, the fluorescence of the reporter molecule increases as products start to accumulate with each successive round of amplification. The point at which the fluorescence appreciably starts to rise above the background level is called the threshold cycle. A standard curve, which can be used for determination of copies of unknown templates using their threshold cycles, can be generated using the log of starting amounts of template and the corresponding threshold cyle. For analysis of the number of viral particles for mPDE3B and mPDE3B604 during these experiments, we used SYBR Green I (intercalating dye) as fluorescent reporter for real-time PCR.

The 50 μL real-time PCR reaction mixture contained 1 μL of sample and 2×25 μL of iQ™ SYBR Green Supermix from Bio-Rad (cat. no. 170-8882) containing 100 mM KCl; 40 mM Tris-HCl, pH 8.4; 0.4 mM of each dNTP (dATP, dCTP, dGTP, and dTTP); 50 U/mL of iTaq DNA polymerase; 6 mM $MgCl_2$; SYBR Green I; 20 nM fluorescein; stabilizers; and forward and reverse primers, 200 nM each (**Table 1**).

Real-time PCR using the iCycler (Bio-Rad) was performed in 96-well real-time PCR format, which included six to ten 10-fold serial dilutions in duplicate of the pJM17 DNA standards (i.e., from 1×10^2 to 1×10^{11} copies per reaction), and 10-fold serial dilutions in duplicate of recombinant mPDE3B (**Fig. 2**) and mPDE3B604 cDNA standards.

Test samples (Ad-mPDE3B, and Ad-mPDE3B604 adenovirus) were assayed in duplicate on each dish, along with duplicate negative controls (i.e., primers without samples) at each dilution. The dishes were covered with a piece of optically clear sealing tape (cat. no. 223-9444; Bio-Rad) and were centrifuged at low speed (e.g., 300g, 5 min) to ensure complete mixing and to bring all reagents to the bottom, and the dishes were placed in the iCycler iQ™ detection system.

The PCR protocol used denaturation of DNA (all at 95°C, 3 min) followed by 40 cycles of denaturation (95°C, 30 s), annealing (58°C, 15 s), and extension (72°C, 45 s). Fluorescent data were collected during the 72°C step (*see* **Note 2**).

Table 1
Real-Time PCR Primers for Amplification of Constructs

	Primer sequence
pJM17	5′-GCAGAACCACCAGCACAGTGT-3′ (sense)
	5′-TCCACGCATTTCCTTCTAAGCTA-3′ (antisense)
β-Gal	5′-CGTTACCCAACTTAATCGCCTT-3′ (sense)
	5′-GCGGGCCTCTTCGCTATTAC-3′ (antisense)
mPDE3B,	5′-ATCGCCTCTTGGTCTGCCAG-3′ (sense)
mPDE3B604	5′-GCCTTCTGTCCATCTCAAATGTAGG-3′ (antisense)

3.4.3. Analysis and Normalization of Real-Time PCR Data

Fluorescent data collected post-real-time PCR was analyzed using the default and variable parameters available in the software provided with the iCycler iQ™ real-time PCR detection system. The real-time PCR threshold cycle number (C_T) for each plasmid DNA standard and test virus sample was calculated at the point where the fluorescence exceeded the threshold limit. The threshold limit was fixed along the linear logarithmic phase of the fluorescence curves at 10 SDs above the average background fluorescence. Average C_T for duplicate standards and samples was calculated. Standard curve equations were calculated by regression analysis of average C_T vs the $\log_{10}$ of the plasmid DNA standard copy number (i.e., the copy number of the plasmid DNA standard equals the initial amplicon DNA copy number). The plasmid DNA standard curve equations (R^2 usually >0.96) were used to calculate the virus copy numbers for test samples.

3.4.4. Plaque Assay

A plaque assay with agarose overlay of virus-producing HEK 293 cells can be used to calculate titers of CVLs and (optional) increase the homogeneity of

---

Fig. 2. *(opposite page)* Amplification plots for reactions on iCycler. The standard curve demonstrated threshold cycles of about 18 for 10^8 copies, approx 100% PCR efficiency, and a correlation coefficient approaching 0.99, indicating good reproducibility and accuracy. Furthermore, no amplification occurred in the zero template control wells. The iCycler platform performed well in terms of specificity and sensitivity. For melting curve analysis, the reactions were cycled at 58°C for 1 min and then ramped to 95°C over a period of 80 cycle repeats (0.5°C/cycle), following amplification of M3B. Melting curve analysis of the reactions revealed that all wells (with the exception of control wells without template) produced a single amplified product at a melting temperature (*Tm*) of 83.5°C (the *Tm* of the intended product), indicating that only the specific product of interest was amplified.

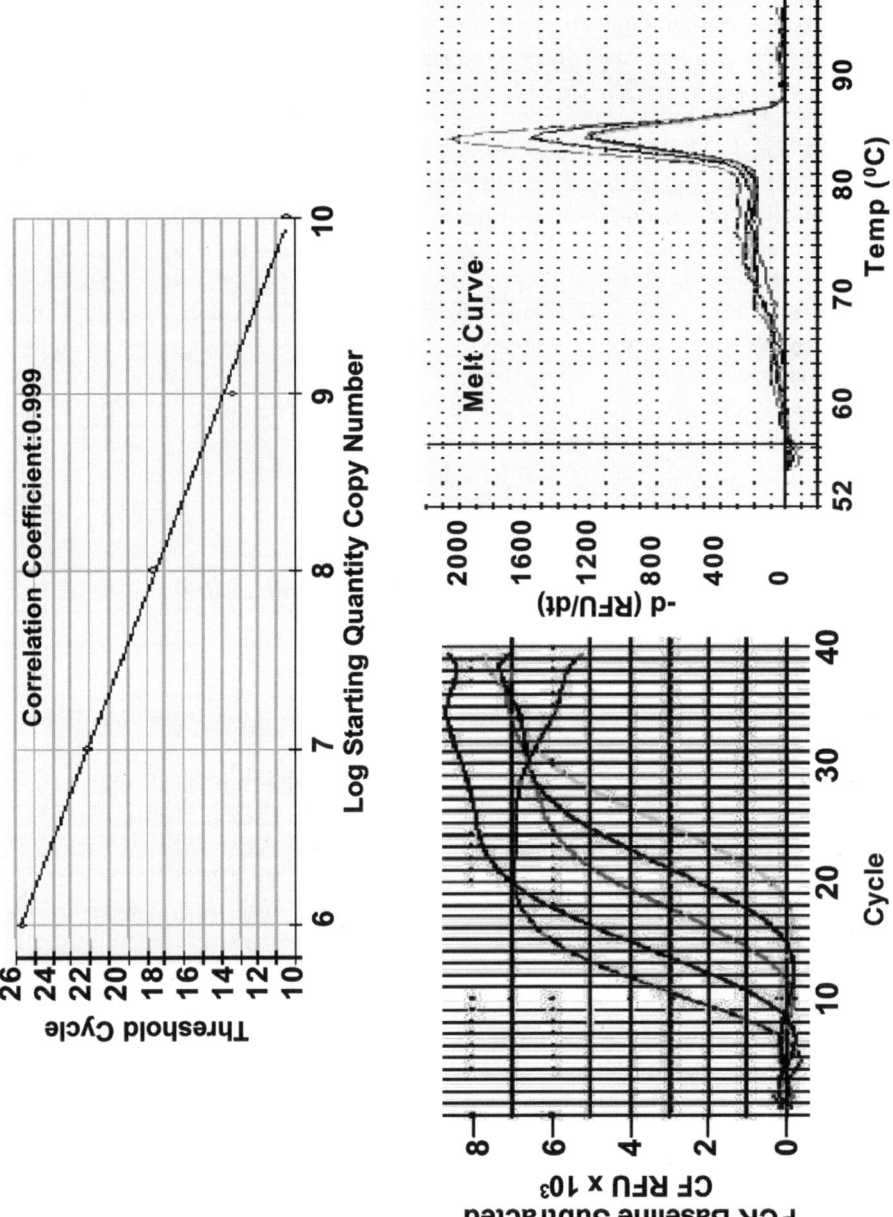

virus stocks by selecting individual recombinant viruses for amplification and/or purification. A protocol modified from Becker et al. *(10)* is used:

1. Grow HEK 293 cells to confluency in 60-mm culture dishes.
2. Prepare serial dilutions (10^{-2}–10^{-9}) of CVLs in 1 mL of DMEM (supplemented with FBS and pen/strep). Infect confluent HEK 293 cells with the viral dilutions (1 mL) for 1 h at 37°C. Aspirate the virus-containing medium and wash the cells with PBS.
3. Mix equal volumes of 1.3% agarose solution (in water) and culture medium (2X MEM) supplemented with FBS and pen/strep, and slowly add to the washed 293 cells (6 mL/dish). It is important that the temperature of the mixed agarose/MEM solution be high enough not to solidify in the tube, while kept sufficiently low not to harm the cells. Before mixing, the agarose solution should therefore be kept in a 42°C water bath and the MEM medium in a 37°C bath. Leave the dishes at room temperature for 30 min before transferring to an incubator at 37°C.
4. Plaques normally appear after 7–9 d. To facilitate counting and/or isolation of plaques, overlay the agarose with 1 mL of a solution of 0.025% neutral red (in PBS) and leave to be absorbed for 2 h. Aspirate the surplus liquid, and keep the dish in a 37°C incubator overnight before evaluating virus titers and/or picking single plaques for further processing.

3.5. Expression of Recombinant mPDE3B and mPDE3B604 in Cultured Cells

3.5.1. Infection of H4IIE Hepatoma Cells With Adenoviruses

1. Propagate H4IIE cells and maintain in DMEM supplemented with 10% FBS and 2 mM glutamine.
2. Amplify viruses containing WT mPDE3B and mutant mPDE3B cDNAs on a preparative scale in HEK 293 cells, and purify using CsCl gradients.
3. Infect exponentially growing H4IIE cells (3×10^6 cells/well in 60-mm Petri dishes) with adenovirus (2–30×10^2 virus particles/cell) in DMEM (without serum for 3 h), and then change to DMEM containing 10% FBS and culture for another 45 h. Also infect H4IIE hepatoma cells with β-gal adenovirus.
4. After 48 h of infection with adenoviruses, suspend the H4IIE cells in buffer A, homogenize (15–25 strokes) in a Dounce homogenizer (Kontes, Vineland, NJ) in buffer A containing 1% Nonidet P-40, sonicate (1×5 pulses, output 2, 30% cycle) at 4°C, and keep for 30 min at 4°C.
5. Centrifuge ($12,000g$, 20 min, 4°C) the homogenates to remove insoluble material.
6. Measure the protein using bicinchoninic acid protein assay reagent with BSA as the standard.
7. Detect the expression of WT and mutant mPDE3B proteins by SDS-PAGE Western blots of H4IIE cell lysates (30–40 µg of proteins) and immunoblotting with anti-PDE3B CT antibody (**Fig. 3**).
8. Assay PDE activities as previously described *(11)*.

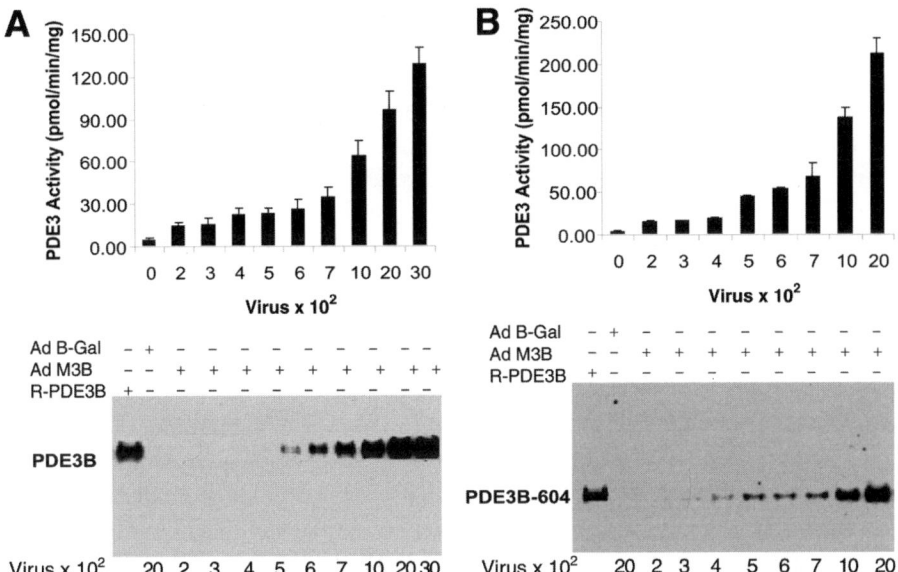

Fig. 3. Adenovirus-mediated gene delivery results in high level of mPDE3B (**A**) and mPDE3B604 protein expression (**B**) in H4IIE hepatoma cells. Viruses containing WT mPDE3B and mPDE3B604 cDNAs were amplified on a preparative scale in HEK 293 cells and purified using CsCl gradients. H4IIE cells were infected with viral concentrations of $2–30 \times 10^2$ virus particles/cell for 48 h. As a control for infection, H4IIE cells were also infected with β-gal adenovirus. Expression of mPDE3B and mPDE3B604 proteins was detected by Western blot of H4IIE cell lysates (30–40 μg of protein/lane) by SDS-PAGE (Novex, San Diego, CA) and immunoblotting with PDE3B-CT antibody; PDE3 activities were assayed in duplicate (mean ± 1/2 range) using ³H-cAMP as substrate (*11*).

3.5.2. Infection of FDCP2 Cells With Adenoviruses

1. Propagate the murine interleukin-3 (IL-3)-dependent hemopoietic cell line FDCP2, and maintain in RPMI-1640 medium supplemented with 10% FBS, 5% WEHI-3B-conditioned medium (which contains IL-3), and 2 m*M* glutamine.
2. Infect exponentially growing FDCP2 cells (2×10^5 cells/mL) with adenovirus ($2–30 \times 10^2$ particles/cell) for 48 h. Also infect FDCP2 cells with β-gal adenovirus to estimate the efficiency of transfection (after staining the cells for the expression of β-gal using a β-gal staining kit from Stratagene) and as a control for cell responses to insulin and forskolin.
3. After 48 h of infection with adenoviruses, collect the FDCP2 cells ($2–5 \times 10^5$ cells/mL), centrifuge (5 min, 1200*g*), and wash twice with PBS; suspend equivalent numbers of cells in serum-free RPMI-1640 medium containing TSB (5 μg/mL of transferrin, 10 μ*M* selenium, and 1 mg/mL of BSA); and incubate (3 mL

of cells/well, approx 7×10^5 cells/mL) in six-well Costar dishes (Cambridge, MA) for 3 h.

4. After 3 h at 37°C, stimulate the quiescent cells with insulin (100 n*M* for 10 min) or forskolin (100 μ*M* for 30 min) using one well each for control and insulin or forskolin and one well for protein analysis and Western blotting. In our experiment, stimulation of FDCP2 cells infected with mPDE3B adenovirus with insulin (100 n*M*, 10 min) or forskolin (100 μ*M* for 30 min) resulted in an approx 1.7-fold increase in endogenous and recombinant mPDE3B activity (*see* Note **3**).

3.5.3. Infection of Pancreatic Islets With Adenoviruses Expressing mPDE3B

Pancreases of Sprague-Dawley rats were digested by collagenase and islets of Langerhans were isolated by handpicking under a stereomicroscope. Islets were then maintained in RPMI-1640 containing 11.1 m*M* glucose supplemented with 10% FBS. Crude stocks of adenoviruses (not CsCl purified) housing mPDE3B or β-gal were titrated using a plaque assay and aliquots of high-titer virus suspensions (1×10^{10} plaque-forming units [PFU]/mL) were used to infect islets for 16 h. Islets were then homogenized in buffer B, followed by SDS-PAGE/ Western blot analysis for detection of PDE3B expression and assays of PDE activities (as previously described). Infection with the mPDE3B virus suspension at 25 μL/100 islets consistently produced a 5- to 10-fold increase in mPDE3B expression and activity (**Fig. 4A**).

Analyses of β-cell function were performed in a parallel set of islets infected for 16 h. Medium was discarded, and islets were washed and preincubated for 2 h in KRBB containing nonstimulatory concentrations of D-glucose (3 m*M*). Groups of three islets were then distributed to 96-well dishes, and insulin secretion was stimulated by exchanging the preincubation buffer for KRBB supplemented with 11.1 m*M* D-glucose in the absence or presence of the secretion-potentiating incretin glucagon-like peptide-1 (GLP-1) (100 n*M*). After stimulatory incubation for 1 h, buffer was withdrawn for determination of insulin concentration using a radioimmunoassay (RIA) system from Linco (St. Charles, MO). Significant blunting of insulin secretion was evident in mPDE3B-overexpressing islets stimulated with high glucose alone or high glucose in combination with GLP-1 (**Fig. 4B**), demonstrating an important role of this enzyme in cAMP-mediated insulin release *(12)* (*see* Note **4**).

3.6. Summary

By employing adenovirus-mediated gene transfer, mPDE3B and mPDE3B604 and β-gal were efficiently expressed, resulting in a 10- to 30-fold increase in PDE3 activity in FDCP2 and H4IIE cells. Adenovirus-mediated gene transfer is useful in studying the regulation of PDE3B in insulin-responsive cells because incubation of infected FDCP2 cells with insulin and forskolin resulted

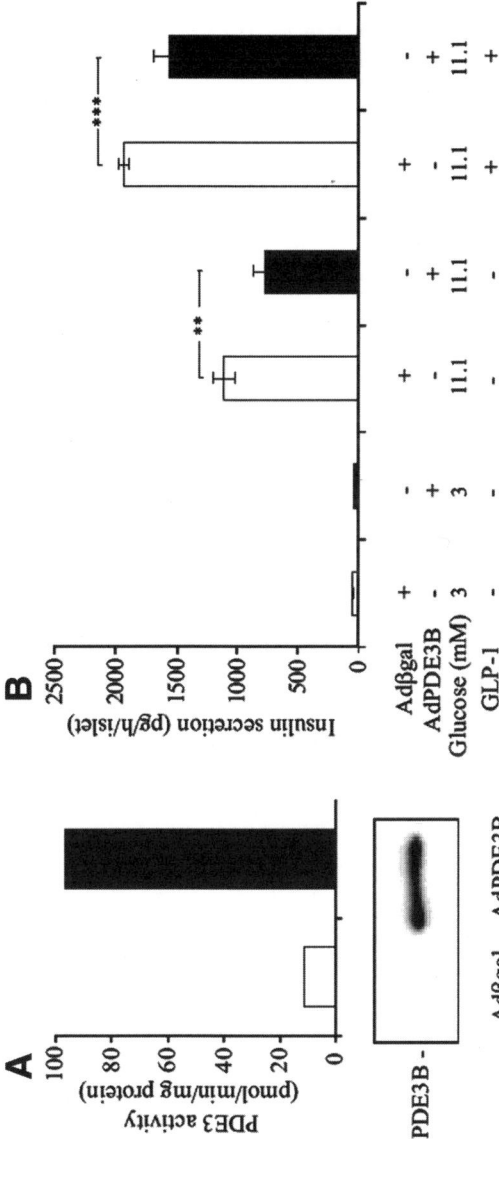

Fig. 4. Adenovirus-mediated overexpression of mPDE3B in rat pancreatic islets results in blunted insulin release. Adenoviruses harboring mPDE3B were amplified in HEK 293 cells, and the CVL was utilized to infect isolated rat islets for 16 h. Lysates were titered in a plaque assay, and 25 µL of high-titer stocks (1×10^{10} PFU/mL) was added to 100 islets. β-Gal virus was used as control of infection. (**A**) Expression of mPDE3B and PDE3 activities was analyzed as described in **Fig. 3**. (**B**) Insulin secretory capacity of mPDE3B-overexpressing β-cells was analyzed in static incubations of groups of three islets stimulated for 1 h with D-glucose at 3 and 11.1 mM, respectively, and D-glucose at 11.1 mM in combination with GLP-1 at 100 nM. Insulin was assayed in duplicate using RIA. The data shown (from **ref. 12**) represent three independent experiments. Statistical significance was evaluated using a student's t-test. **$p < 0.01$; ***$p < 0.001$.

in activation of recombinant mPDE3B. In isolated pancreatic islets infected with adenovirus expressing mPDE3B, a 5- to 10-fold increase in PDE3 activity resulted in significantly blunted insulin secretion. Expression of WT and mutant recombinant PDE3 isoforms should prove useful for structure/function analysis as well for further understanding the molecular mechanisms for the regulation of PDE3 activity and function in intact cells.

4. Notes

1. CVLs usually range from 10^7 to 10^{10} viral particles/mL, and CsCl-purified viral stocks range in titer from 10^{11} to 10^{13} viral particles/mL.
2. Using the intercalation dye SYBR Green I, iCycler iQ™ (Bio-Rad) accurately quantitates PCR products with a wide range of concentrations of viral particles, with a linear response ($R^2 = 0.99$). Melting curve analysis can specifically identify the nonspecific products.
3. Efficiency of gene transfer was assessed by infecting FDCP2 or H4IIE cells with a recombinant adenovirus expressing β-gal using an *in situ* β-gal staining kit (cat. no. 200384-2); greater than 75% of the cells were infected.
4. It is possible that in appropriate target cells, sufficient PDE protein will be synthesized to facilitate analysis of phosphorylation sites by mass spectroscopic analysis.

References

1. Conti, M. and Jin, S. L. (1999) The molecular biology of cyclic nucleotide phosphodiesterases. *Prog. Nucleic Acid Res. Mol. Biol.* **63,** 1–38.
2. Soderling, S. H. and Beavo, J. A. (2000) Regulation of cAMP and cGMP signaling: new phosphodiesterases and new functions. *Curr. Opin. Cell Biol.* **12,** 174–179.
3. Fawcett, L., Baxendale, R., Stacey, P., McGrouther, C., Harrow, I., Soderling, S., Hetman, J., Beavo, J. A., and Phillips, S. C. (2000) Molecular cloning and characterization of a distinct human phosphodiesterase gene family: PDE11A. *Proc. Natl. Acad. Sci. USA* **97,** 3702–3707.
4. Rybalkin, S. D., Yan, C., Bornfeldt, K. E., and Beavo, J. A. (2003) Cyclic GMP phosphodiesterases and regulation of smooth muscle function. *Circ. Res.* **93,** 280–291.
5. Francis, S. H., Turko, I. V., and Corbin, J. D. (2001) Cyclic nucleotide phosphodiesterases: relating structure and function. *Prog. Nucleic Acid Res. Mol. Biol.* **65,** 1–52.
6. Bloom, T. J. (2002) Cyclic nucleotide phosphodiesterase isozymes expressed in mouse skeletal muscle. *Can. J. Physiol. Pharmacol.* **80,** 1132–1135.
7. Wallis, R. M., Corbin, J. D., Francis, S. H., and Ellis, P. (1999) Tissue distribution of phosphodiesterase families and the effects of sildenafil on tissue cyclic nucleotides, platelet function, and the contractile responses of trabeculae carneae and aortic rings in vitro. *Am. J. Cardiol.* **83,** 3C–12C.

8. Degerman, E., Landstrom, T. R., Stenson Holst, L., Goransson, O., Ahmad, F., Shakur, Y., and Kenan, Y. (2000) A role for phosphodiesterase 3B in the antilipolytic action of insulin, in *Diabetes Mellitus: A Fundamental and Clinical Text*, 2nd ed., (LeRoith, D., Olefsky, J., Taylor, S., eds.), Lippincott, Williams, and Wilkins, Philadelphia, PA, pp. 284–291.

9. Shakur, Y., Stenson Holst, L., Landstrom, T. R., Movsesian, M., Degerman, E., and Manganiello, V. (2001) Regulation and function of the cyclic nucleotide phosphodiesterase (PDE3) gene family. *Prog. Nucleic Acid Res. Mol. Biol.* **66,** 241–277.

10. Becker, T. C., Noel, R. J., Coats, W. S., Gomez-Foix, A. M., Alam, T., Gerard, R. D., and Newgard, C. B. (1994) Use of recombinant adenovirus for metabolic engineering of mammalian cells. *Methods Cell. Biol.* **43,** 161–189.

11. Kincaid, R. L. and Manganiello, V. C. (1988) Assay of cyclic nucleotide phosphodiesterase using radiolabeled and fluorescent substrates. *Methods Enzymol.* **159,** 457–470.

12. Härndahl, L., Jing, X. J., Ivarsson, R., Degerman, E., Ahrén, B., Manganiello, V. C., Renström, E., and Stenson Holst, L. (2002) Important role of phosphodiesterase 3B for the stimulatory action of cAMP on pancreatic beta-cell exocytosis and release of insulin. *J. Biol. Chem.* **277,** 37,446–37,455.

9

Identification of Promoter Elements in 5'-Flanking Region of Murine Cyclic Nucleotide Phosphodiesterase 3B Gene

Hanguan Liu, Jing Rong Tang, Eva Degerman, and Vincent C. Manganiello

Summary

We describe techniques for identifying functional promoter elements in the 5'-flanking region of the murine cyclic nucleotide phosphodiesterase 3B (mPDE3B) gene. The 5'-flanking region of the mPDE3B gene was cloned and sequenced, and putative transcription factor binding sites were identified with computational tools. A series of reporter plasmids containing the luciferase gene fused to different fragments of the 5'-flanking region of the mPDE3B gene was constructed and used to transfect 3T3-L1 fibroblasts or differentiating adipocytes. Reporter gene assays showed that there are two promoter regions in the 5'-flanking region in the mPDE3B gene: a distal region located approx 4 kb upstream of the translation initiation site that contains cAMP-response element (CRE) cis-acting elements, and a proximal region that is GC rich and lacks TATA sequences. The distal promoter region induced much higher luciferase activity than did the proximal one. Mutation of the CRE sequences or reversal of the orientation of the CRE-containing region abolished promoter activity of the distal region. Electrophoretic mobility shift assay analysis indicated that binding to CRE elements was greater in nuclear extracts from differentiating adipocytes than from fibroblasts. Mapping of transcription initiation sites suggested that the distal promoter region might function as an enhancer, whereas the proximal promoter drives transcription of the mPDE3B gene.

Key Words

Phosphodiesterase 3B; DNA sequencing; promoter; reverse transcriptase-polymerase chain reaction; mutation; electrophoretic mobility shift assay; transcription initiation.

1. Introduction

Cyclic nucleotide phosphodiesterase 3B (PDE3B) is relatively highly expressed in tissues central to regulation of energy homeostasis, including pancreatic

From: *Methods in Molecular Biology, vol. 307: Phosphodiesterase Methods and Protocols*
Edited by: C. Lugnier © Humana Press Inc., Totowa, NJ

islets, liver, and adipose tissue *(1–3)*. PDE3B is thought to be important in the control of cyclic adenosine monophosphate (cAMP) pools involved in insulin-induced inhibition of lipolysis and glycogenolysis in adipocytes and hepatocytes, respectively, and in modulation of insulin secretion from pancreatic β-cells *(1–3)*. Mechanisms involved in acute activation of PDE3B by insulin and other effectors have been extensively studied *(1,2,4)*. Much less is known regarding long-term regulation of PDE3B gene expression. Identification of promoter regions and key regulatory elements in the PDE3B gene will facilitate the understanding of how PDE3B gene expression can be regulated in different tissues or cells by various effectors, including cAMP, tumor necrosis factor-α, and ceramide *(5–7)*.

To identify promoter elements in the mouse (m)PDE3B gene, we cloned the 5′-flanking region of the gene and constructed a series of luciferase reporter plasmids. These constructs allowed identification of distal (~4 to 5 kb upstream of the translation initiation ATG) and proximal (~0.5 kb upstream of the translation initiation ATG) promoter regions and binding sites for transcription factors. Electrophoretic mobility shift assay (EMSA) and site-directed mutagenesis confirmed the binding of transcription factors to *cis*-elements in the mPDE3B distal promoter region. To map transcription initiation sites (TISs), 5′-rapid amplification of cDNA ends (5′-RACE), cloning and sequencing, reverse transcriptase (RT)(5′-RACE)-polymerase chain reaction (PCR), and ribonuclease protection assays (RPAs) were utilized. Our results indicated that transcription of the mPDE3B gene was initiated at multiple sites within the proximal promoter region (GC rich, lacking TATA sequences); no TISs were detected by 5′-RACE, RT-PCR, and RPA in the distal promoter and its immediate downstream genomic region.

In this chapter, we describe our protocols for conducting these experiments. Because kits from different commercial sources are available for many of the approaches and analyses, we recommend that manufacturer's protocols be followed when commercial kits are used. We only briefly describe portions of the protocols from commercial kits that we used.

2. Materials

1. Tissue culture reagents: Dulbecco's modified Eagle's medium (DMEM), fetal bovine serum (FBS), penicillin/streptomycin, trypsin-EDTA (Invitrogen, Carlsbad, CA).
2. BCA protein assay kit, bovine serum albumin (BSA), and SuperSignal RPAIII® Chemiluminescence Detection Kit (Pierce, Rockford, IL).
3. Enzymes for DNA manipulation (Boehringer Mannheim, Indianapolis, IN).
4. α-^{32}P dATP or γ-^{32}P ATP (Amersham, Piscataway, NJ).
5. Vectors (Basic-pGL3, SV40 promoter-pGL3, and SV40 promoter/enhancer-pGL3, which contain the firefly luciferase gene, and pRL-TK, which contains thymidine kinase [TK] promoter and the Renilla luciferase gene), TransFast™ Reagent, Dual-

    Luciferase™ Reporter Assay System, and Gel Shift Binding Buffer (Promega, Madison, WI).

6. SMART™ RACE cDNA Amplification, AdvanTAge™ PCR Cloning, and Advantage®-GC cDNA PCR kits (Clontech, Palo Alto, CA).
7. Poly(A)Pure™ mRNA isolation kit (Ambion, Austin, TX).
8. Plasmid DNA purification kit (Qiagen, Studio City, CA).
9. ABI PRISM DNA sequence kit (Perkin-Elmer, Foster City, CA).
10. Mouse 3T3-L1 fibroblasts (American Type Culture Collection, Rockville, MD).
11. FastMedia™ LB Agar Amp and FastMedia™ LB Liquid Amp (Fermentas, Hanover, MD).
12. Agarose electrophoresis equipment.
13. MicroSpin™ G-25 columns (Amersham).
14. Tris/borate/EDTA buffer: 50 mM Tris-HCl, 50 mM boric acid, 1 mM EDTA (pH 8.3).

3. Methods

3.1. Cloning of 5′-Flanking Region of mPDE3B Gene

Screening genomic DNA libraries has proven to be a quick and efficient way to clone genes. For identification of promoter regions in the mPDE3B gene, the 5′ end of the gene was cloned and sequenced.

Mouse 129/SvJ and Balb/c genomic libraries were screened using ^{32}P-labeled mPDE3B and rat PDE3B cDNAs as probes. Two sequences from the 5′ end of the mPDE3B gene were cloned (*see* **Note 1**) and submitted to GenBank (GenBank accession nos. AY159890 and AF547434). The 5′ end of the mPDE3B gene was subcloned into pBluescript and used to construct mPDE3B-luciferase reporter vectors. A search of transcription factor–binding sites (TFBSs) in AY159890 at http://www.cbrc.jp/research/db/TFSEARCH.html was conducted. There are binding sites for activator protein 2 (AP-2), specificity protein 1 (Sp-1), and CRE *cis*-acting elements in the *Sal*I-*Xba*I fragment of AY159890 (*see* **Fig. 1A**). The 5′-flanking sequence as well as the ATG and the open reading frame (ORF) of putative exon 1 of the mPDE3B gene are depicted in **Fig. 1A**, and putative TFBSs in **Fig. 1B**.

3.2. Construction of pGL3 Reporter Plasmids With the Luciferase Gene Fused to Different Fragments of the 5′-Flanking Region of mPDE3B Gene

The pGL3-Basic luciferase reporter vector provides a basis for the quantitative analysis of factors that potentially regulate mammalian gene expression. These factors may be *cis*-acting, such as promoters and enhancers, or *trans*-acting, such as various DNA-binding factors. The pGL3-Basic vector contains a modified coding region for firefly luciferase that has been optimized for monitoring transcriptional activity in transfected eukaryotic cells (*see* **Fig. 2**). The

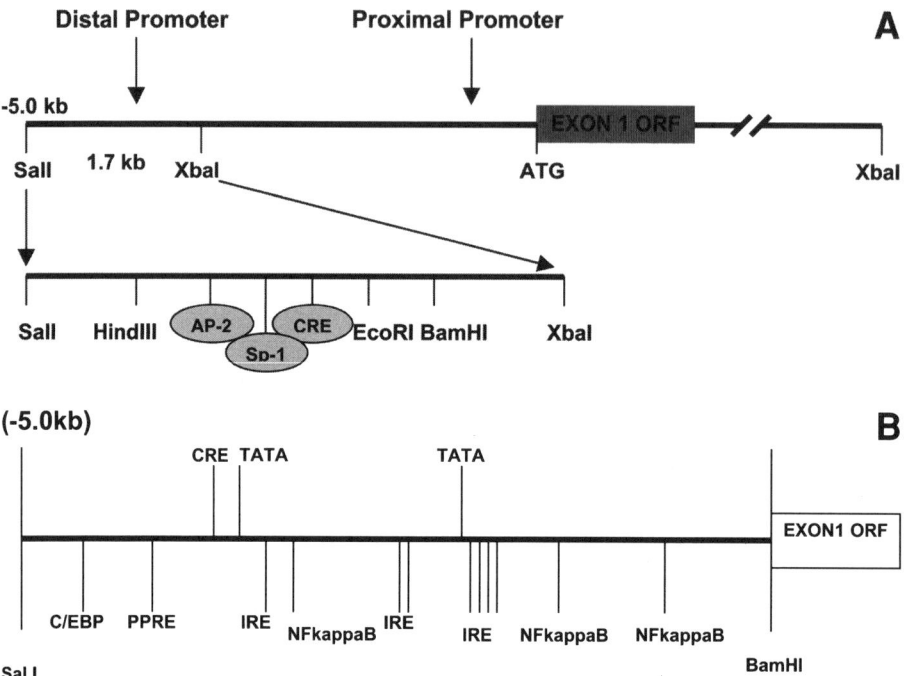

Fig. 1. (**A**) 5′-Flanking region of mPDE3B gene. The 5′-flanking region (~5.0 kb) of the mPDE3B gene was cloned. Arrows identify the proximal and distal promoter regions. The *Sal*I-*Xba*I genomic fragment (~1.7 kb) that contains the distal promoter region is depicted separately. The ATG initiation codon is also shown. (**B**) Putative TFBSs (consensus sequences) in 5′-flanking region of mPDE3B gene. The 5′-flanking regions (AY159890 and AF547434) of mPDE3B gene were checked for putative TFBSs with TFSEARCH (http://www.cbrc.jp/research/db/TFSEARCH.html) and MatInspector (http://www.genomatix.de/cgi-bin/matinspector/matinspector.pl). IRE, insulin-response element; PPRE, peroxisome proliferator-activated receptor response element; C/EBP, CCAAT, enhancer binding protein.

assay of the luciferase activity of this reporter is rapid, sensitive, and quantitative. Because the pGL3-Basic vector lacks promoter and enhancer elements, insertion of foreign genomic DNA into the multiple cloning sites of the vector allows detection of any promoter elements existing in the DNA insert.

mPDE3B reporter plasmids were generated by insertion of mPDE3B genomic fragments and constructs into the pGL3-Basic vector. The size and orientation of all constructs were verified by mapping with corresponding restriction enzymes. Distal and proximal region constructs were generated by restriction enzyme digestion and subsequent ligation into pGL3-Basic vector (**Fig. 2**). Distal-proximal region fusion fragments were constructed by enzymatic restriction and ligation.

A

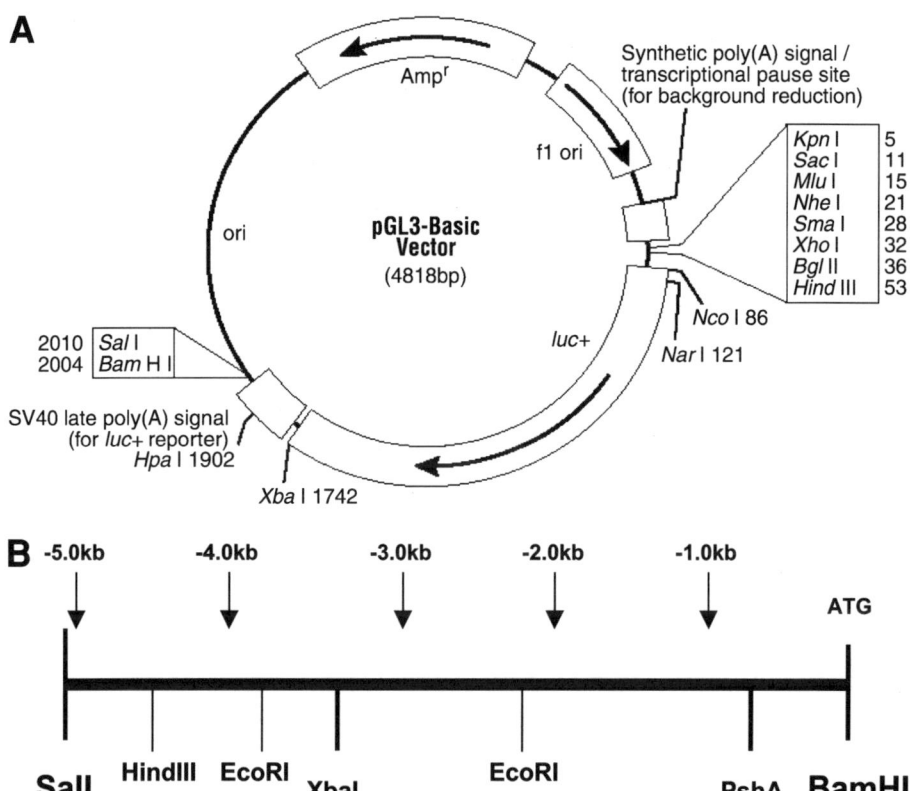

Kpn I	5
Sac I	11
Mlu I	15
Nhe I	21
Sma I	28
Xho I	32
Bgl II	36
Hind III	53

Fig. 2. **(A),(B)** pGL3-mPDE3B luciferase reporter constructs. pGL3-Basic Vector contains neither promoter nor enhancer sequences. A series of DNA fragments of the 5′-flanking region of mPDE3B gene was excised with appropriate restriction enzymes and blunt-ended with Klenow enzyme. The resulting fragments were then inserted into the polylinker region of the pGL3-vector and fused to the downstream firefly luciferase reporter gene. These constructs were used to transfect 3T3-L1 cells and to measure the relative activity of the reporter gene. (Reproduced from Promega.)

Expression vectors containing CRE, Sp-1, and AP-2 sequences were generated from the *Sal*I/*Xba*I fragment of AY159890 (*see* **Fig. 1A**).

A negative control luciferase expression vector containing a mutant CRE element was also created. The sequence and orientation of these pGL3 constructs were also confirmed by automated sequencing.

3.3. Transient Transfection With TransFast™ Reagent (Promega) and Dual Luciferase Assay

Reporter gene assays are widely used to study transcriptional regulation. Transient transfection is often used as a powerful tool in such studies. Among experimental samples, however, inherent variations exist at different stages of the transfection and assay processes. Such variations could be the result of different cell densities at the time of transfection, transfection efficiencies, efficiencies of cell lysis, recovery of samples, and amounts of samples used in the assay. To correct for variables that could confound interpretation of data, a second control reporter under a constitutive promoter is often cotransfected into cells. This control reporter acts as an internal control to normalize experimental variations, such as different transfection efficiencies among samples. Promega's Dual-Luciferase® Reporter (DLR™) Assays (cat. nos. E1910, E1960, E1980) are efficient means of performing dual reporter assays. In the DLR Assay, the activities of firefly (*Photinus pyralis*) and Renilla (*Renilla reniformis*) luciferases are measured sequentially in a single sample. The Renilla luciferase activity serves as an internal control.

3T3-L1 fibroblasts were routinely cultured in DMEM supplemented with 10% calf serum, 100 U/mL of penicillin, and 100 µg/mL of streptomycin at 37°C in a 95% air, 5% CO_2 humidified atmosphere. To induce differentiation, culture medium were removed and replaced with fresh culture medium supplemented with 300 μM isobutylmethylxanthine (M), 1 μM dexamethasone (D), and 5 µg/mL of insulin (I) namely MDI. Preadipocytes were incubated with differentiation agents for 5 d or as otherwise indicated.

Cells were cotransfected with pRL-TK and pGL3 reporter plasmid vectors that expressed Renilla and firefly luciferase activities, respectively. For the different pGL3 reporter plasmids, firefly luciferase expression was driven either by mPDE3B 5′-flanking region genomic fragments or constructs (pGL3-mPDE3B), by the SV40 enhancer and SV40 promoter as positive controls, or by promoter-free pGL3 (pGL3-Basic) as a negative control.

Transfection was performed using TransFast™ Reagent according to the manufacturer's instructions (Promega). On the day before transfection, 400 µL of nuclease-free water was added to the TransFast™ Reagent, and the lipid film was dispersed by vigorous vortexing for 10 s.

After 24 h, a mixture of plasmid DNA, TransFast™ Reagent (3 µL), and serum- and antibiotics-free medium (DMEM, 1 mL) was prepared, incubated for 10–15 min at room temperature, and used to transfect the cells. After removal of the culture medium, 3T3-L1 fibroblasts were washed once with DMEM and cotransfected with the DNA mixture (pRL-TK plasmid vector [0.02 µg] and pGL3 plasmid vector constructs [1.0 µg]) in six-well culture dishes to about 80% confluence. The cells were incubated for 1 h at 37°C and then overlaid with prewarmed DMEM containing 10% FBS (2 mL) and maintained for 48 h at 37°C. For transfection of differentiating 3T3-L1 adipocytes, fibroblasts were grown in DMEM containing 10% FBS to about 80% confluence, when differentiation was induced by the addition of MDI. After 72 h, differentiating 3T3-L1 cells were cotransfected with pRL-TK and with pGL3 containing mPDE3B genomic DNA inserts.

After 48 h at 37°C, the cotransfected cells were washed twice with phosphate-buffered saline (PBS), disrupted in 250 μL of Passive Lysis Buffer (Promega), and centrifuged (1000*g*, 5 min). The supernatants were used for measuring both firefly and Renilla luciferase activities with the Dual-Luciferase Reporter Assay System (Promega) using a luminometer (Lumat LB 9501). Luciferase reagent substrate (100 μL) was mixed with cell lysates (20 μL), and firefly luciferase activity was measured three times. To terminate the reaction and initiate the Renilla luciferase reaction, Stop & GloR Solution (100 μL) was added to the mixture, and Renilla luciferase activity was measured three times. The firefly luciferase activity in lysates from cells transfected with pGL3-mPDE3B reporter constructs, and positive and negative controls, was normalized to Renilla luciferase activity values, and the ratio between firefly and Renilla luciferase activities was referred to as relative luciferase activity (RLA). A representative experiment, shown in **Fig. 3**, demonstrated that the distal promoter region (pGL3-SX construct) induced much higher luciferase activity than did the proximal one (pGL3-PshA construct); the mutated CRE construct, pGL3-mCRE, and the orientation-reversed constructs, pGL3-SX* as well as pGL3-PshA*, exhibited markedly reduced promoter activities.

3.4. Preparation of Nuclear Extracts for EMSA

Nuclear extracts from cells were used to assess specific DNA-protein binding abilities, which can be readily detected by EMSA.

1. Treat 3T3-L1 fibroblasts with or without differentiating agents (MDI) for 5 d. Wash the cells twice with PBS, scrape in PBS, and centrifuge for 5 min at 500*g*.
2. Prepare nuclear extracts on ice from the cells. Briefly, suspend cell pellets in ice-cold hypotonic buffer containing 10 mM HEPES-KOH (pH 7.9), 10 mM KCl, 1.5 mM MgCl$_2$, 1 mM dithiothreitol (DTT), 1 mM phenylmethylsulfonyl fluoride (PMSF), and 2 μg/mL each of aprotinin, pepstatin, and leupeptin.
3. After 15 min on ice, add Nonidet P-40 to a final concentration of 0.6% (v/v). Centrifuge the cell suspension for 5 min at 5000*g*.
4. Resuspend the pelleted nuclei in a high-salt buffer containing 20 mM HEPES-KOH (pH 7.9), 420 mM NaCl, 1.5 mM MgCl$_2$, 0.2 mM EDTA, 25% glycerol (v/v), 1 mM DTT, 1 mM PMSF, and 2 μg/mL each of aprotinin, pepstatin, and leupeptin to solubilize DNA-binding proteins.
5. Gently shake the suspension of nuclei for 30 min at 4°C and then centrifuge for 20 min at 12,000*g*.
6. Divide the resulting supernatants containing nuclear proteins into small aliquots and store at −70°C until use for EMSA. Determine protein concentrations using a Bio-Rad Detergent-compatible Protein Assay Kit with BSA as the standard.

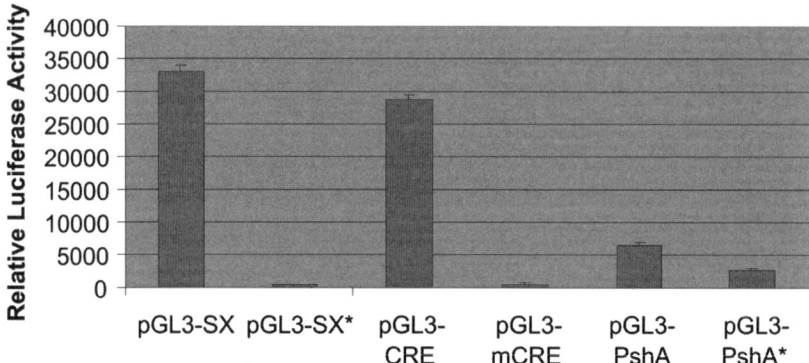

Fig. 3. Promoter activity of CRE *cis*-acting element in mPDE3B gene. Luciferase reporter expression vectors, containing wild-type (WT), orientation-reversed, or mutated CRE *cis*-acting elements in the distal promoter region (pGL3-SX [wild], pGL3-SX* [orientation reversed], pGL3-CRE [wild], pGL3-mCRE [mutated]) and elements in the proximal region (pGL3-PshA [wild], pGL3-PshA* [orientation reversed]) were generated. 3T3-L1 preadipocytes were transfected with the reporter gene constructs. Promoter activity was measured as RLA (mean ± SD; $n = 3$). Promoter-free vector, pGL3-Basic, was also transfected and exhibited little or no promoter activity.

3.5. Electrophoretic Mobility Shift Assay

EMSA is a powerful tool for evaluating DNA-protein interactions. The assay is based on the principle that when subjected to electrophoresis, DNA-protein complexes migrate slower than unbound DNA in a native polyacrylamide or agarose gel, resulting in a "shift" in migration of the labeled DNA band. Labeled DNA containing the binding site of interest is incubated with a nuclear extract or purified factor. The reaction mixture is then subjected to gel electrophoresis on a native polyacrylamide gel and transferred to a nylon membrane for detection/analysis of the labeled DNA. The labeled DNA can be detected based on either radioactive or nonradioactive labeling.

1. Synthesize and purify complementary oligonucleotides derived from the putative sequences in the 5′-flanking region of mPDE3B gene.
2. Denature complementary strands for 5 min at 95°C and then anneal by naturally cooling to room temperature. Use the double-strand DNA (sense and antisense) for making probes or as cold competitors.
3. End-label a probe corresponding to CRE (CTTCCC CGT GAC GTC AAC TCG GCC GAT) elements with γ-^{32}P-ATP using T4 polynucleotide kinase, and purify with MicroSpin G-25 columns.

4. Use unlabeled double-stranded oligonucleotides and mutant oligonucleotides with adenine nucleotide replacement of the CRE-binding sequences as sequence-specific and nonspecific competitors, respectively. Use consensus oligonucleotides with CRE-binding sequences from Promega as positive control probes.

5. To test the specificity of the DNA-protein-binding reactions, add excess unlabeled consensus, WT, or mutated oligonucleotides to the reaction mixture 15 min before adding the labeled probe.

6. Subject oligonucleotide-protein complexes to electrophoresis in 6% DNA retardation gels with Tris/borate/EDTA buffer at 30 mA of constant current at 4°C.

7. Dry the gels before exposing to Kodak-XAR film at –70°C. A representative EMSA experiment, suggesting increased binding to CRE elements in differentiating adipocytes, is shown in **Fig. 4**.

3.6. Mapping of TISs of mPDE3B Gene Using 5′-RACE, RPA, and RT-PCR

TISs can be determined by identifying the 5′ end of the encoded mRNA. It is generally assumed that the sequence at the 5′ end of an mRNA corresponds to the DNA sequence at which transcription initiates. This is not always an accurate assumption, because mRNAs can be cleaved or degraded, which may expose a 5′ end that does not correspond to the authentic start site. Confirmation that the start site has been mapped correctly is provided by additional experiments demonstrating that the surrounding DNA contains a functional promoter with discrete sequence elements that control initiation from the site identified. To identify the 5′ end of mRNA, several methods can be used. Ribonuclease protection, 5′-RACE, and primer extension remain the most common approaches. The following sections describe our approaches to map the TISs in the 5′ end of the mPDE3B gene.

3.6.1. Isolation of mRNA From Mouse 3T3-L1 Fibroblasts

1. Isolate total RNA from differentiated 3T3-L1 adipocytes using Poly(A)Pure™ mRNA isolation kits (Ambion). Trypsinize and harvest differentiated 3T3-L1 adipocytes by centrifuging for 10 min at 600g.

2. Wash the cell pellets (~10^7 cells) by gently dispersing in PBS and centrifuging for 5 min at 1000g.

3. Homogenize the washed cells in 300 μL of the Ambion lysis solution and dilute twofold with dilution buffer according to the manufacturer's instructions.

4. Centrifuge the mixture at 12,000g for 15 min at 4°C to remove cellular debris; add the supernatant to oligo dT cellulose, and incubate for 1 h at room temperature with gentle shaking before centrifuging at 2000g for 3 min at room temperature.

5. Wash the pellet three times with washing buffer, and elute bound poly(A)-mRNA with 100 μL of elution buffer for 10 min at 65°C.

Mutant Probe	-	-	-	-	20	40	-	-	-	-	20	40
Cold Probe	-	-	20	40	-	-	-	-	20	40	-	-
Labeling Probe	+	+	+	+	+	+	+	+	+	+	+	+
Nuclear Extract	-	+	+	+	+	+	-	+	+	+	+	+

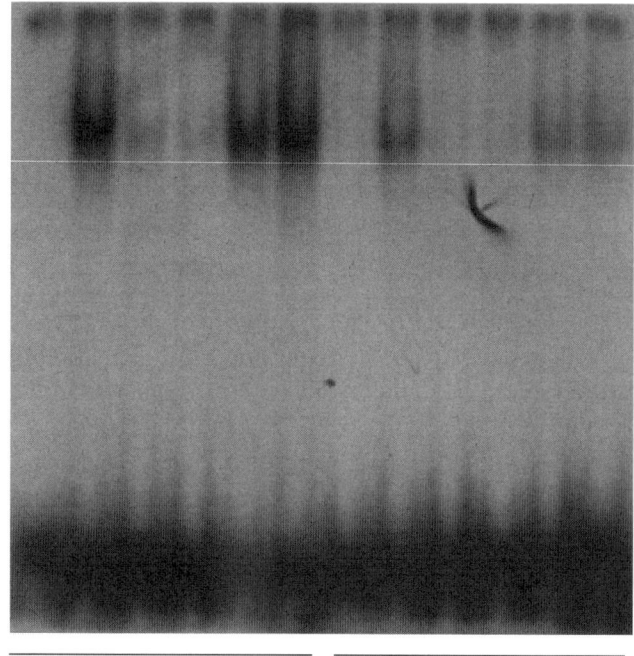

Differentiating Adipocytes   **3T3-L1 Fibroblasts**

Fig. 4. EMSA analysis of *cis*-acting CRE elements in distal promoter region of mPDE3B gene. Double-stranded oligonucleotides with WT CRE sequence (from the distal promoter region of mPDE3B gene) were labeled with ^{32}P and used as labeling probes. Unlabeled WT CRE and its corresponding mutant oligonucleotides with adenine substitutions were used as specific (cold probe) and nonspecific (mutant probe) competitors, respectively. Mobility shift assays were performed as described with 1.75 pmol of ^{32}P-labeled double-stranded CRE oligonucleotides and nuclear extracts (5 µg of protein) from 3T3-L1 fibroblasts and differentiating adipocytes. Sequences of the double-stranded WT CRE oligonucleotide are as follows: CRE (mPDE3B) 5′-CTTCCCCGTGACGTCAACTCGGCCGAT-3′ and 3′-GAAGGGGCACTGCAGT TGAGCCGGCTA-5′. For mutant oligonucleotides, underlined bases were replaced with all As.

6. Pellet the poly(A)-mRNA with glycogen by centrifuging at 12,000*g* for 20 min at 4°C, and dissolve in 10 μL of diethylpyrocarbonate-treated water.

3.6.2. 5′-RACE of Mouse 3T3-L1 Adipocyte mRNA

5′-RACE of mouse 3T3-L1 adipocyte mRNA was conducted using a SMART™ RACE cDNA Amplification kit (Clontech). Briefly, 1 μg of poly(A)-mRNA was used for synthesis of double-stranded cDNA with 5′-GGG-linked universal primers.

PCR amplification of the mPDE3B cDNA from 5′-RACE was performed using an mPDE3B gene-specific antisense primer and Universal Primer Mix. RT-PCR with three other pairs of primers from different regions of mPDE3B cDNA yielded products with different sizes and were used as RT-PCR controls.

The PCR products were directly cloned into pT-Adv vector and transformed into competent *Escherichia coli* using an AdvanTAge PCR Cloning kit (Clontech). After colony selection and nested PCR, positive clones were sequenced, and several TISs were identified and are shown in **Fig. 5B**.

3.6.3. RPA for Mapping TISs of mPDE3B Gene

Because of its high sensitivity and resolution, RPA is well suited for mapping internal and external boundaries in mRNA. The basic requirement for mapping TISs using RPAs is that the probe span the region to be mapped. This usually means that the probe is derived from genomic DNA, as opposed to cDNA. For example, to map the TIS for a given mRNA, a probe is prepared by subcloning and transcribing a genomic fragment that extends from upstream of the gene of interest to some point in the first exon. Probe synthesis, purification, hybridization, and ribonuclease digestion are carried out using standard RPA protocols. The transcription start site is mapped by comparing the size of the protected fragment with the size of the undigested probe.

Poly(A)-mRNA was isolated from 3T3-L1 fibroblasts or differentiated adipocytes treated with MDI for 7 d using an Ambion Poly(A)Pure™ mRNA purification kit. A SuperSignal RPAIII® kit and protocol from Pierce was used for RPA. Briefly, a set of PCR primers containing a bacteriophage T7 promoter for amplification of targeted regions with antisense orientation in the mPDE3B gene was synthesized (*see* **Note 2**). PCR amplification of the 500-bp region upstream of the first ATG codon was performed using mouse genomic DNA as the template. The PCR products were purified by agarose gel.

Biotin-14-CTP, NTP mix, and T7 RNA polymerase from Invitrogen were used to synthesize RNA probes, which were purified by Amersham ProbeQuant G50 Micro Columns (*see* **Note 3**). Control RNA (total RNA from yeast) was from Pierce, and the mRNA used for RPA of the PDE3B gene was isolated from 3T3-L1 preadipocytes or differentiating adipocytes incubated with MDI

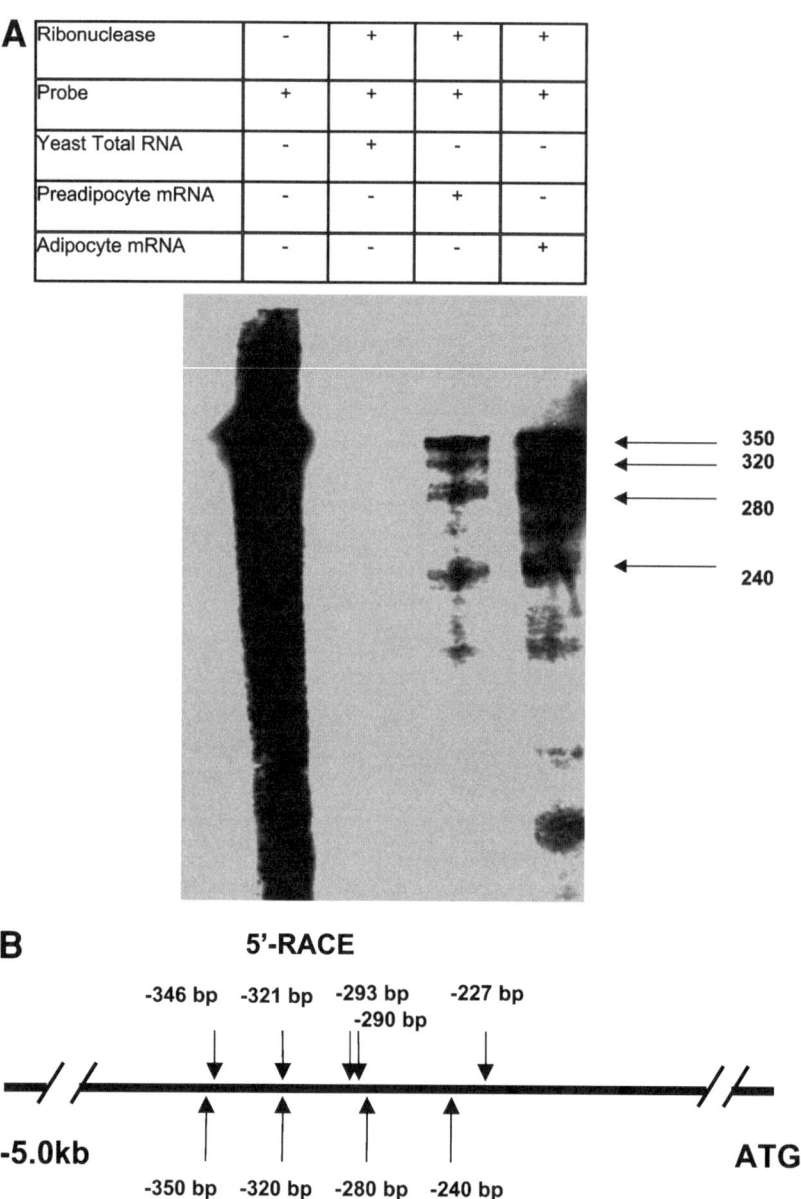

Fig. 5. RPA of TISs of mPDE3B gene. Control RNA (total RNA from yeast) was from Pierce, and the mRNA used for RPA of the mPDE3B gene was isolated from 3T3-L1 preadipocytes or differentiating adipocytes incubated with MDI for 7 d. Twenty micrograms of total RNA or 1 µg of poly(A)-mRNA was hybridized with biotin-labeled RNA probes. The protected RNA fragments were resolved by 6%

for 7 d. Total RNA (~20 μg) or 1 μg of poly(A)-mRNA was hybridized with biotin-labeled RNA probes. The protected RNA fragments were resolved by 6% polyacrylamide TBE-urea gel, transferred to Nylon+ membranes, and detected with Pierce's SuperSignal RPAIII® Chemiluminescence Kit. One representative result shown in **Fig. 5A** demonstrated that there were at least four transcripts protected with sizes of 350, 320, 280, and 240 bp, and that there were multiple TISs in the mPDE3B gene, which is consistent with the results from 5′-RACE. In addition, much greater amounts of mRNA were protected from adipocytes than from fibroblasts, consistent with the marked increase in mPDE3B expression during differentiation of adipocytes.

3.6.4. 5′-RACE(RT)-PCR Amplification of Possible mPDE3B mRNA in Distal Promoter Region

5′-RACE and subsequent PCR cloning as well as clony screening are relatively time-consuming. Direct sequencing of 5′-RACE(RT)-PCR products can quickly detect amplified transcripts. Although 5′-RACE and RPA identified TISs in the proximal promoter region, no TISs were mapped in the distal promoter region using these two methods. To help determine whether the distal promoter in mPDE3B gene actually drives transcription, rather than works as an enhancer in gene transcription, further attempts to identify any TISs in the distal promoter and its immediate downstream region were made using 5′-RACE(RT)-PCR using gene-specific primers.

3T3-L1 cells were cultured in DMEM containing 10% FBS. When they reached confluence, the cells were treated with 300 μM isobutylmethylxanthine for 3 d to increase expression of PDE3B, without inducing significant lipid accumulation. Total RNA was isolated via the TRIzol method, and poly(A)-mRNA was purified using a Qiagen Oligotex mRNA Mini Kit. 5′-RACE of mPDE3B mRNA was performed using a Clontech SMART™ RACE cDNA Amplification kit. PCR amplification between the distal CRE-containing region and ATG translation start codon region (*see* **Notes 4** and **5**) was performed using gene-specific and nested PCR primers (**Table 1**). PCR amplicants were purified and sequenced. Those amplicants that proved difficult to be sequenced were cloned into TOPO-TA vectors from Invitrogen. The TOPO plasmids were

---

Fig. 5. *(continued)* polyacrylamide TBE-urea gel, transferred to Nylon+ membranes, and detected with Pierce's SuperSignal RPAIII® Chemiluminescence Kit. **(A)** One representative result demonstrated that there are multiple TISs in the mPDE3B gene, which is consistent with the data from 5′-RACE-PCR, cloning, and sequencing. **(B)** Similar TISs of the mPDE3B gene were identified by either of two methods for mapping TISs: 5′-RACE and RPA.

Table 1
Primers Used in RT-PCR Amplification of Possible mPDE3B Gene Transcripts and in Mapping Potential Transcription Start Sites in Distal Promoter

Primer	Forward	Reverse 1	Reverse 2
1	−2985/−2962	−1029/−1051	−1025/−1040
2	−2849/−2829	−1029/−1051	−1025/−1040
3	−3206/−3186	−1029/−1051	−1025/−1040
4	−3699/−3679	−3399/−3420	
5	−3422/−3401	−2843/−2863	−2841/−2864
6	−3413/−3393	−2843/−2863	−2841/−2864
7	−4152/−4119	20-1	−3822/−3842
8	−4082/−4046	20-1	−3822/−3842
9	−4057/−4038	20-1	−3822/−3842
10	−2252/−2233	20-1	
11	−1538/−1519	20-1	
12	1501-1520	1601-1580	

transformed into DH5α *E. coli*. Positive clones were selected with amplicillin resistance and further cultured in Luria-Bertani (LB) broth. Vectors containing mPDE3B sequence were identified using PCR methods. *E. coli* harboring pTOPO-mPDE3B were grown overnight with vigorous shaking (240 rpm) in 5 mL of LB broth containing 50 µg/mL of ampicillin; plasmids were purified via the Qiagen Mini-prep vacuum method. Purified plasmid DNA was quantified by measuring the absorbance at 260 nm and was subjected to agarose gel electrophoresis for analysis of the sizes of the DNA fragments. Using Bigdye version 3.1 kits from ABI, the cloned cDNAs were sequenced. The reaction mixture was as follows: 8 µL of Bigdye, 4 µL of 5X buffer, 1 µL of 3.2 pmol primer, 200 ng of double-stranded plasmid DNA, and distilled H_2O to 20 µL. The PCR reaction was run as follows: 96°C for 10 s, 50°C for 20 s, 60°C for 4 min for 25 cycles, and ended at 4°C.

For purification of sequence reaction mixtures, Princeton CentriSep spin columns were soaked in distilled H_2O for 2 h and centrifuged at 720*g* for 2 min, and the sequence reaction mixture was loaded onto the center of the gel bed, centrifuged at 720*g* for 2 min, and kept in a SpeedVac for 30 min to dry out the liquid. Fifteen microliters of high-quality deionized formaldehyde was added to each vial to dissolve the residue, followed by vigorous vortexing and centrifugation. The vials were heated at 95°C for 2 min, chilled on ice for 5 min, and centrifuged. Ten microliters of the samples was used for DNA sequencing. Although control amplicants from the coding region of mPDE3B transcripts were successfully detected with the aforementioned procedure, no transcripts of the mPDE3B gene from the distal promoter region or its immediate downstream region were amplified. These results suggest, but do not definitively prove, that the distal region does not directly drive transcription of the mPDE3B gene.

4. Notes

1. After cloning the 5′-flanking region of mPDE3B gene, it is necessary to check the cloned DNA sequence for possible contamination with vector sequences.
2. To synthesize antisense RNA probes, PCR primers for amplification of targeted regions in the genomic DNA sequence were specifically designed to introduce a promoter, thereby allowing use of the corresponding RNA polymerase for the synthesis of transcript. The orientation of the promoter sequence is critical and should be fused to the 5′ end of the antisense sequence of the targeted region.
3. Synthesized biotin-labeled RNA probes are very stable and under ribonuclease-free conditions can be stored at −20°C for extended periods.
4. To amplify the 5′-flanking, GC-rich region of mPDE3B gene, the annealing temperature for PCR amplification was optimized, because a special design of the PCR primers for high annealing temperatures is required. Primers should have a T_m of approx 70°C to achieve optimal results. The length of primers should be at least 22 nucleotides, and each primer should have a G-C content of 40–60%. In addition, the 3′-terminal ends of each primer should not be complementary to each other and should contain a low G-C content.
5. Because the G-C content of the 5′-flanking region (~500 bp upstream to the ATG start codon) of mPDE3B gene is very high (up to 80%), sometimes multiple PCR bands were observed. Positive and negative controls were therefore included. Furthermore, "Touchdown" PCR was used, which significantly improved the specificity of the PCR reactions.

References

1. Shakur, Y., Holst, L. S., Landstrom, T. R., Movsesian, M., Degerman, E., and Manganiello, V. (2001) Regulation and function of the cyclic nucleotide phosphodiesterase (PDE3) gene family. *Prog. Nucleic Acid Res. Mol. Biol.* **66,** 241–277.
2. Degerman, E., Rahn-Lanström, T., Stenson-Holst, L., Goranssen, O., Harndahl, L., Ahmad, F., Choi, Y.-H., Masciarelli, S., Liu, H., and Manganiello, V. C. (2004) Role for phosphodiesterase 3B in regulation of lipolysis and insulin secretion, in *Diabetes Mellitus: A Fundamental and Clinical Text*, 3rd ed. (LeRoith, D., Taylor, S. I., and Olefsky, J. M., eds.), Lippincott Raven, Philadelphia, pp. 373–381.
3. Pyne, N. J. and Furman, B. L. (2003) Cyclic nucleotide phosphodiesterases in pancreatic islets. *Diabetologia* **46,** 1179–1189.
4. Degerman, E., Smith, C. J., Tornqvist, H., Vasta, V., Belfrage, P., and Manganiello, V. C. (1990) Evidence that insulin and isoprenaline activate the cGMP-inhibited low-Km cAMP phosphodiesterase in rat fat cells by phosphorylation. *Proc. Natl. Acad. Sci. USA* **87,** 533–537.

5. Rahn, L. T., Mei, J., Karlsson, M., Manganiello, V., and Degerman, E. (2000) Down-regulation of cyclic-nucleotide phosphodiesterase 3B in 3T3-L1 adipocytes induced by tumour necrosis factor alpha and cAMP. *Biochem. J.* **346(Pt.2),** 337–343.
6. Mei, J., Holst, L. S., Landstrom, T. R., Holm, C., Brindley, D., Manganiello, V., and Degerman, E. (2002) C(2)-ceramide influences the expression and insulin-mediated regulation of cyclic nucleotide phosphodiesterase 3B and lipolysis in 3T3-L1 adipocytes. *Diabetes* **51,** 631–637.
7. Zhang, H. H., Halbleib, M., Ahmad, F., Manganiello, V. C., and Greenberg, A. S. (2002) Tumor necrosis factor-alpha stimulates lipolysis in differentiated human adipocytes through activation of extracellular signal-related kinase and elevation of intracellular cAMP. *Diabetes* **51,** 2929–2935.

10

Purification of PDE6 Isozymes From Mammalian Retina

Dana C. Pentia, Suzanne Hosier, Rachel A. Collupy,
Beverly A. Valeriani, and Rick H. Cote

Summary

The photoreceptor phosphodiesterase (PDE6) is the central effector of visual transduction in vertebrate retinal photoreceptors. Distinct isozymes of PDE6 exist in rods and cones. Mammalian retina serves as an abundant source of tissue for PDE6 purification. Methods are described for the isolation and purification of membrane-associated PDE6 from rod outer segment membranes. Purification of cone PDE6 from the soluble fraction of retinal extracts is also described. Several procedures that can purify the rod and cone isozymes to homogeneity, including anion exchange, hydrophobic interaction, gel filtration, hydroxyapatite, and immunoaffinity chromatography, are presented. A method to activate PDE6 by limited proteolysis of its inhibitory γ-subunit is also provided.

Key Words

Photoreceptor; phosphodiesterase; retina; phototransduction; anion-exchange chromatography; cone and rod; hydrophobic interaction chromatography.

1. Introduction

The cyclic nucleotide phosphodiesterase (PDE) that is abundantly expressed in retinal photoreceptor cells constitutes the PDE6 family of PDEs. Features of PDE6 that distinguish it from other PDE families include primary localization in photoreceptive cells, a catalytic mechanism operating at the diffusion-controlled rate, association with a low molecular weight protein inhibitor of catalysis (γ-subunit), and attachment to cellular membranes via isoprenylated C-termini.

Rod and cone photoreceptors express distinct isoforms of PDE6. The rod PDE6 enzyme (PDE6R) is the only PDE that is a catalytic dimer of nonidentical catalytic subunits, α and β (P$\alpha\beta$). PDE6R forms a holoenzyme on binding of two identical γ-subunits (Pγ) to the catalytic $\alpha\beta$ dimer: $\alpha\beta\gamma_2$. Cone PDE6

From: *Methods in Molecular Biology, vol. 307: Phosphodiesterase Methods and Protocols*
Edited by: C. Lugnier © Humana Press Inc., Totowa, NJ

(PDE6C) is a catalytic dimer of the cone-specific α'-subunit, to which two cone-specific γ'-subunits bind to form $\alpha'_2\gamma'_2$. PDE6C, as well as a fraction of the total PDE6R, is found in a soluble form in mammalian retinal homogenates in association with a prenyl-binding protein, formerly defined as the δ-subunit of PDE6), which is believed to be responsible for solubilizing PDE6 from photoreceptor membranes *(1)*.

This chapter presents a set of methods for purification of each of the iso-forms of PDE6 from mammalian retina.

2. Materials

All solutions used to isolate and purify PDE6 are supplemented just before use with 1 mM dithiothreitol (DTT) and 0.3 mM phenylmethylsulfonyl fluoride (PMSF) or protease inhibitor cocktail (following the manufacturer's recommendations). All chromatography buffers are filtered with a 0.45-µm membrane under vacuum immediately before use to remove particulates and degas the solvent.

1. Bovine retinas (W. Lawson, Lincoln, NE).
2. Chromatography columns and media: Mono Q and Superdex prepacked columns, butyl-Sepharose, and Q-Sepharose (Amersham Biosciences), ceramic hydroxy-apatite (Type I, 40-µm particle size) (Bio-Rad, Hercules, CA), Sulfolink coupling gel and Immunopure (G) IgG Purification Kit (Pierce, Rockford, IL).
3. Membrane filtration devices (Millipore, Bedford, MA).
4. ROS1 monoclonal antibody (MAb) cell line: This was a kind gift from Dr. R. L. Hurwitz *(2)*.
5. Miscellaneous stock solutions: 1 M DTT in water, 100 mM PMSF in 95% ethanol, Mammalian Protease Inhibitor Cocktail (P8340; Sigma Chemical Corporation, St. Louis, MO).
6. Solution A: 20 mM 3-morpholinopropane-1-sulfonic acid (MOPS), pH 7.2; 2.0 mM MgCl$_2$, 60 mM KCl; 30 mM NaCl.
7. Solution B: 50% (w/v) sucrose in solution A. It is used to prepare the following sucrose density gradient solutions, which are checked for the correct density with a hydrometer at 4°C:
   a. Solution B1: 51 mL of B diluted to 100 mL with A (ρ = 1.105 g/mL).
   b. Solution B2: 54.25 mL of B diluted to 100 mL with A (ρ = 1.115 g/mL).
   c. Solution B3: 64.5 mL of B diluted to 100 mL with A (ρ = 1.135 g/mL).
8. Solution C: 45% (w/v) sucrose in solution A.
9. ROS membrane solutions:
   a. High Mg^{2+} hypotonic buffer: 5 mM Tris-HCl (pH 7.5 at 4°C), 10 mM MgCl$_2$, 10 mM DTT.
   b. PDE6R extraction buffer: 5 mM Tris-HCl (pH 7.5 at 4°C), 5 mM DTT.
10. Mono Q chromatography solutions:
   a. MQ-A buffer: 10 mM Tris-HCl, pH 7.5, 100 mM NaCl, 2 mM MgCl$_2$.
   b. MQ-B buffer: MQ-A containing 1.0 M NaCl.

11. Hydrophobic interaction chromatography (HIC) solutions:
    a. HIC-A: 400 mM (NH$_4$)$_2$SO$_4$, 5 mM Tris-HCl, pH 7.5.
    b. HIC-B: 5 mM Tris-HCl, pH 7.5.
12. Gel filtration chromatography (GFC) buffer: 10 mM Tris-HCl, pH 7.5, 300 mM NaCl, 2 mM MgCl$_2$.
13. Q-Sepharose chromatography solutions:
    a. Q-A buffer: 10 mM Tris-HCl, pH 7.5, 1 mM MgCl$_2$, 100 mM NaCl.
    b. Q-B buffer: 10 mM Tris-HCl, pH 7.5, 1 mM MgCl$_2$, 350 mM NaCl.
14. Hydroxyapatite (HAP) chromatography solutions:
    a. 0.5 M sodium phosphate solution, pH 7.2.
    b. HAP-A buffer: 75 mM sodium phosphate, pH 7.2, 50 mM NaCl.
    c. HAP-B buffer: 150 mM NaH$_2$PO$_4$, 150 mM K$_2$HPO$_4$, pH 7.2.
15. Immunoaffinity purification solutions:
    a. TMN buffer: 50 mM Tris-HCl, pH 7.5, 140 mM NaCl, 0.5 mM MgCl$_2$.
    b. pH 9.0 wash buffer: 25 mM 3-cyclohexylaminopropane-1-sulfonic acid (CAPS), 200 mM NaCl, 2 mM MgCl$_2$, pH 9.0.
    c. pH 10.8 elution buffer: 25 mM CAPS, 200 mM NaCl, 2 mM MgCl$_2$, 10% glycerol, pH 10.8.
    d. Neutralizing buffer: 1.0 M Tris-HCl, pH 6.7.
16. PDE storage buffer: 100 mM NaCl, 10 mM Tris-HCl, pH 7.5, 2 mM MgCl$_2$, 2 mM DTT, sterile filtered.
17. Solutions for trypsin activation of PDE6:
    a. 2X Proteolysis buffer: 20 mM Tris-HCl, pH 7.5, 200 mM NaCl, 4 mM MgCl$_2$, 40% glycerol.
    b. 2X Proteolysis stop solution: 20 mM Tris-HCl, pH 7.5, 200 mM NaCl, 4 mM MgCl$_2$, 0.5 mg/mL of soybean trypsin inhibitor (T9128; Sigma Chemical Corporation), 0.4 mg/mL of bovine serum albumin, 0.4 mM Pefabloc, 4 mM DTT.

3. Methods

An overview of the purification scheme for rod and cone PDE6 is shown in **Fig. 1**. Following the general procedure of McDowell *(3)*, photoreceptors are detached from bovine retina by mechanical disruption, and the rod outer segments (ROSs) are separated from soluble proteins by centrifugation. Following removal of soluble ROS proteins, the membrane-associated PDE6R is extracted from the membrane with a hypotonic buffer lacking magnesium *(4)* and then chromatographically purified. PDE6C and a soluble form of PDE6R found in the soluble retinal extract are separated by strong anion-exchange chromatography *(5)* and then further purified by several chromatographic techniques. Purified PDE6R and PDE6C can be stored at –20°C for several months with minimal loss of activity. The final method in this chapter is a procedure to proteolytically activate the PDE6 holoenzyme by digesting the inhibitory Pγ-subunits *(6)*, leaving the PDE6 catalytic dimer fully activated.

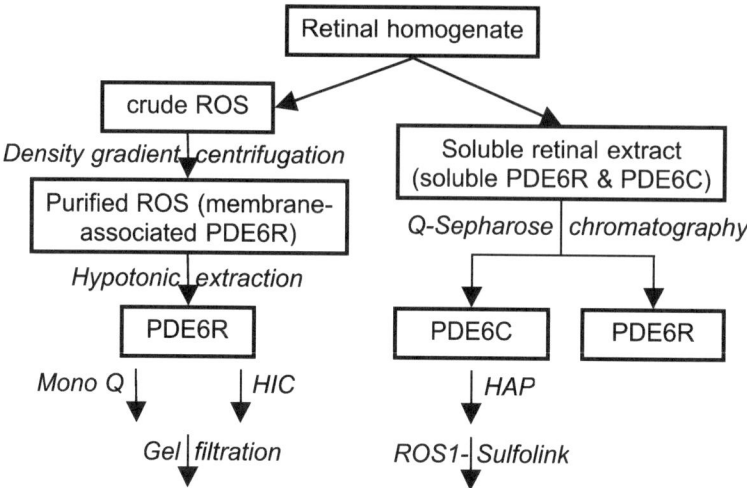

Fig. 1. Flow chart showing major strategies used to purify PDE6R and PDE6C from bovine retina. Once the initial PDE6 isozyme has been isolated, various combinations of chromatographic techniques can be used to purify the enzyme to homogeneity. Mono Q, anion-exchange chromatography using Mono Q resin; HIC, hydrophobic interaction chromatography on butyl-Sepharose resin; HAP, hydroxyapatite chromatography; ROS1-Sulfolink, immunoaffinity purification using the ROS1 antibody coupled to Sulfolink resin.

3.1. Initial Isolation of PDE6 Isoforms From Bovine Retina and Purification of ROS

In this section, mechanical disruption of photoreceptor cells from the neural retina is the starting point for isolating both the rod photoreceptor PDE6 associated with the outer segment membranes, and the soluble rod and cone PDE6 that are recovered in the soluble portion of the retinal extract. For the case of the membrane-associated PDE6R, sucrose density gradient centrifugation results in purified ROS in which rhodopsin (~70% of total protein), transducin (~10% of total protein), and PDE6R (1 to 2% of total protein) are membrane bound. Because of the relative purity of the membrane-associated PDE6R, we use only the soluble retinal extract for purification of PDE6C and discard the soluble PDE6R, which is difficult to purify to homogeneity.

To prevent activation of the components of the phototransduction pathway, the following procedures should be performed in a darkroom with infrared (IR) illumination and IR viewers. All solutions should be ice cold throughout the ROS purification process.

3.1.1. Preparation of Retinal Homogenate From Previously Frozen Bovine Retinas

1. Quickly thaw 50 frozen bovine retinas and keep on ice once thawed.
2. Add 45 mL of solution C (supplemented with 200 μL of Mammalian Protease Inhibitor Cocktail) to a beaker containing the 50 retinas.
3. Use a magnetic stir bar in the bottom of the beaker at low speed to mechanically disrupt photoreceptors from the retinas for 1 h in the dark (*see* **Note 1**).
4. Transfer the solution containing the disrupted retinas to 50-mL centrifuge tubes and centrifuge at $3000g_{max}$ for 3 min to pellet the retinal debris. (Save the retinal debris that has pelleted if purification of PDE6C is desired [*see* **Subheading 3.1.3.**]).

3.1.2. Fractionation of Retinal Extract to Separate ROS From Soluble Retinal Proteins

1. Pour off the retinal extract (containing ROS) through a nylon sock into a cold beaker.
2. Add 1.5 vol of solution A to the supernatant to dilute the sucrose concentration. Mix well.
3. Centrifuge the retinal extract in 50-mL tubes for 30 min at $23,000g_{max}$ in a fixed-angle rotor.
4. Pour off the supernatant (containing soluble retinal proteins, including PDE6C and soluble PDE6R, to be used for PDE6C purification; *see* **Subheading 3.6.**). Further purify the pellet (containing ROS and its membrane-associated PDE6R) (*see* **Subheading 3.1.4.**).

3.1.3. Re-Extraction of Retinal Debris Pellet to Recover Additional Soluble PDE6

If the primary purpose is to purify PDE6C, then the pelleted retinal debris should be resuspended to extract additional soluble PDE6.

1. Recover debris trapped by filtering the retinal extract (*see* **Subheading 3.1.2.**, **step 1**) by rinsing with 100 mL of solution A, and combine this with the retinal debris pellets (*see* **Subheading 3.1.2.**, **step 4**).
2. Vortex vigorously to release additional soluble PDE6.
3. Centrifuge the solution for 20 min at $30,000g_{max}$. Combine this supernatant with the soluble retinal proteins obtained previously in **Subheading 3.1.2.**, **step 4**. This is the starting material for purification of PDE6C in **Subheading 3.6.**

3.1.4. Purification of ROS

1. Resuspend the ROS-containing pellets from **Subheading 3.1.2.** in 15 mL of solution B1 ($\rho = 1.105$ g/mL).
2. Prepare discontinuous sucrose gradients just before use in 18-mL centrifuge tubes (Beckman) by layering 5 mL of solution B3, then 5 mL of solution B2. Layer the resuspended ROS pellets (in solution B1) to within 1 cm of the top of the tubes.

3. Centrifuge the sucrose gradients for 60 min at $116,000g_{max}$ in a swinging-bucket rotor at 4°C.
4. Remove the interface of solutions B2 and B3 containing the purified ROS with a 15-gage needle attached to a 5-mL syringe.
5. Dilute the ROS with 2 vol of solution A, and then centrifuge for 60 min at $30,000g_{max}$ to pellet the ROS (*see* **Note 2**).

3.2. Extraction of PDE6R From ROS Membranes

In this section, the soluble proteins present in the ROS are removed by disrupting the plasma membrane of the ROS in a moderate ionic strength buffer in the dark, and separating ROS membranes from the soluble proteins by centrifugation *(7)*. The resulting ROS membranes contain integral membrane proteins (predominantly rhodopsin), along with peripheral membrane proteins (notably PDE6R and a reduced amount of transducin). Because transducin undergoes a light-dependent binding to photoactivated rhodopsin *(4)*, the ROS membranes are exposed to light just before releasing PDE6R with a low ionic strength buffer.

3.2.1. Preparation of ROS Membranes Enriched in PDE6R

1. Resuspend the ROS pellets in 15 mL of solution A, and then homogenize 10–12 times with a tight-fitting Teflon pestle homogenizer driven by a Talboy Model 134-1 overhead stirrer.
2. Spin the ROS homogenate for 45 min at $110,000g_{max}$, and discard the soluble ROS proteins in the supernatant.
3. Resuspend the pellets in 15 mL of solution A and recentrifuge to remove residual soluble proteins.
4. Resuspend the ROS membranes in 15 mL of high Mg^{2+} hypotonic buffer to deplete the ROS membranes of some peripheral membrane proteins, and centrifuge as in **step 2**. Discard the supernatant following centrifugation (*see* **Note 3**).

3.2.2. Extraction of PDE6R From Washed ROS Membranes With Hypotonic Buffer

1. Expose the ROS membrane pellets to room light for 1 min at 4°C to photoactivate rhodopsin, thereby inducing tight binding of transducin to the ROS membranes.
2. Resuspend the light-exposed ROS membranes in a hypotonic PDE6R extraction buffer.
3. Homogenize the ROS membranes in a Dounce tissue grinder at 4°C using a pestle with a tight clearance (~0.025 mm).
4. Centrifuge the hypotonic extract for 60 min at $30,000g_{max}$, and recover the hypotonic supernatant, which contains solubilized PDE6R.
5. Repeat the hypotonic extraction procedure (without homogenization) two additional times.
6. Pool the hypotonic extracts, and clarify the solution by ultracentrifugation at more than $100,000g_{max}$ for 30 min (*see* **Note 4**).

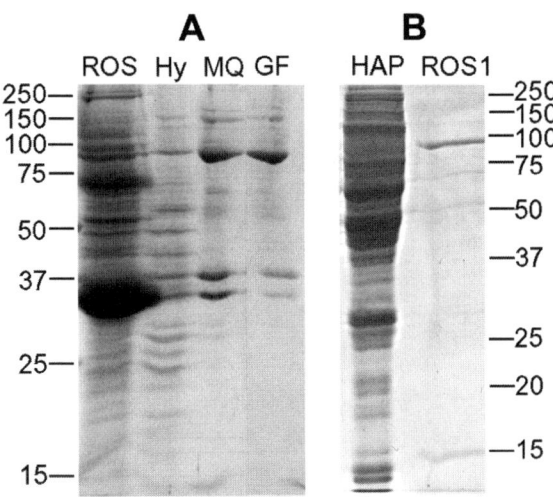

Fig. 2. Sodium dodecyl sulfate-polyacrylamide gel electrophoresis (SDS-PAGE) of PDE6 at various stages of purification. Samples were applied to 12% acrylamide gels and electrophoresed and the gel was stained with Coomassie Blue. (**A**) Membrane-associated PDE6R purification. Purified ROS contain high concentrations of rhodopsin, which are removed by hypotonic extraction of PDE6R from ROS membranes (Hy). Mono Q chromatography (MQ) removes most impurities except for major bands at 35–37 kDA (transducin subunits). Gel filtration chromatography (GF) yields a PDE6R preparation that is more than 95% pure. (**B**) Soluble PDE6C purification. PDE6C that has been resolved from soluble PDE6R by Q-Sepharose chromatography (not shown) is further purified and concentrated by hydroxyapatite chromatography (HAP). A greater than 100-fold purification of the PDE6C to more than 95% purity is achieved by immunoaffinity purification on the ROS1-Sulfolink column (ROS1).

3.3. Purification of PDE6R by Mono Q Anion-Exchange Chromatography

Anion-exchange chromatography of the PDE6 family has traditionally utilized the weak anion exchanger, diethylaminoethyl (DEAE), as the functional group *(5,7)*. We find that greater reproducibility and better resolution are achieved for PDE6 isozymes when a strong anion exchanger (quaternary ammonium [Q]) is used. The purification of hypotonically extracted PDE6R on a Mono Q column eliminates most other proteins, with transducin subunits remaining the predominant impurity following chromatography (**Fig. 2A**).

1. Equilibrate a Mono Q column with 10 column vol of MQ-A buffer prior to loading the sample.

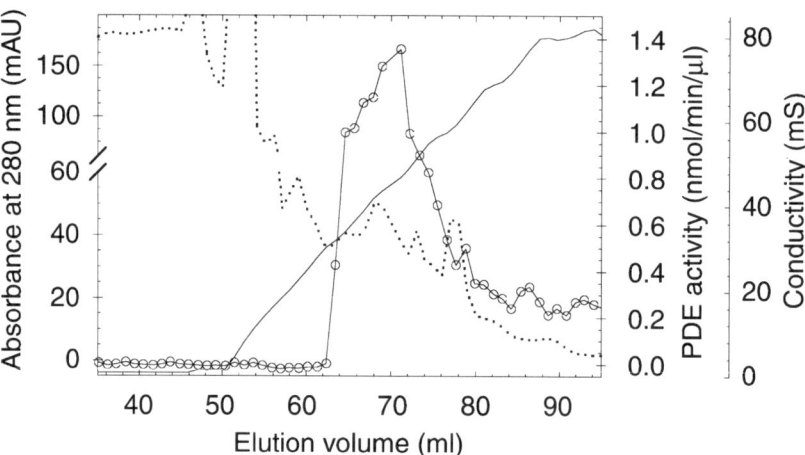

Fig. 3. Mono Q chromatography of hypotonically extracted PDE6R. Forty-five milli-liters of hypotonically extracted PDE6R was loaded onto a Mono Q column followed by 5 mL of MQ-A. PDE6R was eluted with a linear gradient from 0 to 100% MQ-B. For this particular experiment, the buffers were different from the standard procedure: MQ-A lacked NaCl, and MQ-B was 800 m*M* NaCl. Absorbance (dotted line) and conductivity (continuous line) were recorded, and PDE activity (—O—) was assayed for each fraction collected.

2. Adjust the hypotonic PDE6R sample to the approximate ionic strength of MQ-A buffer by adding 0.10 vol of MQ-B buffer. Filter the sample with a low-protein-binding 0.22-µm filter to remove particulates.
3. Load the PDE6R sample onto the Mono Q column at a flow rate of 0.5 mL/min and wash the column with 5 column vol of MQ-A buffer.
4. Perform a linear salt gradient from 0% MQ-B (100 m*M* NaCl) to 100% B (1.0 *M* NaCl) in a total volume of 40 mL, and collect 1-mL fractions.
5. After the elution is completed, wash the Mono Q column with 5 column vol of MQ-B, and then store as directed by the manufacturer.
6. Identify fractions containing PDE6R by a colorimetric PDE activity assay *(8)*. See **Fig. 3** for a typical result.

3.4. Removal of the PDE6R-Binding Protein, Glutamic Acid–Rich Protein-2, by Hydrophobic Interaction Chromatography

Hydrophobic Interaction Chromatography(HIC) separates proteins based on their relative hydrophobicity. This method is useful for the purification of PDE6R free of a high-affinity PDE6R-binding protein, glutamic acid-rich protein-2 (GARP2) *(9)*, that is found associated with Mono Q-purified PDE6R.

1. Adjust hypotonically extracted PDE6R (*see* **Subheading 3.3.**) or Mono Q–purified PDE6R (*see* **Subheading 3.3.**) to 400 mM $(NH_4)_2SO_4$ by the addition of solid $(NH_4)_2SO_4$ to the solution (*see* **Note 5**).
2. Wash a 30-mL butyl-Sepharose column with 5 column vol of HIC-A prior to loading the PDE6R sample at a flow rate of 0.5 mL/min.
3. Remove unbound proteins (including PDE6R) with 4 column vol of HIC-A.
4. Perform concentration and buffer exchange of the recovered PDE6R by ultrafiltration (*see* **Subheading 3.9.**).
5. To elute GARP2 from the butyl-Sepharose column, perform stepwise elution with 2 column vol of 65% HIC-B and then 2 vol of 100% HIC-B.
6. Regenerate the HIC column following the manufacturer's instructions.

3.5. Gel Filtration Chromatography of PDE6

Gel filtration (size exclusion) chromatography is the most suitable final step for PDE6 purification. It not only purifies PDE6 from other proteins based on size, but it also equilibrates the enzyme in a buffer more suitable for long-term storage.

1. Reduce the PDE6 sample to be purified in volume by ultrafiltration (*see* **Subheading 3.9.**) to ≤2% of the total volume of the gel filtration column in order to obtain maximum resolution of the PDE6R peak (*see* **Note 6**).
2. Equilibrate the gel filtration column with 2 column vol of GFC buffer prior to injecting the PDE6 sample on the column.
3. Operate the column at a flow rate of 0.4 mL/min and collect 0.4-mL fractions.
4. Analyze the absorbance at 280 nm and the activity of each fraction (*see* **Fig. 4**).

3.6. Separation of PDE6C From Soluble PDE6R by Q-Sepharose Anion-Exchange Chromatography

The purification of PDE6C is hampered by the low abundance of cone photoreceptors in most retinas, the lack of procedures for isolating intact cone photoreceptor cells, and the occurrence of PDE6C in the soluble retinal extract in conjunction with a more than 10-fold excess of soluble PDE6R. Gillespie and Beavo *(5)* exploited differences in the surface charges of PDE6C and PDE6R to effect a separation of the rod and cone enzyme using DEAE-cellulose anion-exchange chromatography. Using a strong anion exchanger attached to a rigid bead (Q-Sepharose) and a batch adsorption method, we have enhanced the resolution and reduced the time required to separate PDE6C from PDE6R.

1. Equilibrate 45 mL of Q-Sepharose resin with 3 column vol of Q-A buffer at a flow rate of 3 mL/min.
2. Before mixing the Q-Sepharose with the soluble retinal proteins, check the conductivity of the protein solution to ensure that it is less than 16 mS.

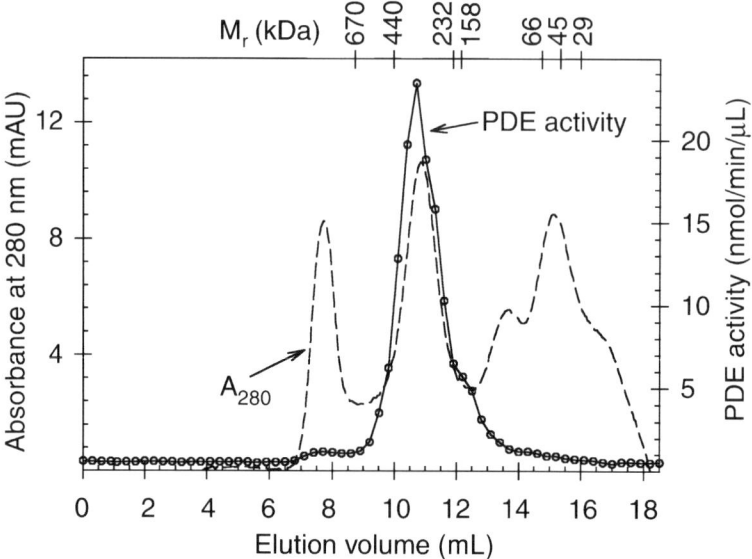

Fig. 4. Gel filtration chromatography of PDE6R on Superdex 200. A 0.5-mL sample of Mono Q-purified PDE6R was injected onto a Superdex 200 HR 10/30 column at 0.4 mL/min, and 0.3-mL fractions were collected. The 280-nm peak at 7.5 mL represents material in the void volume, with other impurities eluting at less than 100 kDa. The protein and PDE activity peak at 11.0 mL has an apparent mol mass of 300 kDa.

3. Mix the Q-Sepharose and the soluble retinal proteins (*see* **Subheadings 3.1.2.** and **3.1.3.**) by gently stirring with a cross-shaped magnetic stir bar. Incubate for 45 min.

4. Pour the suspension into 250-mL centrifuge bottles, and centrifuge for 1 min at $1000g_{max}$. Decant the supernatant and discard.

5. Add the Q-Sepharose slurry to a wide-diameter chromatography column (e.g., Bio-Rad 2.5-cm Econo column or Amersham XK 26 column), attach the flow adapter, and hook up to the chromatography system.

6. Wash the packed column with 3 column vol of Q-A buffer at 3 mL/min, and then perform a linear gradient from 0% Q-B to 100% Q-B over 5 column vol, collecting 5-mL fractions.

7. Continuously monitor the absorbance at 280 nm and the conductivity, and perform a colorimetric PDE activity assay on every other fraction to identify the PDE6C (~24 mS) and PDE6R (~32 mS) (*see* **Note 7**).

8. Regenerate the Q-Sepharose resin by washing with 2 vol of 1 *M* NaCl, 3 vol of 0.5 *M* NaOH, and then equilibrate and store in 10 m*M* NaOH containing 20% ethanol.

3.7. Purification of PDE6C by HAP Chromatography

HAP is a crystalline form of calcium phosphate suitable for chromatographic separations based on the interaction of the Ca^{2+} and PO_4^{3-} ions of the column matrix with oppositely charged ionic groups on the surface of proteins. Because of its unique adsorptive properties, HAP can remove contaminants from PDE6 samples that copurify by other methods. HAP has been previously used to purify hypotonically extracted PDE6R *(10)*, and it is utilized here as a prelude to immunoaffinity purification of PDE6C (*see* **Subheading 3.8.**).

1. Pack a 1×10 cm column with 1.5 mL of swollen HAP, wash with 10 column vol of HAP-B buffer, and then equilibrate with 10 vol of HAP-A.
2. Prepare a pooled Q-Sepharose-purified PDE6C sample by adding 0.10 vol of 0.5 *M* sodium phosphate.
3. Load the PDE6C onto the column at a flow rate of 0.2 mL/min, and then wash the column with 10 column vol of HAP-A.
4. Elute the PDE6C with a linear gradient of 0–100% HAP-B over 20 column vol, and monitor the absorbance at 280 nm and the conductivity.
5. Collect 1-mL fractions, assay for PDE activity, and pool the peak fractions. HAP-purified PDE6C has severalfold higher specific activity than Q-Sepharose PDE6C but remains highly impure (**Fig. 2B**) (*see* **Note 8**).
6. Regenerate the HAP column following the manufacturer's recommendations.

3.8. Immunoaffinity Chromatography of PDE6

Immunoaffinity purification is the single most effective chromatographic technique used in protein purification. Because antibodies are raised against unique epitopes of a particular protein, they afford the capability of enriching for the protein of interest more than 1000-fold in a single step.

The murine MAb ROS1 was originally raised against the PDE6R holoenzyme *(2)*. The epitope recognized by the ROS1 antibody has not been defined, but the affinity of the antibody for PDE6R and PDE6C is very high *(2)*. Elution of active PDE6 from ROS1 has required extensive manipulation of buffer conditions, with high pH and the inclusion of glycerol serving to elute the antibody with the least loss of biological activity. Inclusion of a pH 9.0 wash step just prior to elution serves to eliminate proteins that would otherwise contaminate PDE6 preparation.

3.8.1. Coupling of ROS1 Antibody to Sulfolink Beads

1. Purify the ROS1 antibody from ascites fluid on a Pierce Immunopure Protein G-agarose column using the manufacturer's protocols.
2. Concentrate the ROS1 antibody eluted from the Protein G column to less than 1 mL using a Centricon Plus-20 (*see* **Subheading 3.9.**), dilute 10-fold with the coupling buffer, and reconcentrate to ≥10 mg/mL in a volume of ≤2.5 mL.

3. Couple the ROS1 antibody to the Sulfolink beads at a coupling density of 5 mg of antibody/mL of resin following the manufacturer's instructions.

3.8.2. Purification of PDE6C by Adsorption to ROS1-Sulfolink Column and Elution of Active Enzyme

1. Wash the ROS1-Sulfolink column with 10 column vol of pH 10.8 elution buffer at 1 mL/min, and then equilibrate with 10 column vol of TMN buffer.
2. Dilute the PDE6C sample in TMN buffer (*see* **Note 9**).
3. Load PDE6C onto the column at a flow rate of 0.5 mL/min.
4. Recover the eluate and reapply it to the ROS1 column to ensure that all PDE6C has bound.
5. Wash the beads with 5 column vol of TMN at 1 mL/min.
6. Wash with 5 column vol of pH 9.0 wash buffer to remove contaminating proteins that nonspecifically bind to the ROS1-Sulfolink beads.
7. Elute the PDE6C with 10 column vol of pH 10.8 elution buffer.
8. Collect 1.0-mL fractions in tubes containing 0.1 mL of neutralization buffer.
9. Assess fractions for PDE activity using standard activity assays *(8)*.
10. Pool fractions containing PDE activity and concentrate prior to long-term storage (*see* **Subheading 3.9.**).
11. Regenerate the ROS1-Sulfolink resin by washing with 5 column vol of pH 10.8 elution buffer, then 10 column vol of TMN buffer containing 0.05% NaN_3 (*see* **Note 10**).

3.9. Concentration of PDE6 by Ultrafiltration

Centrifugal ultrafiltration of dilute PDE6 samples serves three purposes: concentration of the PDE6 prior to storage, removal of low molecular weight impurities, and exchange of the purification buffer with the PDE storage buffer. In our experience, the Amicon/Millipore devices with the Ultracel PL membrane (Centricon and Centricon Plus-20) offer the best rate of concentration and highest recoveries for PDE6 isozymes.

1. Before adding a PDE6 sample to the centrifugal filtration unit, add PDE storage buffer, and centrifuge briefly to prewet the membrane. Discard the buffer in the filtrate and retentate compartments.
2. For PDE6 samples with volumes <2 mL, load the Centricon YM-100 with the PDE6, and centrifuge at $1000g_{max}$ until the desired volume is achieved (*see* **Note 11**).
3. For PDE6 samples with volumes of 2–20 mL, load the Centricon Plus-20 with PL-30 membrane (30,000 mol wt cutoff) with the PDE6 sample and then centrifuge at $4000g_{max}$ until the volume is reduced sufficiently.
4. To carry out buffer exchange, resuspend the concentrated PDE6 sample to the full volume capacity of the device with PDE storage buffer and reconcentrate the sample. Repeat the process until the contaminating solutes are reduced to an acceptable level (typically <5% of the original concentration).

5. Recover the concentrated PDE6 from the retentate chamber following the manufacturer's directions.
6. For long-term storage of PDE6, mix the concentrated enzyme 1:1 (v/w) with molecular biology–grade glycerol, and keep at –20°C.

3.10. Preparation of Activated PDE6 Lacking Inhibitory Pγ-Subunit

Because of the very high affinity with which Pγ binds to the catalytic dimer of PDE6 *(11)*, the most effective way to prepare fully activated PDE6 is by limited proteolysis. Under controlled conditions, trypsin can effectively degrade the Pγ-subunits (and relieve their inhibition of the active site) without harming the properties of the catalytic dimer of PDE6 *(6,12,13)*.

3.10.1. Determining Optimal Conditions for Trypsin Activation of PDE6

In this procedure, PDE6 is proteolyzed with trypsin at 4°C for various times to determine the minimum time necessary to activate the enzyme fully. After quenching the reaction with soybean trypsin inhibitor, samples are assayed for PDE activity.

1. Dilute PDE6 to a concentration of 200 n*M* in 2X proteolysis buffer, and add an equal volume of 100 µg/mL of TPCK-treated trypsin (T1426; Sigma Chemical Corporation) in 10 m*M* Tris-HCl (pH 7.5) at 4°C.
2. At various intervals, remove portions and quench with an equal volume of 2X proteolysis stop solution (final PDE6 concentration is 50 n*M*).
3. Assay the quenched samples for PDE6 catalytic activity using a colorimetric PDE activity assay *(8)*.

3.10.2. Removal of Proteolytic Fragments of Pγ
From Trypsin-Activated PDE6

Although limited proteolysis with trypsin effectively activates PDE6 without adversely affecting the catalytic subunits, a large proteolytic fragment of Pγ consisting of the C-terminal half of the protein is generated (**Fig. 5**). This peptide has low affinity for binding to the active site of the enzyme *(14,15)*, but at high concentrations Pγ C-terminal peptides can act as competitive inhibitors of catalysis *(15)*. This section provides a procedure to remove this Pγ fragment to prepare purified Pαβ dimer lacking Pγ or Pγ fragments (**Fig. 5**).

1. Concentrate trypsin-activated PDE6 (*see* **Subheading 3.10.1.**) by ultrafiltration (*see* **Subheading 3.9.**) using a 30-kDa mol mass cutoff filter, and resuspend in MQ-A buffer and reconcentrate to remove trypsin, trypsin inhibitor, and some Pγ fragments from the PDE6 sample.
2. Load the concentrated PDE6 sample onto a Mono Q column as described in **Subheading 3.3.**

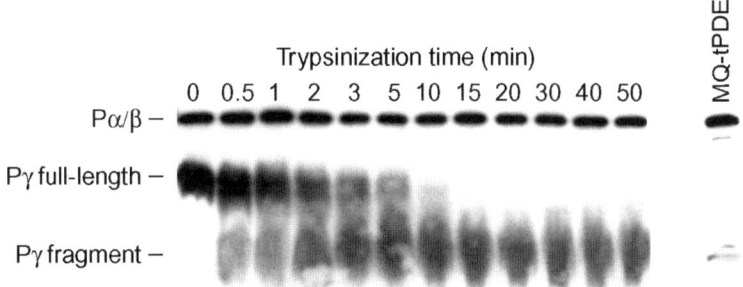

Fig. 5. Proteolytic digestion of PDE6R holoenzyme preferentially destroys the Pγ-subunit. PDE6R (50 n*M*) purified by Mono Q and gel filtration chromatography was incubated with 50 μg/mL of trypsin at 4°C as described in **Subheading 3.10.1.** At the indicated times, samples were mixed with soybean trypsin inhibitor and run on SDS-PAGE. After transfer to a nitrocellulose membrane at 60 V for 1 h, the membrane was probed with a mixture of catalytic and inhibitory subunit antibodies. The time course of disappearance of the 11-kDa Pγ-subunit correlates with the appearance of an approx 5 kDa Pγ fragment (amino acids 45–87; *see* **refs.** *13* and *14*). The α and β catalytic subunits are not degraded by this treatment. Following purification of the trypsin-activated PDE6R on Mono Q (MQ-EPDE), most of the Pγ fragment is removed.

3. Concentrate PDE6-containing fractions from the Mono Q column by ultrafiltration (*see* **Subheading 3.9.**) to a volume of less than 500 μL and store at –20°C with 50% glycerol (*see* **Note 12**).

4. Notes

1. To optimize recovery of soluble rod and cone PDE6 instead of the membrane-associated PDE6R, the magnetic stir bar is replaced with a four-blade propeller paddle attached to an overhead electric stirrer. The use of the propeller blade substantially reduces the yield of ROS (and hence membrane-associated PDE6R) from the sucrose gradient centrifugation step.
2. At this stage, the pelleted ROS can be frozen at –70°C (wrapped in aluminum foil to maintain a dark-adapted state) and PDE6R extracted at a later time. The yield of purified ROS can be estimated by spectrophotometric determination of the rhodopsin concentration (*16*).
3. The ROS and ROS membrane preparations should be kept in a dark-adapted state until this step is completed.
4. At this stage of purification, a typical yield of hypotonically extracted PDE6R from 50 retinas is 3 nmol when the ROS are mechanically disrupted with the magnetic stir bar. PDE6R constitutes approx 10% of the total protein in the hypotonic extract.
5. The high concentration of ammonium sulfate disrupts the high-affinity interaction of GARP2 with PDE6R, thus allowing their separation on the HIC column.

6. A Superdex 200 HR 10/30 column with a total volume of 24 mL allows 0.5 mL of PDE6 to be loaded. We use the Superdex 200 beads for good separation of PDE6 elution from the void volume of the column as well as high resolution over the mol mass range of 10,000–600,000 Daltons.

7. The typical yield of PDE6C from 50 bovine retinas is 40 μg at this stage. The PDE6C is quite impure at this step and should be further purified as soon as possible.

8. The phosphate buffer used for HAP chromatography interferes with the colorimetric PDE activity assay unless precautions are taken. Assaying 1-μL portions of 1/20 diluted fractions will reduce the phosphate concentration sufficiently to avoid interference with the colorimetric PDE assay. Alternatively, a radiotracer PDE activity assay *(8)* can be used.

9. The PDE6C buffer conditions are not critical to the success of ROS1 immunopurification. However, we routinely dilute the PDE6C 1:1 with TMN buffer to adjust the pH and ionic strength to be similar to the equilibration conditions. It is important to determine the effective binding capacity of the ROS1-Sulfolink column by testing each batch of ROS1 beads for the maximum amount of PDE6 that will bind per milliliter of resin.

10. ROS1-Sulfolink columns can be reused for many years, but a slow decline in the maximum PDE6-binding capacity is observed. To avoid cross-contamination, separate ROS1 columns are reserved for PDE6R and for PDE6C purifications.

11. The Centricon YM-100 membrane (100,000 molecular mass cutoff) retains all of the PDE6 activity in the retentate chamber when the holoenzyme is being concentrated. However, for trypsin-activated PDE6 (*see* **Subheading 3.10.**), significant amounts of PDE6 activity pass through into the filtrate, and the YM-50 membrane is recommended in this case.

12. The use of ultrafiltration and diafiltration of the trypsin-treated PDE6 prior to Mono Q chromatography causes the Pγ fragments to dissociate from the catalytic dimer, and the Pγ fragments are then able to pass through the membrane in the subsequent concentration step. An additional gel filtration step can be added to this procedure if residual Pγ fragments must be removed.

Acknowledgment

This work was supported by National Eye Institute (National Institues of Health) grant EY 05798 and is Scientific Contribution Number 2195 from the New Hampshire Agricultural Experiment Station.

References

1. Cote, R. H. (2003) Structure, function, and regulation of photoreceptor phosphodiesterase (PDE6), in *Handbook of Cell Signaling* (Bradshaw, R. and Dennis, E., eds.), Academic, San Diego, pp. 453–457.
2. Hurwitz, R. L., Bunt Milam, A. H., and Beavo, J. A. (1984) Immunologic characterization of the photoreceptor outer segment cyclic GMP phosphodiesterase. *J. Biol. Chem.* **259,** 8612–8618.

3. McDowell, J. H. (1993) Preparing rod outer segment membranes, regenerating rhodopsin, and determining rhodopsin concentration, in *Methods in Neuroscience, vol. 15* (Hargrave, P. A., ed.), Academic, San Diego, pp. 123–130.

4. Kuhn, H. (1982) Light-regulated binding of proteins to photoreceptor membranes and its use for purification of several rod cell proteins. *Methods Enzymol.* **81,** 556–564.

5. Gillespie, P. G. and Beavo, J. A. (1988) Characterization of a bovine cone photoreceptor phosphodiesterase purified by cyclic GMP-Sepharose chromatography. *J. Biol. Chem.* **263,** 8133–8141.

6. Hurley, J. B. and Stryer, L. (1982) Purification and characterization of the gamma regulatory subunit of the cyclic GMP phosphodiesterase from retinal rod outer segments. *J. Biol. Chem.* **257,** 11,094–11,099.

7. Baehr, W., Devlin, M. J., and Applebury, M. L. (1979) Isolation and characterization of cGMP phosphodiesterase from bovine rod outer segments. *J. Biol. Chem.* **254,** 11,669–11,677.

8. Cote, R. H. (2000) Kinetics and regulation of cGMP binding to noncatalytic binding sites on photoreceptor phosphodiesterase. *Methods Enzymol.* **315,** 646–672.

9. Körschen, H. G., Beyermann, M., Müller, F., Heck, M., Vantler, M., Koch, K. W., Kellner, R., Wolfrum, U., Bode, C., Hofmann, K. P., and Kaupp, U. B. (1999) Interaction of glutamic-acid-rich proteins with the cGMP signalling pathway in rod photoreceptors. *Nature* **400,** 761–766.

10. Malinski, J. A. and Wensel, T. G. (1992) Membrane stimulation of cGMP phosphodiesterase activation by transducin: comparison of phospholipid bilayers to rod outer segment membranes. *Biochemistry* **31,** 9502–9512.

11. Wensel, T. G. and Stryer, L. (1986) Reciprocal control of retinal rod cyclic GMP phosphodiesterase by its gamma subunit and transducin. *Protein Struct. Funct. Genet.* **1,** 90–99.

12. Catty, P. and Deterre, P. (1991) Activation and solubilization of the retinal cGMP-specific phosphodiesterase by limited proteolysis—role of the C-terminal domain of the β-subunit. *Eur. J. Biochem.* **199,** 263–269.

13. Mou, H., Grazio, H. J., Cook, T. A., Beavo, J. A., and Cote, R. H. (1999) cGMP binding to noncatalytic sites on mammalian rod photoreceptor phosphodiesterase is regulated by binding of its γ and δ subunits. *J. Biol. Chem.* **274,** 18,813–18,820.

14. Artemyev, N. O. and Hamm, H. E. (1992) Two-site high-affinity interaction between inhibitory and catalytic subunits of rod cyclic GMP phosphodiesterase. *Biochem. J.* **283,** 273–279.

15. Mou, H. and Cote, R. H. (2001) The catalytic and GAF domains of the rod cGMP phosphodiesterase (PDE6) heterodimer are regulated by distinct regions of its inhibitory γ subunit. *J. Biol. Chem.* **276,** 27,527–27,534.

16. Bownds, D., Gordon-Walker, A., Gaide Huguenin, A. C., and Robinson, W. (1971) Characterization and analysis of frog photoreceptor membranes. *J. Gen. Physiol.* **58,** 225–237.

11

Cyclic Guanosine 5′-Monophosphate Binding to Regulatory GAF Domains of Photoreceptor Phosphodiesterase

Rick H. Cote

Summary

Of the 11 families of mammalian cyclic nucleotide phosphodiesterases (PDEs), 5 contain regulatory domains capable of binding cyclic guanosine 5′-monophosphate (cGMP). The best understood of the GAF-containing PDEs is the family of rod (PDE6R) and cone (PDE6C) photoreceptor PDEs. Binding of cGMP to the rod PDE6 catalytic dimer ($\alpha\beta$) allosterically regulates the affinity of the inhibitory subunits of PDE6 (γ) for the enzyme. Two nonidentical, high-affinity cGMP-binding sites exist on the nonactivated mammalian PDE6R holoenzyme ($\alpha\beta\gamma\gamma$). One of the sites does not readily exchange with free cGMP when the catalytic dimer is complexed with Pγ. On dissociation of γ from the catalytic dimer, one of the two cGMP-binding sites undergoes a transition from high to low affinity. This chapter describes techniques to quantify cGMP binding to PDE6 in order to study the regulatory significance of the GAF domains. For high-affinity cGMP binding sites on PDE6, membrane filtration is the method of choice because of its speed, simplicity, and sensitivity. However, lower-affinity cGMP-binding sites require a method that does not perturb the equilibrium between bound and free ligand. The use of ammonium sulfate solutions during filtration extends to lower-binding affinities the useful range of membrane filtration. However, a centrifugal separation technique that minimizes perturbation of the cGMP-binding equilibrium is also presented for measuring lower-affinity cGMP-binding sites. These methods are applicable to understanding the regulatory mechanisms regulating other GAF-containing PDEs as well.

Key Words

Photoreceptor; phosphodiesterase; cyclic guanosine 5′-monophosphate; GAF domain; transducin; membrane filtration; ligand binding.

1. Introduction

Binding of a ligand to its cognate receptor induces conformational changes in the protein receptor that alter its structure and function. Changes in the

From: *Methods in Molecular Biology, vol. 307: Phosphodiesterase Methods and Protocols*
Edited by: C. Lugnier © Humana Press Inc., Totowa, NJ

intracellular concentration of cyclic guanosine 5′-monophosphate (cGMP) can cause changes in cGMP binding to specific cGMP receptor proteins, including the cyclic nucleotide-gated ion channel, cGMP-dependent protein kinase, and cGMP-binding PDEs. In the former case, the relatively low affinity of cGMP ($K_D > \mu M$) for this class of ion channel precludes the use of traditional binding assays, and electrical recordings of channel activity are the preferred approach for studying this class of cGMP receptor (e.g., *see* **ref. 1**). For cGMP-binding kinases and PDEs, a number of cGMP-binding assays have been used to characterize these regulatory cGMP-binding sites. Quantitation of cGMP-binding to receptor proteins can be carried out using equilibrium or nonequilibrium methods.

The choice of the optimal method for assaying cGMP binding depends on several factors, the most important of which is the intrinsic dissociation rate of cGMP from its receptor protein. If the rate of cGMP dissociation is slow relative to the time needed to separate bound from free ligand in the binding assay, then nonequilibrium binding assays (e.g., membrane filtration) are usually the method of choice, because they are rapidly performed and exhibit high sensitivity and low nonspecific binding. If significant dissociation of bound cGMP can occur during the separation of free from bound cGMP, then an equilibrium binding method (e.g., equilibrium dialysis, ultrafiltration) is required to prevent underestimating the extent of cGMP binding.

This chapter presents both nonequilibrium (membrane filtration) and equilibrium (centrifugal separation) methods for quantitating cGMP binding to rod PDE6.

2. Materials

1. Photoreceptor cell extracts containing PDE6 or purified PDE6 (*see* Chapter 10 on PDE6 purification).
2. Recombinant bovine rod inhibitory γ-subunit expressed in *Escherichia coli* harboring the pET11a/γ expression plasmid (*2*).
3. PDE6 storage buffer: 10 mM Tris-HCl, 100 mM NaCl, 2 mM MgCl$_2$, pH 7.5, supplemented with 2 mM dithiothreitol and 0.1 mM Pefabloc.
4. 100 mM zaprinast (cat. no. Z0878; Sigma Chemical Corporation, St. Louis, MO) dissolved in 1-methyl-2-pyrrolidinone or dimethyl sulfoxide.
5. Millipore MF membrane, 0.45-μm pore size, 25-mm diameter (cat. no. HAWP 025 00; Millipore, Bedford, MA).
6. Hoefer FH 225V 10-place filter manifold or Millipore 1225 sampling manifold.
7. Gast vacuum pump.
8. Buffer A: 60 mM KCl, 30 mM NaCl, 20 mM HEPES, 2 mM MgCl$_2$, 0.1 mM EDTA, pH 7.5.
9. [^{3}H]cGMP (cat. no. NET-554; Perkin Elmer) prepared as a 10X stock solution in buffer A containing 100 μM zaprinast or 10 mM EDTA. Adjust the cGMP

concentration and specific activity with nonradioactive cGMP as needed for particular experiments.

10. Ultima Gold and Ultima Gold XR scintillation fluids (Packard).
11. 95% Buffered ammonium sulfate (BAS): First dissolve 353.3 g of ammonium sulfate in 500 mL of water chilled to 4°C, and then mix with 35 mL of 200 mM Tris-HCl, pH 7.5.
12. 2% Sodium dodecyl sulfate (SDS): Add 10 g of SDS to 500 mL of distilled water.
13. Scintillation counter with automatic quench correction and disintegration-per-minute calculations.
14. Silicone oils: Dow Corning 550 fluid (density: 1.07 g/mL) and Dow Corning 200 fluid (density: 0.818 g/mL, viscosity: 1.0 cP) (William F. Nye, New Bedford, MA).
15. Tabletop air-driven ultracentrifuge (Beckman Airfuge) with A-110 rotor.
16. Airfuge centrifuge tubes (5 × 20 mm, polyallomer, no. 342630).

3. Methods

The methods described outline preparation of various forms of PDE6 for assays of cGMP binding, a general membrane filtration assay for cGMP binding to photoreceptor cell extracts or purified PDE6, the use of ammonium sulfate to stabilize cGMP bound to PDE6 during the membrane filtration assay, a centrifugal separation assay that sediments membrane-associated PDE6 through a silicone oil layer to separate cGMP bound to PDE from free cGMP, and guidelines for the analysis of ligand-binding data to extract equilibrium binding constants and kinetic rate constants.

3.1. Preparation of PDE6 for cGMP-Binding Assays

3.1.1. Purification of PDE6 From Photoreceptor Cells

Methods for isolating rod and cone PDE6 from retinal photoreceptors are described in Chapter 10. Because PDE6 is the only high-affinity cGMP-binding protein detected in rod and cone cells *(3–5)*, assays of cGMP binding to PDE6 can be performed in crude photoreceptor cell extracts as well as with highly purified PDE6 with similar results (*see* **Note 1**). For storage periods of more than 2 mo, purified PDE6 is stored in PDE storage buffer containing 50% glycerol and kept at –20°C. For longer storage, PDE6 is frozen in this solution at –80°C.

3.1.2. Depletion of Endogenous cGMP From GAF Domains

Because of the high affinity with which PDE6 binds cGMP, care must be taken to remove endogenous, bound cGMP prior to addition of radiolabeled cGMP. Repeated washing of rod cell membranes, or chromatographic purification of soluble PDE6, will remove unbound cGMP. Bound cGMP is allowed to dissociate from the GAF domains by incubating nonactivated PDE6 for 1 to 2 h at 30°C in the absence of PDE6 catalytic site inhibitors *(4,6)*. In this way, cGMP that dissociates is hydrolyzed and thus prevented from rebinding to the GAF domains. PDE6 catalytic dimers lacking bound γ-subunits are more readily depleted of endogenous cGMP (0.5 h at 30°C) because the cGMP dissociation rate is accelerated when γ is not present. The state of occupancy of PDE6 can be ascertained either by radioimmunoassay measurements of PDE6 cGMP content or by empirically determining the maximum stoichiometry of cGMP binding (B_{max}) at a saturating (1 μM) concentration of [³H]cGMP (*see* **Subheading 3.2.1.**).

3.1.3. Inhibition of Enzymatic Activity to Prevent Ligand Destruction

Complicating cGMP-binding assays is the fact that the ligand of the binding assay is also the substrate for the active site of PDE6. Unless precautions are taken to prevent catalysis by inclusion of active-site inhibitors, added cGMP will be consumed and reduced cGMP binding will be observed. The extent to which [³H]cGMP is broken down should be ascertained under the experimental conditions of the binding assay using a standard cyclic nucleotide radiotracer assay method *(7)*. Either of the following conditions is effective in preventing the breakdown of cGMP during the time needed for PDE6 to reach cGMP-binding equilibrium at its GAF domains: (1) 10 mM EDTA preincubated for 10 min prior to nucleotide addition to PDE6 $\alpha\beta$ catalytic dimers (*see* **Note 2**); (2) 0.1 mM zaprinast, a PDE5/6-selective inhibitor (*see* **Note 3**).

3.2. Membrane Filtration Assay of cGMP Binding

Receptor binding studies often use membrane filtration to separate ligand bound to its receptor protein from unbound ligand. Membranes composed of mixed esters of cellulose (typically nitrate and acetate) are the most popular, because they are hydrophilic, have a high capacity for binding proteins, and usually exhibit low levels of nonspecific binding of the ligand. For multiple samples, a vacuum filtration manifold should be used. High-throughput receptor-binding assays rely increasingly on 96-well dishes to which various membrane filters are attached.

The following section describes the general membrane filtration assay for quantitating cGMP binding to PDE6. The validity of the method depends on meeting the following criteria: (1) quantitative retention of protein on the filter, (2) low nonspecific binding of the ligand to the membrane, and (3) no ligand dissociation from the receptor protein caused by the separation of free from bound ligand.

3.2.1. Membrane Filtration Assay for cGMP Binding to PDE6

1. Incubate nucleotide-depleted PDE6 with [³H]cGMP in buffer A (supplemented with zaprinast or EDTA) until binding equilibrium is attained (*see* **Note 4**).

2. Apply membrane filter disks to the filter manifold.
3. Just before applying the sample to the membrane, prewet the filter disks with 1 mL of cold filter wash buffer, and open the vacuum to draw the wash solution through the filter.
4. Apply the [^{3}H]cGMP-PDE6 mixture (typically 20–100 µL) directly to the membrane with a micropipet.
5. Immediately rinse the filter three times with 1 mL of cold wash buffer with a repetitive dispensing pipet (e.g., Eppendorf Repeater Pipette). This process should be completed as quickly as possible to minimize cGMP dissociation from PDE6-binding sites.
6. Include samples that will allow determination of the extent of nonspecific binding of [^{3}H]cGMP to the filter (*see* **Subheading 3.2.3.**). In addition, determine the total disintegrations per minute applied to the filter by adding known volumes of the incubation mixture directly to scintillation vials.
7. After drying the filter under vacuum for approximately one additional minute, place the filter in a scintillation vial, and add 4 mL of scintillation cocktail.
8. Vigorously shake the vials until the membrane filter becomes translucent.
9. Set up a scintillation counter to automatically calculate disintegrations per minute, and count each sample until sufficient disintegrations per minute have been recorded to reduce the counting error to less than 1%.

3.2.2. Assessment of Quantitative Recovery of PDE6 on Membrane Filter

Failure to retain all of the applied PDE6 on the MF-Millipore membrane can be assessed by comparing the enzyme activity of the PDE6 before filtration with the enzyme activity recovered in the filtrate. Because of the high sensitivity of the filter-binding method and the high protein-binding capacity of the MF-Millipore membranes, even rod photoreceptor extracts in which PDE is <1% of the total protein show no evidence for less than full recovery of PDE6 on the membrane. Quantitative binding of PDE6 to the membrane does not require PDE6 attachment to photoreceptor membranes; purified PDE6 αβ dimers are also quantitatively retained on the filter membranes.

3.2.3. Determination of Nonspecific Binding of cGMP to Filter Membrane

One of the advantages of membrane filtration over other ligand-binding methods is the low level of nonspecific binding generally observed. Since the nonspecific binding component of the disintegrations per minute recovered on the membrane depends on the total disintegrations per minute applied to the filter, nonspecific binding controls should be performed in each membrane filtration experiment.

1. To test for nonspecific binding of the [^{3}H]cGMP to the filter, perform a filter-binding assay in which the PDE6 is omitted from the incubation (*see* **Subheading 3.2.1.**, **step 1**), but otherwise the sample is treated identically to the PDE6-containing samples.

2. To test for nonspecific binding of [³H]cGMP to PDE6, incubate the PDE6 with [³H]cGMP to which a >1000-fold excess of unlabeled cGMP has been added (i.e., 0.5 μM [³H]cGMP premixed with 1.0 mM cGMP). Elimination of unbound [³H]cGMP from the membrane filter can be optimized by varying the volume and number of washes of the filters.

3.2.4. Assessment of Whether cGMP Dissociation From PDE6 Occurs During Membrane Filtration

During membrane filtration, the equilibrium between bound and free cGMP is perturbed, and the washing step can potentially cause bound cGMP to dissociate. The rate of dissociation of cGMP from its bound state on PDE6 is defined by the dissociation rate constant, k_{-1}, which can be experimentally measured (*4*). Furthermore, the k_{-1} value depends on several factors, including the activation state of PDE6 (*5,6*), the amount of exogenous γ-subunit (*8*), and the temperature. If k_{-1} is known, then the time needed to complete membrane filtration and washing with a loss of <10% of the bound nucleotide can be calculated as described originally by Bennett and Yamamura (*9*): $T_{10\%} = 0.14/k_{-1}$. If k_{-1} is not known, it can be evaluated by an isotopic dilution experiment in which a large excess of nonradioactive cGMP is added to an equilibrium mixture of [³H]cGMP-PDE6 and the kinetics of cGMP dissociation are monitored:

1. Incubate nucleotide-depleted PDE6 with 1 μM [³H]cGMP to saturate all of the cGMP-binding sites. Allow the binding reaction to reach equilibrium.
2. Following **steps 1–5** in **Subheading 3.2.1.**, determine the extent of bound [³H]cGMP using the standard protocol.
3. To the remaining PDE6 sample, add 1 mM nonradioactive cGMP (1/50 dilution of a 50 mM cGMP solution).
4. Immediately remove portions and perform the standard membrane filtration, noting the time at which each sample is filtered and washed.
5. Perform nonspecific binding controls as in **Subheading 3.2.3., step 2**.

The initial time course of [³H]cGMP dissociation from PDE6 should follow an exponential decay. k_{-1} can be estimated from standard curve-fitting programs and used to calculate the $T_{10\%}$ in the previous equation. If significant cGMP dissociation from PDE6 is occurring faster than the filtration process can be completed, lowering the temperature to 4°C may retard the dissociation reaction sufficiently.

3.3. Ammonium Sulfate Stabilization of cGMP Binding to PDE6

Under conditions in which cGMP dissociation from the PDE6 GAF domains is occurring during the standard membrane filtration procedure, high concentrations of ammonium sulfate can be used to effectively retain bound cGMP

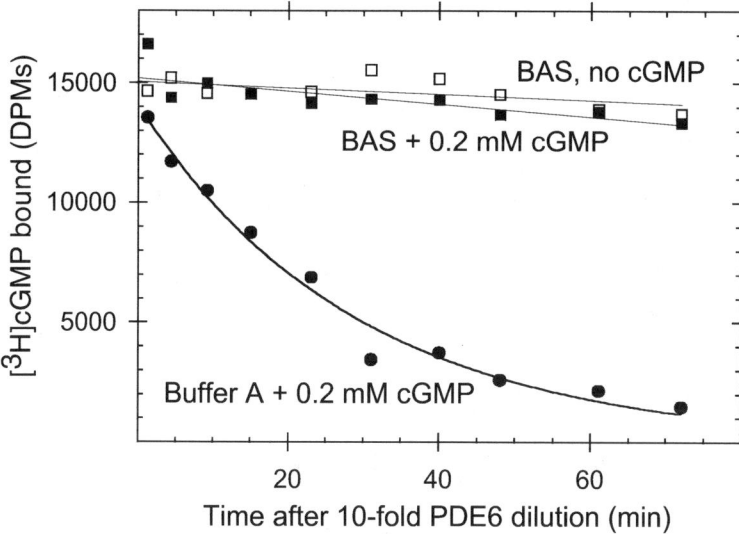

Fig. 1. Ammonium sulfate stops cGMP exchange at the GAF domain of rod PDE6. Rod PDE6 (24 n*M*) was incubated with 200 n*M* [³H]cGMP for 30 min at 4°C. At time zero, portions (200 µL) were added to 2.0 mL of the following solutions (at 4°C): an isotonic solution (buffer A) containing 0.2 m*M* nonradioactive cGMP (filled circles), buffered 95% ammonium sulfate (BAS) either lacking (open squares) or containing (filled squares) 0.2 m*M* cGMP. Aliquots were removed for membrane filtration at the indicated times. Even in the presence of a large excess of cold cGMP, the ammonium sulfate solution essentially halted cGMP exchange at the GAF domains of PDE6. In contrast, dissociation of [³H]cGMP from PDE6 followed an exponential decay when cold cGMP was added in Buffer A.

during filtration (*see* **Fig. 1**). The use of ammonium sulfate is also appropriate for dissociation kinetics experiments, in which the dissociation reaction can be effectively halted on addition of PDE6 to the ammonium sulfate solution. This approach has previously been used for studying cyclic nucleotide-dependent protein kinases *(10–12)*, PDE2 *(13)*, and PDE5 *(14)*.

3.3.1. Halting of cGMP Dissociation from PDE6 GAF Domains Using Buffered Ammonium Sulfate Stop Solution

1. Chill microcentrifuge tubes containing 200 µL of 95% BAS to 4°C in an ice water bath.
2. Prepare PDE6 (≥5 n*M*) as described in **Subheading 3.1.**, and incubate with [³H]cGMP in buffer A (containing zaprinast or EDTA to block cGMP hydrolysis). Permit the binding reaction to reach equilibrium.
3. Add a 20-µL portion of the PDE-[³H]cGMP mixture to the 200-µL BAS solution and mix well.

4. After adding each of the PDE samples to the BAS stop solution, keep samples ice cold until filtration is performed.
5. Include samples to determine the extent of nonspecific binding (*see* **Subheading 3.2.3.**) and to quantify the total disintegrations per minute added to the BAS.
6. Prewet Millipore filters with 2 mL of BAS just before use. Allow the filters to become fully wetted before applying a vacuum.
7. Apply the 220-μL PDE-containing sample to the prewet filters.
8. Rinse immediately with three 1-mL portions of ice-cold BAS.
9. Allow the filters to dry partially under a vacuum for approx 1 min, and then place the filters in 20-mL scintillation vials with 2.0 mL of 2% SDS (*see* **Note 5**). Shake the vials for 10 min.
10. Add 3.5 mL of Ultima Gold XR, and mix well.
11. Count the samples in a liquid scintillation counter.

3.4. Centrifugal Separation of Membrane-Associated PDE6 Through Silicone Oil

A major drawback of membrane filtration described in **Subheading 3.2.** is that low-affinity cGMP-binding sites might not be detected, because of rapid dissociation during the filtration and washing steps. True equilibrium binding methods such as equilibrium dialysis are well suited to studying low-affinity ligand-binding sites but often have high nonspecific binding, require large quantities of binding protein, and are laborious to perform (*15*). This section describes centrifugation of PDE6 through silicone oil, in which radiolabeled cGMP bound to membrane-associated PDE6 is sedimented through the oil, while leaving unbound [^{3}H]cGMP in the upper aqueous layer (*16*). This method minimizes disruption of the cGMP-binding equilibrium, while reducing greatly the nonspecific entrapment of unbound [^{3}H]cGMP with the pelleted PDE6. This centrifugal separation method has been successfully used to validate the membrane filtration method, to identify a class of lower-affinity cGMP-binding sites on PDE6, and to correlate the dissociation of bound cGMP with the dissociation of the inhibitory γ-subunit from the PDE6 holoenzyme (*8,16*).

3.4.1. Preparation of Centrifuge Tubes Containing Silicone Oil

1. Mix the higher (1.07 g/mL)- and lower (0.818 g/mL)-density silicone oils to obtain a final density of 1.02 g/mL. The refractive index of the oils can be used to estimate the density of the oil mixture.
2. Add 100 μL of the 1.02 g/mL silicone oil mixture to a 5 × 20 mm centrifuge tube (*see* **Note 6**).
3. Briefly spin the centrifuge tubes to ensure that no oil adheres to the side of the tube.

*3.4.2. Preparation of Rod Outer Segment Membranes Containing
Bound PDE6 and Incubation With [³H]cGMP*

The procedures for preparing disrupted rod outer segment (ROS) membranes and removing soluble proteins are described in **Chapter 10, Subheading 3.2.1.**

1. Treat the PDE6-containing ROS membranes as described in this chapter, **Subheading 3.1.2.** to deplete endogenous cGMP from the PDE6 GAF domains (*see* **Note 7**).
2. Incubate nucleotide-depleted ROS membranes containing 60 n*M* PDE6 with [³H]cGMP in buffer A supplemented with catalytic site inhibitors (as described in **Subheading 3.1.3.**). Allow the binding reaction to come to equilibrium at room temperature.

*3.4.3. High-Speed Centrifugation of Membrane-Associated PDE6
Through Silicone Oil and Quantitation of Bound [³H]cGMP*

1. Layer portions (10 μL) of ROS homogenates preincubated with [³H]cGMP on top of the silicone oil in centrifuge tubes.
2. Centrifuge the tubes at 90,000 rpm (110,000g_{av}) for 3 min at room temperature. The PDE6-containing ROS membranes are sedimented through the silicone oil layer, while the aqueous layer remains above the oil (*see* **Note 8**).
3. Immerse the tubes in a dry ice–ethanol bath to freeze the aqueous layer.
4. Cut the tubes with a razor blade through the oil layer just above the pelleted membranes, and discard the frozen top layer.
5. Transfer the contents of the bottom portion of the tubes to a scintillation vial. Resuspend the pellets at the bottom of the tube with 50 μL of 2% SDS, and then transfer to the scintillation vial.
6. Add 4 mL of Ultima Gold XR to the vial, mix thoroughly, and quantify the radioactivity in a scintillation counter.

*3.4.4. Verification of Efficacy of Centrifugal Separation
of ROS Membranes From Aqueous Layer*

Quantitative sedimentation of the ROS membranes can be estimated by spectroscopic determination of the rhodopsin concentration *(17)* in the pellet fraction. This requires that the ROS membrane preparation and centrifugation be carried out under infrared illumination conditions. Nonspecific entrapment of unbound [³H]cGMP can be estimated exactly as described in **Subheading 3.2.3.** In addition, ^{14}C-labeled compounds that do not bind to ROS membranes (e.g., sorbitol) can be included with the [³H]cGMP-binding mixture to assess the efficacy of the separation procedure (*see* **Note 9**).

3.5. Analysis and Interpretation of Radiolabeled cGMP-Binding Data

Because the type of data analysis depends on the nature of the experiment, this section presents some general guidelines that are relevant to all analyses of radiolabeled cGMP-binding experiments.

Transformation of ligand-binding data for graphic interpretation of the results using simple linear regression (e.g., Scatchard plot) should be avoided *(18)*. With the ready availability of powerful nonlinear curve-fitting applications (e.g., Sigmaplot, Graphpad Prism, Origin), the experimental results (corrected for non-specific binding) can be fit to an exact mathematical model, and a statistical analysis of the goodness of fit can be reviewed. One specialized program, KELL (Biosoft), uses an iteratively weighted, nonlinear curve-fitting routine to specifically analyze radioligand-binding experiments *(19)*. This program is particularly useful when the variance of the data as a function of ligand concentration is known, or when ligand depletion owing to receptor binding is significant *(20)*. KELL also offers the advantage of several built-in statistical tests that help assess the degree to which the chosen model fits the actual data; this is important when multiple classes of ligand-binding sites may exist. However, in most instances, we have found that judicious use of standard curve-fitting programs such as Sigmaplot works well for interpreting the results of [^{3}H]cGMP binding to PDE6.

For saturation-binding studies in which the extent of [^{3}H]cGMP binding to PDE6 is measured as a function of the total cGMP concentration added, the binding affinity (K_D) and the maximum binding (B_{max}) can be estimated from fitting the data to a two-parameter hyperbola (one class of sites) or four-parameter double hyperbola (two classes of binding sites). Various statistical tests can be applied to the results to decide which model is most appropriate for the data *(21)*. In practice, six to eight data points covering 10–90% of B_{max} are needed to determine accurately the affinity and density of a single class of sites. To resolve two classes of binding sites whose affinity differs by less than 100-fold, our experience is that 15–20 data points that span the range of 10–90% of B_{max} are needed to discern accurately the individual classes of sites. A final caution is that low-affinity cGMP-binding sites are sometimes difficult to demonstrate because relatively small errors in estimating the magnitude of nonspecific binding can be mistaken for a low-affinity binding site.

Graphic analysis of the kinetics of [^{3}H]cGMP dissociation from PDE6 GAF domains following the addition of a large excess of nonradioactive cGMP (cold chase) can be fit to a two-parameter (one class of sites) or four-parameter (two sites) exponential decay model to estimate the dissociation rate constant (k_{-1}) and the concentration for each class of cGMP-binding sites. Similar considerations to those described for statistical analysis and experimental design of saturation-binding studies apply equally to kinetic measurements of cGMP dissociation.

4. Notes

1. The state of activation, membrane attachment, and species from which PDE6 is isolated all influence the cGMP-binding properties of the enzyme. Differences also exist between rod and cone PDE6 from a given species.

a. The state of activation of PDE6 greatly influences the affinity and stoichiometry of cGMP binding. Nonactivated PDE6 holoenzyme ($\alpha\beta\gamma\gamma$) exhibits a single class of high-affinity binding sites that exchange slowly their bound cGMP with cGMP in solution *(6,22)*. PDE6 activated by transducin or by physical removal of the inhibitory γ-subunits has two classes of nonidentical binding sites *(5,8)*. Rod PDE6 catalytic dimers ($\alpha\beta$) can be reconstituted with recombinant γ or with peptide fragments of γ containing a GAF-stimulatory domain (amino acids 18–41 of the γ sequence), thereby restoring the binding properties of the PDE6 holoenzyme *(6,23)*.

b. The exchange rates of cGMP with the GAF domains of rod PDE6 are slower for PDE6 attached to cell membranes than is observed for PDE6 that is hypotonically extracted from the membrane. It is unclear whether this is owing to an effect of the isoprenyl moieties that anchor PDE6 to the membrane, or whether the procedure to release PDE6 from the membrane alters the interaction of a PDE6-binding protein that modulates cGMP affinity to PDE6.

c. Species differences in the cGMP-binding properties of PDE6 exist. In general, mammalian rod PDE6 has higher affinity for cGMP *(6,22)* than amphibian rod PDE6 *(4,5,8)*.

d. Although not as well characterized as the rod enzyme, cone PDE6 appears to have the same binding stoichiometry for cGMP as rod PDE6 but binds cGMP with lower affinity *(24,25)*.

2. EDTA acts by chelating divalent cations that are required for the catalytic mechanism of cGMP hydrolysis. EDTA treatment is most effective for PDE6 catalytic dimers lacking bound γ-subunits. In the case of PDE6 holoenzyme, EDTA is much less effective, presumably because the γ-subunit binds in the active site and prevents dissociation of the divalent cations responsible for catalysis *(26)*.

3. Previous studies have shown that the cGMP-binding GAF domains are highly specific for cGMP *(27)* and bind PDE5/6 inhibitors very poorly *(26)*. Thus, drugs acting on the catalytic site do not greatly influence cGMP-binding properties at the GAF domains.

4. The exact incubation conditions depend on the purpose of the experiment. To give one specific example, to measure the maximum extent to which cGMP can bind to a nonactivated PDE6 preparation (B_{max}), incubate 25 μL of 6.0 n*M* PDE6 with 500 n*M* [^{3}H]cGMP (specific activity = 3.5×10^{16} disintegrations per min, per mol) for 30 min at 4°C. Under these conditions, high-affinity cGMP-binding sites will be saturated, and specific binding will be more than 10-fold higher than the level of nonspecific binding and easily quantified in a scintillation counter.

5. The use of 2% SDS is unnecessary in the standard membrane filtration procedure but is helpful to solubilize rapidly the radioactive cGMP when ammonium sulfate is used. Alternatively, the dried filters can be shaken for 48 h in scintillation fluid (in the absence of SDS) with similar results.

6. To facilitate removal of the pelleted ROS membranes following centrifugation, 5 μL of 50% glycerol can be added to the bottom of the tube prior to the addition of the oil.

7. Bovine rod PDE6 in its membrane-associated state is not easily depleted of all endogenous bound cGMP. One of the two high-affinity sites is much more easily exchangeable than the second cGMP site *(6)*. By contrast, membrane-associated amphibian rod PDE6 can be depleted of all endogenous cGMP using the method in **Subheading 3.1.2.** *(4,8)*.
8. The method used to disrupt intact ROS and to prepare soluble protein–depleted ROS membranes can influence the final density of the ROS membranes used for these experiments. If inversion of the aqueous and oil layers is observed, then the density of the oil may need to be increased up to 1.028 g/mL. Incomplete sedimentation of the ROS membranes (determined by <100% recovery of rhodopsin in the pellet) will require lowering the oil density.
9. With this method, no radioactivity can be detected in the silicone oil layer itself, and less than 0.04% of the total [^{14}C]sorbitol is found in the pellet fraction using this procedure.

Acknowledgments

I am grateful to the past and present members of my laboratory for developing the procedures and conducting the experiments reported in this chapter. This work was supported by the National Institutes of Health (National Eye Institute, EY-05798). This is Scientific Contribution #2191 from the New Hampshire Agricultural Experiment Station.

References

1. Taylor, W. R. and Baylor, D. A. (1995) Conductance and kinetics of single cGMP-activated channels in salamander rod outer segments. *J. Physiol. (Lond.)* **483,** 567–582.
2. Artemyev, N. O., Arshavsky, V. Y., and Cote, R. H. (1998) Photoreceptor phosphodiesterase: interaction of inhibitory γ subunit and cyclic GMP with specific binding sites on catalytic subunits. *Methods* **14,** 93–104.
3. Yamazaki, A., Sen, I., Bitensky, M. W., Casnellie, J. E., and Greengard, P. (1980) Cyclic GMP–specific, high affinity, noncatalytic binding sites on light-activated phosphodiesterase. *J. Biol. Chem.* **255,** 11,619–11,624.
4. Cote, R. H. and Brunnock, M. A. (1993) Intracellular cGMP concentration in rod photoreceptors is regulated by binding to high and moderate affinity cGMP binding sites. *J. Biol. Chem.* **268,** 17,190–17,198.
5. Cote, R. H., Bownds, M. D., and Arshavsky, V. Y. (1994) cGMP binding sites on photoreceptor phosphodiesterase: role in feedback regulation of visual transduction. *Proc. Natl. Acad. Sci. USA* **91,** 4845–4849.
6. Mou, H., Grazio, H. J., Cook, T. A., Beavo, J. A., and Cote, R. H. (1999) cGMP binding to noncatalytic sites on mammalian rod photoreceptor phosphodiesterase is regulated by binding of its γ and δ subunits. *J. Biol. Chem.* **274,** 18,813–18,820.
7. Kincaid, R. L. and Manganiello, V. C. (1988) Assay of cyclic nucleotide phosphodiesterase using radiolabeled and fluorescent substrates. *Methods Enzymol.* **159,** 457–470.

8. Norton, A. W., D'Amours, M. R., Grazio, H. J., Hebert, T. L., and Cote, R. H. (2000) Mechanism of transducin activation of frog rod photoreceptor phosphodiesterase: allosteric interactions between the inhibitory γ subunit and the noncatalytic cGMP binding sites. *J. Biol. Chem.* **275,** 38,611–38,619.

9. Bennett, J. P. Jr. and Yamamura, H. I. (1985) Neurotransmitter, hormone, or drug receptor binding methods, in *Neurotransmitter Receptor Binding*, 2nd ed. (Yamamura, H. I., ed.), Raven, New York, pp. 61–89.

10. Doskeland, S. O., Ueland, P. M., and Haga, H. J. (1977) Factors affecting the binding of [³H] adenosine 3′:5′-cyclic monophosphate to protein kinase from bovine adrenal cortex. *Biochem. J.* **161,** 653–665.

11. Corbin, J. D. and Doskeland, S. O. (1983) Studies of two different intrachain cGMP-binding sites of cGMP-dependent protein kinase. *J. Biol. Chem.* **258,** 11,391–11,397.

12. Doskeland, S. O. and Ogreid, D. (1988) Ammonium sulfate precipitation assay for the study of cyclic nucleotide binding to proteins. *Methods Enzymol.* **159,** 147–151.

13. Miot, F., Van Haastert, P. J. M., and Erneux, C. (1985) Specificity of cGMP binding to a purified cGMP stimulated phosphodiesterase from bovine adrenal tissue. *Eur. J. Biochem.* **149,** 59–65.

14. Coquil, J. F., Franks, D. J., Wells, J. N., Dupuis, M., and Hamet, P. (1980) Characteristics of a new binding protein distinct from the kinase for guanosine 3′-5′-monophosphate in rat platelets. *Biochim. Biophys. Acta* **631,** 148–165.

15. Hulme, E. C. (1990) Receptor binding studies, a brief outline, in *Receptor-Effector Coupling, A Practical Approach* (Hulme, E. C., ed.), Oxford University Press, UK, pp. 203–215.

16. Forget, R. S., Martin, J. E., and Cote, R. H. (1993) A centrifugal separation procedure detects moderate affinity cGMP binding sites in membrane-associated proteins and permeabilized cells. *Anal. Biochem.* **215,** 159–161.

17. Bownds, D., Gordon-Walker, A., Gaide Huguenin, A. C., and Robinson, W. (1971) Characterization and analysis of frog photoreceptor membranes. *J. Gen. Physiol.* **58,** 225–237.

18. Leatherbarrow, R. J. (1990) Using linear and non-linear regression to fit biochemical data. *Trends Biochem. Sci.* **15,** 455–458.

19. McPherson, G. A. (1985) Analysis of radioligand binding experiments on a microcomputer system. *J. Pharmacol. Meth.* **14,** 213–228.

20. Munson, P. J. (1983) LIGAND: a computerized analysis of ligand binding data. *Methods Enzymol.* **92,** 543–576.

21. Munson, P. J. and Rodbard, D. (1980) LIGAND: a versatile computerized approach for the characterization of ligand binding systems. *Anal. Biochem.* **107,** 220–239.

22. Gillespie, P. G. and Beavo, J. A. (1989) cGMP is tightly bound to bovine retinal rod phosphodiesterase. *Proc. Natl. Acad. Sci. USA* **86,** 4311–4315.

23. Mou, H. and Cote, R. H. (2001) The catalytic and GAF domains of the rod cGMP phosphodiesterase (PDE6) heterodimer are regulated by distinct regions of its inhibitory γ subunit. *J. Biol. Chem.* **276,** 27,527–27,534.

24. Gillespie, P. G. and Beavo, J. A. (1988) Characterization of a bovine cone photo-receptor phosphodiesterase purified by cyclic GMP–Sepharose chromatography. *J. Biol. Chem.* **263,** 8133–8141.

25. Cote, R. H., Daly, A. E., Valeriani, B. A., and Vardi, N. (2002) Regulation of cone photoreceptor phosphodiesterase (PDE6C) by its inhibitory γ′ subunit and by cGMP binding. *Invest. Ophthalmol. Vis. Sci.* **43,** ARVO E-Abstract 1960.

26. D'Amours, M. R., Granovsky, A. E., Artemyev, N. O., and Cote, R. H. (1999) The potency and mechanism of action of E4021, a PDE5-selective inhibitor, on the photoreceptor phosphodiesterase depends on its state of activation. *Mol. Pharmacol.* **55,** 508–514.

27. Hebert, M. C., Schwede, F., Jastorff, B., and Cote, R. H. (1998). Structural features of the noncatalytic cGMP binding sites of frog photoreceptor phosphodiesterase using cGMP analogs. *J. Biol. Chem.* **273,** 5557–5565.

12

Renaturation of the Catalytic Domain of PDE4A Expressed in *Escherichia coli* as Inclusion Bodies

Wito Richter, Thomas Hermsdorf, and Dietrich Dettmer

Summary

Owing to simplicity, speed, cost advantage, and a generally high product yield, expression in *Escherichia coli* is the method of choice for the production of large amounts of protein. However, because of the high expression level, proteins often accumulate within the cells as insoluble aggregates called inclusion bodies. The inclusion body protein is misfolded and biologically inactive and, thus, needs to be refolded into its native conformation. There is no universal method for refolding inclusion bodies and optimal conditions have to be determined empirically for any given protein. Here, we describe a simple and efficient refolding protocol for the catalytic domain of type 4 cyclic nucleotide phosphodiesterases (PDE4s). This method has the potential for adaptation to other PDE subtypes.

Key Words

Refolding; renaturation; inclusion bodies; recombinant expression; *Escherichia coli;* phosphodiesterase; PDE4; PDE7; cAMP.

1. Introduction

Overexpression in *Escherichia coli* is the fastest and most economic and efficient method for producing large amounts of recombinant protein. However, this expression system has two major drawbacks when compared to alternatives such as overexpression in mammalian cell lines, yeast, or insect cells. First, proteins expressed in *E. coli* lack many posttranslational modifications, such as glycosylation, phosphorylation, and disulfide bond formation, that might be generated by an eukaryotic cell as the expression host. Second, the high-level expression in *E. coli* often results in accumulation of the recombinant protein in the form of large insoluble aggregates of misfolded protein known as inclusion bodies.

From: *Methods in Molecular Biology, vol. 307: Phosphodiesterase Methods and Protocols*
Edited by: C. Lugnier © Humana Press Inc., Totowa, NJ

Several techniques to avoid the accumulation of inclusion bodies, thus allow-ing the recombinant protein to be expressed in its soluble and biologically active form, are described in the literature including the following:

1. Modification of cell culture conditions (e.g., a decrease in incubation tempera-ture, an increase in osmotic pressure on the cells, and the use of culture medium additives *[1,2]*).
2. Expression of the protein of interest as a fusion protein with highly soluble tags such as glutathione-*S*-transferase, thioredoxin, or the maltose-binding protein to improve its solubility *(3,4)*.
3. Coexpression of chaperones that assist and support the proper folding of the protein within the cell such as members of the heat-shock protein family (e.g., Hsp60, 70, and 90) and the *E. coli* GroE family of proteins *(5,6)*.
4. The use of secretion tags to export the recombinant protein into the periplasm or medium to prevent its accumulation within the cells (for a comprehensive review on optimizing recombinant expression in *E. coli, see* **ref. 7**).

Alternatively, the expression of recombinant protein in the form of inclusion bodies can have the following advantages:

1. Proteins often accumulate to a higher yield as compared to expression in a soluble form.
2. Proteins that are toxic to the host cell can be highly overproduced in the form of these biologically inactive protein aggregates.
3. Expressed proteins are well protected from degradation by proteases while locked inside these dense particles.
4. Inclusion bodies can be easily purified from the cells using simple centrifugation protocols often resulting in preparations with a purity of the recombinant protein of 50–80% (*see* **Fig. 1**).

When expressed in the form of inclusion bodies, the recombinant protein subsequently must be refolded to regain its native conformation and biological activity. Protein refolding, however, is a complex process. There are no general rules and an extensive trial-and-error approach is often required. The develop-ment of a simple and effective refolding procedure is critical because a differ-ent method of expression may be the only alternative. Many factors influence the outcome of the refolding reaction such as the purity of the inclusion bodies, choice of the initial inclusion body denaturant, protein concentration, temper-ature, pH, ionic strength, and redox conditions. Furthermore, the addition of protein-specific coenzymes and substrates or metal ions and ligands, the use of low molecular weight additives that function as folding enhancers and aggre-gation suppressors, and the application of folding catalysts and chaperones also may have an effect *(8–14)*. However, expression in *E. coli* and the subsequent

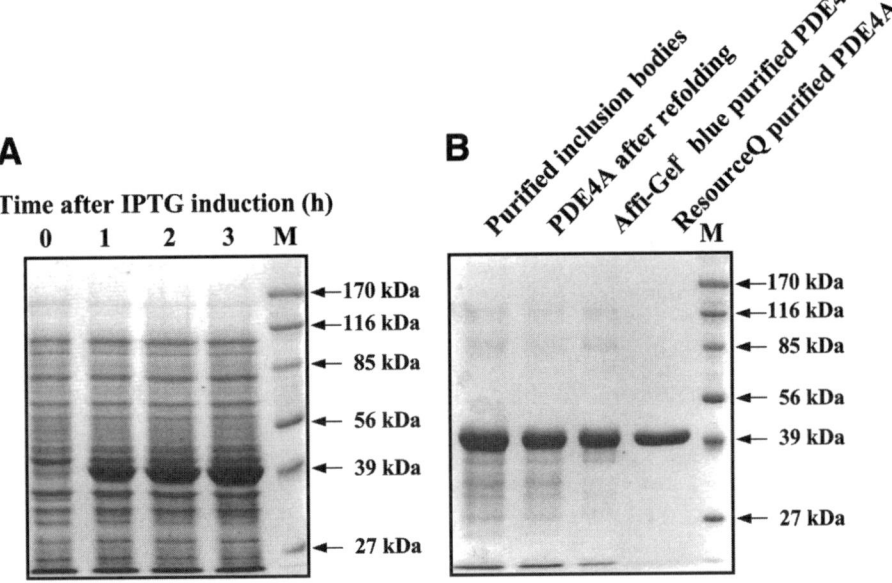

Fig. 1. Time-dependent expression, refolding, and purification of catalytic domain of human PDE4A (PDE4A$_{342-704}$). **(A)** The Coomassie gel shows the cell homogenate of BL21(DE3) cells carrying the pET21a-PDE4A expression plasmid at the time of IPTG addition (0 h; *see* **Subheading 3.1., step 3**) and after further cultivation for 1, 2, and 3 h. **(B)** Shown in a Coomassie-stained gel are from left to right: purified inclusion bodies of PDE4A$_{342-704}$ (*see* **Subheading 3.2., step 7**), the construct after refolding (*see* **Subheading 3.3., step 3**), the enzyme after Affi-Gel blue purification (*see* **Subheading 3.4., step 3**), and after subsequent ResourceQ anion-exchange chromatography (*see* **Subheading 3.4., step 4**). M, molecular weight markers.

refolding of inclusion bodies seems a rather promising strategy in the case of cyclic nucleotide phosphodiesterases (PDEs) for the following reasons:

1. The catalytic domain of these enzymes does not contain disulfide bonds.
2. They do not require any posttranslational modifications.
3. With the exception of metal ions, they do not depend on coenzymes and ligands.
4. PDEs do not have a complex multisubunit quaternary structure.
5. These proteins are enzymes, thus, permitting the use of a simple activity assay to screen a large variety of refolding conditions at the same time *(15,16)*.

These factors might, to some extent, simplify the refolding process and improve the chances of finding efficient refolding conditions for PDEs.

Previously, we developed a refolding assay for PDE4A expressed in the form of inclusion bodies *(17)*. Based on this method, this chapter describes a step-by-step procedure for the large-scale expression, refolding, and purification of the catalytic domain of PDE4A. In addition, suggestions are made on how to adapt this protocol to the refolding of other PDE subtypes.

2. Materials

1. Luria Bertani–ampicillin (LB-Amp) medium and agar dishes: 10 g/L of bacto-tryptone, 5 g/L of yeast extract, 10 g/L of NaCl (pH 7.0), 50 µg/mL of ampicillin, and 15 g/L of agar (only LB agar dishes).
2. Isopropyl-β-D-thiogalactopyranoside (IPTG).
3. French Press® (originally SLM Instruments Aminco, presently ThermoSpectronic, Pittsford, NY).
4. Lysis buffer: 0.1 M Tris-HCl (pH 7.0) containing 1 mM EDTA.
5. Extraction buffer: 60 mM EDTA (pH 7.0) containing 6% Triton X-100 and 1.5 M NaCl.
6. Washing buffer: 0.1 M Tris-HCl (pH 7.0) containing 20 mM EDTA.
7. Solubilization buffer: 0.1 M Tris-HCl (pH 8.0) containing 6 M guanidine hydrochloride (*see* **Note 1**), 100 mM dithiothreitol (DTT) (*see* **Note 2**), and 1 mM EDTA.
8. PDE4A refolding buffer: 1 M Tris-HCl (pH 7.0) (*see* **Note 1**) containing 40 µM ZnSO$_4$, 20 mM MgCl$_2$, and 10 mM DTT.
9. Affi-Gel® blue (Bio-Rad, Munich, Germany), ResourceQ anion-exchange column (Amersham Pharmacia Biotech, Freiburg, Germany).

3. Methods
3.1. Expression of PDE4A in E. coli

A cDNA encoding amino acids 342–704 of human PDE4A4B (Genbank accession no. L20965) was cloned into the *Nde*I-*Hin*dIII restriction enzyme sites of the pET21a vector (Novagen, Bad Soden, Germany) to generate a PDE4A expression plasmid. The resulting construct encoding the catalytic domain of PDE4A fused to a C-terminal (His)$_6$-tag was transformed in competent BL21(DE3) cells (Novagen), the cultures of which were stored as glycerol stocks at –80°C.

1. Streak a small volume of glycerol stock of BL21(DE3) cells carrying the pET21a-PDE4A vector on an LB-Amp agar dish and incubate overnight at 37°C. Inoculate 20 mL of LB-Amp medium with one of the resulting *E. coli* colonies, and shake this preculture overnight at 37°C and 225 rpm.
2. The following day, dilute the preculture into a 1-L batch of LB-Amp medium (*see* **Note 3**), and incubate further at 30°C (*see* **Note 4**) until the culture reaches an OD$_{600nm}$ of 0.6–0.8.
3. Add IPTG to a final concentration of 0.1 mM, and continue shaking at 30°C and 225 rpm for 3 h (*see* **Note 4**).
4. Harvest the cells with a 10-min centrifugation at 3500g, and then combine and weigh the cell pellets (*see* **Note 5**). About 5 g of *E. coli* (wet cell pellet) will be

obtained (*see* **Fig. 1A**). The cells can be used immediately for preparation of inclusion bodies or stored at –80°C.

3.2. Purification of PDE4A Inclusion Bodies (see Note 6)

1. Resuspend the bacterial cell pellet (~5 g) in 30 mL of lysis buffer (*see* **Subheading 2., item 4**) using a glass homogenizer.
2. Add 7.5 mg of lysozyme, transfer the mixture into an Erlenmeyer flask, and incubate at 4°C for 30 min while stirring the suspension using a magnetic stirrer.
3. Break up the cells by passing the suspension three times at 1200 psi through a French Press.
4. Transfer the solution back into the Erlenmeyer flask, add 300 μg of DNase and MgCl$_2$ to a final concentration of 3 mM, and stir at room temperature for another 30 min.
5. Add approx 18 mL of extraction buffer (*see* **Subheading 2., item 5**, 50% of the sample volume at the previous step), and incubate for an additional 30 min at 4°C.
6. Pellet the inclusion bodies by a 10-min centrifugation at 4°C and 30,000g.
7. Resuspend the pellet in 30 mL of washing buffer (*see* **Subheading 2., item 6**) using a glass homogenizer, and centrifuge at 30,000g at 4°C for 10 min. Repeat this step three times. This protocol should yield about 500 mg of purified inclusion bodies (*see* **Fig. 1B** and **Note 5**). Inclusion body pellets can be stored at –20°C for several weeks.

3.3. Solubilization and Refolding of PDE4A Inclusion Bodies

1. Solubilize the inclusion body pellets to approx 80 mg/mL in 6 mL of solubilization buffer (*see* **Subheading 2., item 7**) by pipetting up and down several times. Incubate the mixture for 2 h at room temperature in a rotating mixer.
2. Centrifuge at 30,000g for 10 min, transfer the supernatant into a new tube (*see* **Note 7**), and determine the protein concentration using a standard protein assay such as Bradford (*see* **Note 5**).
3. Prepare 1 L of refolding buffer (*see* **Subheading 2., item 8** and **Note 8**), precool to 10°C, and stir in a beaker at high speed using a magnetic stirrer. Pipet the solubilized PDE4A inclusion bodies (*see* **step 2**) directly into the stirring solution to a final protein concentration of 50 μg/mL (*see* **Note 9**), and further incubate the reaction for 16 h at 10°C.

3.4. Purification of Refolded PDE4A (see Notes 8 and 10)

1. After refolding, dialyze the solution (~1 L from Subheading 3.3., step 3) for 6 h at 4°C against 4 L of 50 mM Tris-HCl (pH 7.4). Then centrifuge for 30 min at 10,000g at 4°C to remove any aggregates or impurities.
2. Preequilibrate an Affi-Gel blue column with a bed volume of 40 mL with 50 mM Tris-HCl (pH 7.4), and then load the sample onto the column at a flow rate of 1 mL/min. Wash the column with 300 mL of 50 mM Tris-HCl (pH 7.4) containing 1 M KCl, 1 mM EDTA, and 0.2 mM DTT. Elute the PDE with 120 mL of 50 mM Tris-HCl (pH 10.6) containing 1 M KCl, 0.2 mM DTT, and 1 mM adenosine while collecting fractions of 4 mL.

3. Pool the fractions with the highest protein concentration, adjust the pH of the solution to 8.0 using 1 M HEPES (pH 4.7), and dialyze overnight at 4°C against 2 L of 50 mM Tris-HCl (pH 7.4) containing 0.2 mM DTT.
4. After a 20-min centrifugation at 20,000g and 4°C, load the supernatant at a flow rate of 1 mL/min onto a ResourceQ anion-exchange column with a bed volume of 6 mL. Wash the column with 20 mL of 50 mM Tris-HCl (pH 7.4), and subsequently elute the PDE using a 50-mL gradient from 0 to 400 mM NaCl in 50 mM Tris-HCl (pH 7.4). Pool fractions containing the highest specific PDE activity, and concentrate to about 10 mg/mL by ultrafiltration using an Ultrafree-15 centrifugal filter unit (Millipore, Bedford, MA).

Using this standard protocol, we typically obtain approx 7.5 mg of highly purified PDE (*see* **Fig. 1B**) from an initial 1 L of *E. coli* culture. Enzyme preparations usually exhibit k_m values of about 3 µM cAMP and specific activities of approx 5 µmol/(min•mg) at 1 µM cAMP.

3.5. Finding Efficient Refolding Conditions

The refolding conditions developed for the catalytic domain of the PDE4A also were applied effectively to the refolding of longer PDE4A constructs *(17)* as well as to the refolding of other PDE4 subtypes (unpublished results). However, application of these conditions to the refolding of the catalytic domain of PDE7A was not successful despite the high sequence homology of approx 40% between both enzymes, and new refolding conditions had to be identified for PDE7 (*[18]*; the refolding conditions of PDE4A and PDE7A are compared in **Table 1**). This indicates that the conservation of residues that are crucial for catalysis and substrate binding does not guarantee a similar folding behavior of the respective proteins. On the contrary, the modification of any part of the amino acid sequence or even the substitution of a single residue (especially the exchange of hydrophilic for hydrophobic residues or vice versa or the introduction of cysteine residues) can significantly change the ratio of successfully renatured vs misfolded and aggregated protein and might, consequently, require the adaptation of the refolding conditions (*see* **Note 11** for the experimental design of a renaturation screening).

Finding effective refolding conditions can prove to be a difficult task because usually a large number of reagents and additives (often over a wide range of concentrations) and various other refolding conditions need to be tested (*see* **Table 1** and **refs. 8** and *19* for lists of previously applied reagents and conditions). Additionally, refolding parameters have to be cross-checked against each other. For example, particular concentrations of both reagent A and reagent B could be critical for efficient refolding whereas either substance separately or

Table 1
Comparison of Refolding Conditions for PDE4A and PDE7A

Refolding parameter	Effect on PDE4A refolding (optimal concentration)	Effect on PDE7A refolding (optimal concentration)
Buffer substance[a]	Tris	Tris
pH	7.0	8.5
Temperature	10°C	10°C
Protein concentration	50 µg/mL	100 µg/mL
Incubation time	~16 h	~16 h
Refolding solution additive[b]		
Tris	Important (1 M)	Supportive as buffer (50 mM)
Reductant	Important (10 mM DTT)	Important (5 mM DTT)
Zn^{2+}	Crucial (40 µM)	No effect
Mg^{2+}/Mn^{2+}	Minor effects (20 mM Mg^{2+})	Crucial (200 mM Mg^{2+})
Arginine[c]	No effect	Important (1 M)
Ethylene glycol	No effect	Important (50%)

[a]Tris-HCl was found to be superior to phosphate, acetate, and HEPES buffers.

[b]In addition to the substances listed, the following additives were tested but were found to have minor or no effect on PDE4A and PDE7A refolding (shown in parentheses are the highest concentrations tested): guanidine hydrochloride (<0.5 M); glycerol (<20%); polyethylene glycol (<10%); cyclodextrin (<10 mM); sodium dodecyl sulfate (<1%); Triton X-100 (<1%); rolipram (<1 µM, only PDE4); NaCl (<1 M); and the metal ions Co^{2+}, Ni^{2+}, Cu^{2+}, and Mn^{2+} (0.002–200 mM).

[c]Arginine improved refolding at low Tris concentrations. At a Tris concentration of 1 M, however, arginine had no additional effect.

161

combined at suboptimal concentrations might prove ineffective. In this context, comparing the refolding conditions for PDE4A and PDE7A may provide an advantage when establishing a refolding assay for a different PDE subtype. As these parameters are similar for PDE4 and PDE7 refolding, an incubation temperature of 10°C, a pH of 7.0 to 8.0, a protein concentration of approx 50 μg/mL, a Tris-based buffer, and a reductant such as DTT at 5–10 mM could be used as fixed conditions when starting the screening process, while various concentrations of metal ions and additives such as Tris, arginine, and ethylene glycol are tested for their refolding efficiency. Subsequently, other additives or parameters could be explored for further optimization.

4. Notes

1. When utilized at high concentrations, reagents of the highest purity should be used to prevent artifacts caused by the accumulation of trace contaminants.
2. The addition of a high concentration of reductant is necessary to disrupt interchain disulfide bonds and to allow the complete solubilization and denaturation of the inclusion bodies.
3. The culture volume should not exceed a quarter of the volume of the culture flask to ensure sufficient aeration.
4. Decreasing the incubation temperature to 30°C reduces the proteolytic activity encountered during protein purification of longer PDE4A constructs *(17)*. Most likely, cultivation at higher temperatures induces proteases that aggregate together with the PDE protein and are trapped within the inclusion bodies. During the PDE refolding procedure, these proteases are released and renatured as well and can cause protein degradation during the subsequent protein purification steps.
5. The protocol described in this chapter is standardized regarding the volume of buffers and column materials used at every step based on the following experiences: A bacteria culture of 1 L produces a cell pellet of approx 5 g (wet cell weight; **step 4** in **Subheading 3.1.**). About 500 mg of inclusion bodies (wet weight; **step 7** in **Subheading 3.2.**) can be extracted from this cell pellet, and the solubilized inclusion bodies will contain approx 50 mg of protein (**step 2** in **Subheading 3.3.**). After refolding and subsequent Affi-Gel blue and ResourceQ protein purification, 7.5 mg of highly purified PDE4A is typically obtained *(see* **Subheading 3.4.**). The buffer volumes should be adapted proportionally if different amounts of cell pellet, inclusion bodies, or protein are used.
6. High purity of the inclusion body preparations is critical for refolding efficiency and reduces the necessity of subsequent purification steps. Therefore, cell disruption and solubilization of cell debris should be as efficient as possible to eliminate contamination of the inclusion bodies. According to the protocol of Rudolph et al. *(8)*, this is achieved with the following techniques:
    a. Lysozyme treatment to degrade the bacterial cell wall, making the cells more vulnerable to cell disruption (**step 2** in **Subheading 3.2.**).

    b. The use of a French Press as the most effective way to break up the cells (**step 3** in **Subheading 3.2.**).

    c. The digestion of DNA that otherwise would be pelleted during centrifugation and contaminate the inclusion body pellet (**step 4** in **Subheading 3.2.**).

    d. The treatment of the cell homogenate with high concentrations of detergent and salt to solubilize membrane fragments and debris (**step 5** in **Subheading 3.2.**)

    e. An extensive washing of the inclusion bodies (**step 7** in **Subheading 3.2.**).

7. At this step, the centrifugation supernatant should be a clear solution, indicating that the inclusion bodies are completely solubilized and denatured. For best renaturation results, solubilize inclusion bodies shortly before starting the refolding assay and do not use samples stored for longer than a day.

8. The catalytic domain of PDE4A (PDE4A4B$_{342-704}$) appears to be quite stable against proteolysis, thus, the addition of protease inhibitors was unnecessary throughout the entire protocol. However, protease inhibitors should be used through the refolding and purification procedure if proteolytic degradation is encountered during the preparation of other PDE constructs.

9. Other refolding methods such as one-step or stepwise dialysis, buffer exchange by gel filtration, and diafiltration are described in the literature *(10)*. However, rapid dilution into the refolding buffer was found to be superior by far to several other methods tested for the refolding of PDE4.

10. During refolding, the renaturation of the protein competes with other reactions such as misfolding and aggregation. Therefore, a certain percentage of misfolded, aggregated, or partially folded protein is typically produced even under highly optimized refolding conditions. For subsequent protein purification, methods with the potential to selectively enrich the properly folded protein should be chosen over purification techniques that can not distinguish between correctly folded and misfolded protein. For example, the Affi-Gel blue column used in this study will bind only PDE4 containing a proper nucleotide-binding pocket but not misfolded protein. On the other hand, a metal chelate column would unspecifically bind all proteins that contain a (His)$_6$-tag. We have used metal chelate columns successfully, however, to remove proteolytic fragments and to enrich full-length (His)$_6$-tagged protein *(17)*.

11. To determine optimal refolding conditions for a PDE, a vast number of buffers of diverse compositions were prepared. One-milliliter aliquots of these refolding solutions were pipetted into 1.5-mL Eppendorf tubes and precooled to 10°C. The refolding reaction was then started one sample at a time by diluting the solubilized inclusion bodies into the tube to the desired protein concentration (typically between 10 and 200 µg/mL) and vortexing the tube briefly at high speed. After a further incubation for 16 h at 10°C, all reactions were subjected to PDE activity assays. Based on the enzyme activity measured, it was determined whether the modification of the refolding buffer (e.g., the addition of a certain reagent, a different protein concentration or pH value) had improved PDE refolding or not, and if so, which concentration or which combination of buffer additives had been the most effective and resulted in the highest PDE activity.

Acknowledgment

We are indebted to Caren Spencer for editorial work on the manuscript.

References

1. Chalmers, J. J., Kim, E., Telford, J. N., Wong, E. Y., Tacon, W. C., Shuler, M. L., and Wilson, D. B. (1990) Effects of temperature on Escherichia coli overproducing beta-lactamase or human epidermal growth factor. *Appl. Environ. Microbiol.* **56,** 104–111.
2. Kopetzki, E., Schumacher, G., and Buckel, P. (1989) Control of formation of active soluble or inactive insoluble baker's yeast alpha-glucosidase PI in Escherichia coli by induction and growth conditions. *Mol. Gen. Genet.* **216,** 149–155.
3. Kapust, R. B. and Waugh, D. S. (1999) Escherichia coli maltose-binding protein is uncommonly effective at promoting the solubility of polypeptides to which it is fused. *Protein Sci.* **8,** 1668–1674.
4. Kovala, T., Sanwal, B. D., and Ball, E. H. (1997) Recombinant expression of a type IV, cAMP-specific phosphodiesterase: characterization and structure-function studies of deletion mutants. *Biochemistry* **36,** 2968–2976.
5. Gatenby, A. A., Viitanen, P. V., and Lorimer, G. H. (1990) Chaperonin assisted polypeptide folding and assembly: implications for the production of functional proteins in bacteria. *Trends Biotechnol.* **8,** 354–358.
6. Machida, S., Yu, Y., Singh, S. P., Kim, J. D., Hayashi, K., and Kawata, Y. (1998) Overproduction of beta-glucosidase in active form by an Escherichia coli system coexpressing the chaperonin GroEL/ES. *FEMS Microbiol. Lett.* **159,** 41–46.
7. Makrides, S. C. (1996) Strategies for achieving high-level expression of genes in Escherichia coli. *Microbiol. Rev.* **60,** 512–538.
8. Rudolph, R., Böhm, G., Lilie, H., and Jaenicke, R. (1997) Folding proteins, in *Protein Function, A Practical Approach* (Creighton, T. E., ed.), IRL, Oxford, UK, pp. 57–99.
9. Rudolph, R. and Lilie, H. (1996) In vitro folding of inclusion body proteins. *FASEB J.* **10,** 49–56.
10. Tsumoto, K., Ejima, D., Kumagai, I., and Arakawa, T. (2003) Practical considerations in refolding proteins from inclusion bodies. *Protein Expr. Purif.* **28,** 1–8.
11. Lilie, H., Schwarz, E., and Rudolph, R. (1998) Advances in refolding of proteins produced in E. coli. *Curr. Opin. Biotechnol.* **9,** 497–501.
12. Middelberg, A. P. (2002) Preparative protein refolding. *Trends Biotechnol.* **20,** 437–443.
13. Guise, A. D., West, S. M., and Chaudhuri, J. B. (1996) Protein folding in vivo and renaturation of recombinant proteins from inclusion bodies. *Mol. Biotechnol.* **6,** 53–64.
14. Chaudhuri, J. B. (1994) Refolding recombinant proteins: process strategies and novel approaches. *Ann. NY Acad. Sci.* **721,** 374–385.
15. Francis, S. H., Turko, I. V., and Corbin, J. D. (2001) Cyclic nucleotide phosphodiesterases: relating structure and function. *Prog. Nucleic Acid Res. Mol. Biol.* **65,** 1–52.

16. Xu, R. X., Hassell, A. M., Vanderwall, D., et al. (2000) Atomic structure of PDE4: insights into phosphodiesterase mechanism and specificity. *Science* **288,** 1822–1825.
17. Richter, W., Hermsdorf, T., Lilie, H., Egerland, U., Rudolph, R., Kronbach, T., and Dettmer, D. (2000) Refolding, purification, and characterization of human recombinant PDE4A constructs expressed in Escherichia coli. *Protein Expr. Purif.* **19,** 375–383.
18. Richter, W., Hermsdorf, T., Kronbach, T., and Dettmer, D. (2002) Refolding and purification of recombinant human PDE7A expressed in Escherichia coli as inclusion bodies. *Protein Expr. Purif.* **25,** 138–148.
19. Armstrong, N., de Lencastre, A., and Gouaux, E. (1999) A new protein folding screen: application to the ligand binding domains of a glutamate and kainate receptor and to lysozyme and carbonic anhydrase. *Protein Sci.* **8,** 1475–1483.

13

Determining the Subunit Structure of Phosphodiesterases Using Gel Filtration and Sucrose Density Gradient Centrifugation

Wito Richter

Summary

Size-exclusion chromatography (gel filtration) is a widely used method to determine the molecular weight of a protein. Often, the elution volume of several standard proteins is plotted against their known molecular weight to generate a standard curve, which is then used to determine the molecular weight of the protein of interest by its elution volume. However, gel filtration does not measure the mass of a particle as such, but the Stokes radius (R_s), a property dependent on mass, shape, and hydration of a protein. Thus, this method works well only if the protein of interest has a spherical symmetrical shape and an average hydration level. For all other proteins, the use of gel filtration as the sole means to determine the molecular weight will be misleading. The molecular weight of any given protein can be calculated, however, using the method of Siegel and Monty. This method combines Stokes radii obtained from gel filtrations and sedimentation coefficients derived from density gradient centrifugations to calculate the mass of a protein independently of its shape or hydration. It has been shown previously that PDE4D3, a representative of the long PDE4 splice forms, behaves as a dimer, whereas PDE4D2, a prototype of the short PDE4 splice forms, is a monomer. Both proteins exhibit an anomalous behavior on gel filtration columns. For this reason, they are used in this study to demonstrate the necessity of performing both gel filtration and density gradient centrifugation to determine the molecular weight of a protein.

Key Words

Dimerization; oligomerization; subunit structure; gel filtration; sucrose density gradient centrifugation; sedimentation coefficient; Stokes radius; PDE4.

1. Introduction

The mammalian cyclic nucleotide phosphodiesterases (PDEs) comprise a large group of isoenzymes that are divided into 11 PDE families on the basis of their substrate specificity, inhibitor sensitivity, and ultimately, their amino acid

From: *Methods in Molecular Biology, vol. 307: Phosphodiesterase Methods and Protocols*
Edited by: C. Lugnier © Humana Press Inc., Totowa, NJ

2V sequence homology *(1)*. All PDEs possess a conserved carboxy-terminal catalytic domain of about 270 amino acids fused to N-termini harboring unique regulatory domains such as Ca^{2+}-calmodulin-binding sites (PDE1), GAF domains (domain named for its presence in cyclic guanosine 5′-monophosphate [cGMP]-regulated PDEs, adenylyl cyclases, and the *Escherichia coli* FhlA protein; PDE2,5,6,9), PAS domains (domain named for its presence in Per, ARNT, and Sim proteins; PDE8), or UCR domains (upstream conserved region[s]; PDE4). Modification of these N-terminal domains by binding of messenger molecules such as Ca^{2+}-calmodulin, cGMP (GAF domains), and phosphatidic acid (UCR domains) *(2–4)*, or by phosphorylation by a variety of kinases including the Ca^{2+}-calmodulin-dependent kinase, the extracellular signal-regulated kinase 2, and the protein kinases A, B, and G *(5–9)* typically results in PDE activation.

Most of the PDEs thus far characterized regarding their quaternary structure behave as dimers. The dimerization domains were further mapped for PDE2, 5, and 6 *(10–14)* and PDE4 *(15)* and were found to be mediated by the conserved amino-terminal GAF and UCR domains, respectively. Of the PDE4 isoenzymes, only the so-called long PDE4 splice forms, which have a complete UCR1/UCR2 module, are dimers, whereas the so-called short splice forms, which only partially contain the UCR½ module, are monomers. That dimerization is mediated by the highly conserved UCR domains indicates that their subunit structure is an evolutionarily conserved structural property of PDE4 enzymes. This in turn leads to the speculation that dimerization is conserved because it is critical for key functions of PDE4. However, these possibilities remain to be investigated.

Up until now, a variety of methods have been applied to investigate the quaternary structure of PDE4 isoenzymes including gel filtration, analytical ultracentrifugation, crosslinking experiments, dynamic light scattering, and sucrose density gradient centrifugation *(15–21)*. Indeed, even crystal structures have been analyzed regarding PDE4 oligomerization *(22,23)*. A major limitation of these studies is the tendency of the PDE4 proteins to form larger aggregates during purification. For this reason, truncated PDE4 constructs encoding only the catalytic domain were usually used in these studies. For the analysis of full-length PDE4 proteins, analytical ultracentrifugation and dynamic light scattering, the only two methods that can determine directly the molecular weight of a protein in solution, cannot be used because they depend on pure enzyme preparations. Alternatively, gel filtration and sucrose density gradient centrifugation, using crude enzyme extracts, can be applied. Both methods, however, do not measure the molecular weight of a protein *per se* but measure properties known as the Stokes radius (R_s) and the sedimentation coefficient (S), respectively, both of which depend not only on the mass of a protein but also on its shape and hydration. The sole use of either method to determine the molecular

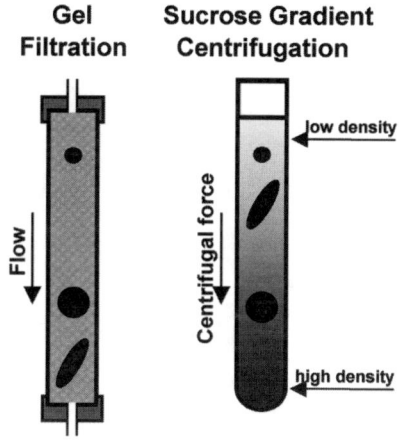

Molecule 1
(globular symmetric molecule)

Molecule 2
(globular symmetric molecule with a higher molecular weight than molecule 1)

Molecule 3
(asymmetric molecule with the same molecular weight as molecule 2)

Fig. 1. Behavior of a symmetric vs an asymmetric molecule in gel filtration and density gradient centrifugation. A protein with an anomalous (asymmetrical) shape or an above-average hydration level will elute faster from a gel filtration column and migrate slower in density gradient centrifugation than a globular (symmetrical) protein of the same mass.

weight of a protein will, therefore, be misleading if the protein does not have a globular symmetrical shape and an average hydration level (**Fig. 1**). According to Siegel and Monty *(24)*, both methods can be combined, however, to calculate the molecular weight of a protein, independent of its shape and hydration, as follows: The sedimentation coefficient (*S*) derived from density gradient centrifugation is given as:

$$S = M{\bullet}(1 - v_2{\bullet}\rho)/(N_A{\bullet}f) \qquad (1)$$

in which *S* is the sedimentation coefficient (Svedberg units), *M* is the molecular weight (kDa), v_2 is the partial specific volume of a protein (0.73 cm³/g), ρ is the density of the medium (1 g/cm³), N_A is Avogadro's constant ($6.022{\bullet}10^{23}$ mol⁻¹), and *f* is the frictional coefficient.

The frictional coefficient (*f*) is the property that expresses the shape and hydration level of a protein. In addition to the sedimentation behavior, the frictional coefficient is correlated with the Stokes radius (R_s), which can be measured by gel filtration as shown by **Eq. 2**:

$$R_s = f/(6 \bullet \pi \bullet \eta) \qquad \text{or} \qquad f = R_s \bullet 6 \bullet \pi \bullet \eta \qquad (2)$$

in which R_s is the Stokes radius (nm) (*see* **Note 1**), and η is the viscosity of the medium (0.01 g/[cm$\bullet$s]). Substituting f in **Eq. 1** with its term in **Eq. 2** leads to

$$S = M \bullet (1 - v_2 \bullet \rho)/(N_A \bullet R_s \bullet 6 \bullet \pi \bullet \eta) \qquad (3)$$

or rearranged for M as

$$M = S \bullet N_A \bullet R_s \bullet 6 \bullet \pi \bullet \eta \bullet /(1 - v_2 \bullet \rho) \qquad (4)$$

Equation 4 can be further simplified by inserting all fixed quantities as listed to

$$M = S \bullet R_s \bullet 4.2 \qquad (5)$$

Thus, by combining R_s derived from gel filtrations and S obtained from density gradient centrifugations, f cancels out of the equation, and the mass of a protein can be calculated independently of its shape or hydration level (*see* **Note 2**).

This chapter describes a step-by-step procedure for gel filtration and sucrose density gradient centrifugation using a long and a short PDE4 splice form that behave as a dimer and a monomer, respectively. Both methods can be applied similarly for determining the subunit structure of other PDEs or any other suitable protein.

2. Materials

2.1. Reagents

1. Homogenization buffer: 50 mM Tris-HCl (pH 7.4) containing 150 mM NaCl, 1 mM EDTA, 0.2 mM EGTA, 5 mM β-mercaptoethanol, a protease inhibitor mixture (Roche Molecular Biochemicals, Indianapolis, IN), and 1 mM 4-(2-aminoethyl)-benzenesulfonyl fluoride hydrochloride (AEBSF, Roche Molecular Biochemicals). The buffer must be prepared fresh on the day of the experiment, because the protease inhibitors are unstable in this solution.
2. External molecular weight standards: These can be purchased separately or as pre-mixed sets from several companies such as Sigma-Aldrich (St. Louis, MO), Bio-Rad (Hercules, CA), or Amersham Pharmacia Biotech (Piscataway, NJ). The range of the molecular weights of the standard proteins should include the anticipated molecular weight of the proteins of interest. Unless otherwise indicated by the manufacturer, the standard proteins are solubilized in homogenization buffer to approx 1 mg/mL and can be stored in this form at –20°C. The external marker solutions are mixed with approx 5000 cpm of [^{14}C]-methylated bovine serum albumin (BSA) that is used as an internal standard and are filtered through 0.2-μm filters (*see* **Note 3**) using a 1-mL syringe before they are loaded onto gel filtration columns or density gradients. The following are several standards used in this chap-

ter together with their molecular weights (kDa), their Stokes radii (Å), and their sedimentation coefficients (Svedberg, S): bovine thyroglobulin (670 kDa; 85 Å; 19S), horse ferritin (440 kDa; 61 Å; 17.6 S), bovine catalase (250 kDa; 52 Å; 11.3 S), bovine gamma globulin (158 kDa), rabbit aldolase (158 kDa; 48.1 Å; 7.3 S), bovine serum albumin (67 kDa; 35.5 Å; 4.6 S), chicken ovalbumin (44 kDa; 30.5 Å; 3.5 S), horse myoglobin (17 kDa; 20.4 Å), and cytochrome-*c* (12 kDa; 17 Å; 1.86 S). Additionally, Blue-dextran and deuterium-labeled water can be used to determine the void and the total volume of the gel filtration column, respectively.

3. Internal molecular weight standard: [^{14}C]-methylated BSA purchased from Sigma-Aldrich.
4. Gel filtration running buffer (*see* **Note 4**): 10 mM sodium acetate buffer (pH 6.5) containing 150 mM NaCl, 1 mM EDTA, 0.2 mM EGTA, 5 mM β-mercaptoethanol, and 20% ethylene glycol. The buffer should be filtered through 0.2-μm filters and degassed before use. The buffer can be used for several days if stored at 4°C.
5. Sucrose density gradient stock solutions: To obtain 5% (or 40%) sucrose stock solutions, dissolve 10 g (or 80 g) of sucrose to a final volume of 200 mL in 50 mM Tris-HCl (pH 7.4) containing 150 mM NaCl, 1 mM EDTA, 0.2 mM EGTA, and 5 mM β-mercaptoethanol.

2.2. Equipment for Running Gel Filtrations

Required is a high-performance liquid chromatography (HPLC) system comprising of an HPLC pump (e.g., the series 410 LC pump from Perkin Elmer, Boston, MA), a gel filtration column (e.g., a TSK-G3000SW from Tosoh Bioscience, Montgomeryville, PA; *see* **Note 5**), an optional guard column (e.g., the TSK-Guard-SW; Tosoh Bioscience), an ultraviolet (UV) detector (e.g., the LC-95UV/VIS from Perkin Elmer; *see* **Note 6**), a chart recorder, and a fraction collector (e.g., the Pharmacia LKB HeliFrac).

2.3. Equipment for Running Sucrose Density Gradients

1. Gradient maker such as the Jule gradient former (Jule, New Haven, CT).
2. Appropriate centrifugation vials (e.g., 14 × 89 mm Polyallomer centrifuge tubes from Beckman Coulter, Fullerton, CA), to prepare the gradients.
3. Ultracentrifuge (e.g., the Beckman L-70) equipped with a swinging-bucket rotor fitting centrifugation vials such as the Beckman SW41-TI, to run the samples.
4. Peristaltic pump (e.g., the Bio-Rad Econo Pump).
5. Fraction collector (e.g., the Bio-Rad model 2110), to collect the gradient fractions from the tube after centrifugation.

3. Methods

3.1. Preparation of Samples

1. Harvest one tissue culture dish of MA10 cells overexpressing PDE4D2 or PDE4D3 into 1 mL of homogenization buffer (*see* **Subheading 2.1.1.**), and lyse the cells with 30 strokes in a glass homogenizer.

2. Centrifuge the homogenate at 14,000g for 20 min and subsequently at 100,000g for 30 min at 4°C. If the supernatant is not clear after the first centrifugation, filter it through a 0.2-µm filter (*see* **Note 3**) using a 1-mL syringe before continuing with the second centrifugation step.
3. Mix the sample with approx 5000 cpm of [^{14}C]-methylated BSA that is used as an internal standard.

3.2. Gel Filtration

1. Equilibrate the gel filtration column with running buffer (*see* **Subheading 2.1.4.**) at a flow rate of 0.5 mL/min until the backpressure of the column reaches a constant level and the UV recording at 280 nm achieves a baseline (*see* **Note 7**). Set the UV detector to zero.
2. Load 500 µL of enzyme sample or the same volume of external standard solution onto the column and simultaneously start the timer for the fraction collector and the UV chart recorder to monitor the elution of the molecular weight standards.
3. Elute the samples or standards from the column with running buffer at a constant speed of 0.5 mL/min while collecting eluate fractions of 500 µL and recording the UV profile at 280 nm. Once the samples are completely eluted, the UV reading should reach baseline again and the next sample can be separated (*see* **Note 8**).
4. Analyze the gel filtration fractions obtained using a PDE activity assay and/or Western blotting to determine the elution volume of the enzymes.
5. Use 200 µL of each fraction obtained from the separation of PDE samples or external molecular weight standards for scintillation counting in order to determine the elution volume of the internal marker [^{14}C]-methylated BSA (*see* **Note 9**).
6. Plot the Stokes radii (or molecular weights) of the molecular weight standards vs their elution volume, generate a smooth curve through these points, and determine the Stokes radius (or the apparent molecular weight) of the protein of interest by its elution volume on this standard curve (*see* **Note 10**).

3.3. Sucrose Density Gradient Centrifugation

1. Assemble the gradient maker according to the manufacturer's manual or as shown in **Fig. 2A**.
2. Close the B-seals, open the A-seals, and fill the storage and mixing chambers to the desired volume (*see* **Note 11**) with the 40 and 5% sucrose solutions, respectively (*see* **Note 12**). Switch on the magnetic stirrer in the mixing chamber.
3. Close the A-seals, open the B-seals, place the end of the tubing into the centrifugation tube (*see* **Note 13**), and push down the pistons at a slow and constant speed. The highly concentrated sucrose solution will stream slowly into the mixing chamber, where it is constantly mixed with the low-sucrose concentration, thereby slowly increasing the concentration of the solution that flows from the mixing chamber into the centrifugation tube. To obtain linear gradients, both pistons must be pushed down at exactly the same pace. Once the final gradient volume is reached, carefully remove the end of the tubing from the centrifugation vial without disturbing the gradient. For the remainder of the procedure, it is

A

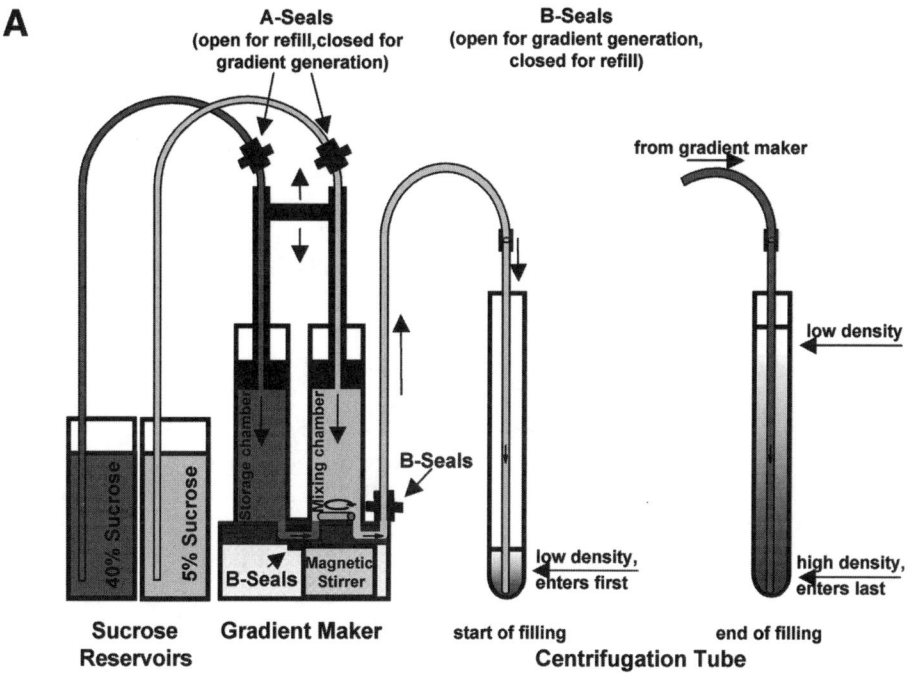

B

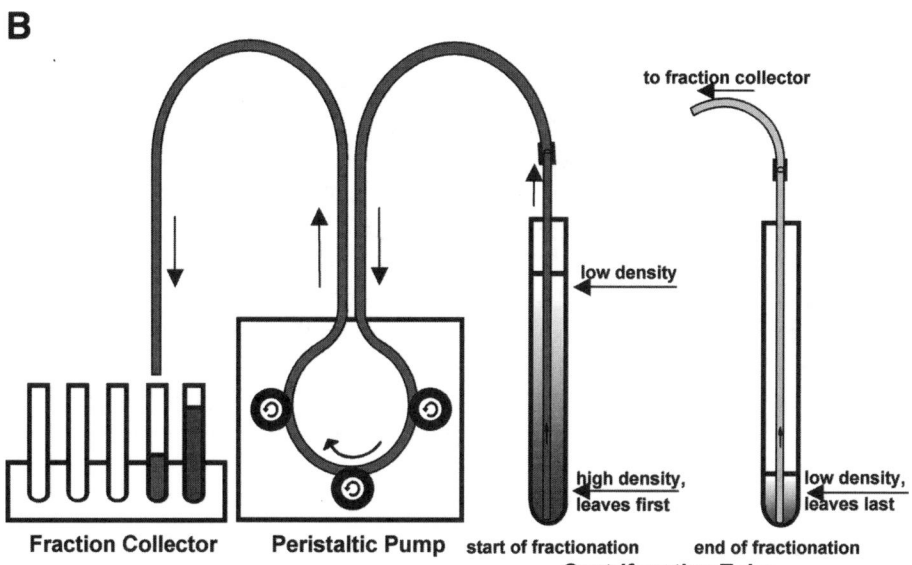

Fig. 2. Experimental design for (**A**) generation and (**B**) fractionation of sucrose density gradients.

important to handle the filled centrifugation tubes very gently, keeping them in the upright position and avoiding any vibration that might disturb the density gradient.

4. Before preparing the next sucrose gradient, wash the mixing chamber several times with the 5% sucrose solution by filling and emptying it repeatedly.

5. Store all gradients at 4°C overnight (*see* **Note 14**).

6. Load 200 µL of sample or external standard solution onto the gradient by slowly pipetting it against the wall of the tube just above the top of the gradient (*see* **Note 15**).

7. Place the tubes in a swinging-bucket rotor and centrifuge at 28,000 rpm for 36 h at 4°C (*see* **Notes 12** and **16**).

8. After centrifugation, carefully place a thin rigid tubing connected to a peristaltic pump into the centrifugation tube (**Fig. 2B**), and then remove the gradient starting from the bottom of the tube with a constant flow of 0.5 mL/min. Collect fractions of approx 225 µL using a fraction collector (*see* **Note 17**).

9. Analyze the resulting fractions for protein concentration (molecular weight markers) or PDE activity (recombinant PDE samples; **Fig. 3A**). In addition, use approx 150 µL of all the fractions for scintillation counting in order to determine the migration point of the internal marker [^{14}C]-methylated BSA (*see* **Notes 9** and **18**).

10. Plot the sedimentation coefficients (or molecular weights) of the external standard proteins vs their migration point, draw a smooth curve through these points, and determine the sedimentation coefficient (or apparent molecular weight) of the protein of interest by its migration position on this standard curve (**Fig. 3B**).

3.4. Calculation of Molecular Weight

By applying gel filtration and density gradient centrifugation as described in **Subheadings 3.2.** and **3.3.** the apparent molecular weights, Stokes radii, and sedimentation coefficients in **Table 1** were determined for PDE4D3 and PDE4D2 *(15)*. For proteins of a globular symmetrical shape and an average hydration level, the apparent molecular weights determined by gel filtration and sucrose density gradient centrifugation should be very similar. As shown in **Table 1**, this is not the case for PDE4D3 and PDE4D2. Both exhibit the behavior of proteins with either an asymmetrical shape or an unusually strong hydration resulting in higher apparent molecular weights as determined in gel filtration when compared to density gradient centrifugation (*see* **Fig. 1**). Using either method by itself would have resulted in an erroneous estimate of the molecular weight. The correct molecular weight of both enzymes can be calculated, however, using Eq. 4 established in **Subheading 1.** As a result, the long splice form PDE4D3 was found to be a dimer, and the short splice form PDE4D2, a monomer. This is consistent with the apparent molecular weight of PDE4D3 being twice the apparent molecular weight of PDE4D2 in both gel filtration and density gradient centrifugation, indicating that PDE4D3 is an oligomer one magnitude higher than PDE4D2.

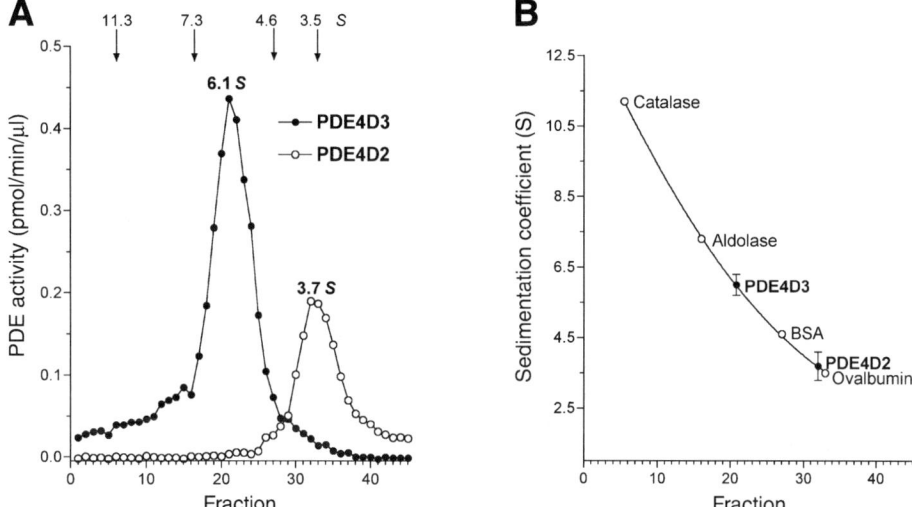

Fig. 3. Sedimentation of PDE4D2 and PDE4D3 in sucrose density gradients. **(A)** PDE activity profile for the migration of PDE4D2 and PDE4D3. Arrows indicate the migration positions of external standard proteins. **(B)** Generation of a standard curve based on the migration of several standard proteins. The migration of PDE4D2 and PDE4D3 is also indicated.

4. Notes

1. It is important to remember that the Stokes radius used in this equation must be in nanometers and not in Angstrom. The conversion is 1 nm = 10 Å.
2. For an excellent background presentation under the heading "How big is a protein molecule?" visit the website of Dr. Harold P. Erickson (http://www.cellbio.duke.edu/Faculty/Erickson, Department of Cell Biology, Duke University Medical Center, Durham, NC).
3. Retention of fluid in larger filters can cause a significant loss of sample. We routinely use 13-mm-wide syringe filters from Nalgene.
4. The composition of the running buffer (pH, buffer substance, and concentration) can be widely modified unless the properties of the gel filtration column or the protein samples analyzed are limiting. For example, higher concentrations of chelators or detergent as well as pH values over 8.0 would decrease the life-span of the silica-based column used in our study.
5. Gel filtration columns are optimized for the separation of different molecular weight ranges. The anticipated molecular weight of the protein of interest should be considered when selecting the column.
6. Although continuous recording of the absorbance at 280 nm using a UV detector simplifies determination of the elution profiles of the external standard proteins, the elution of these markers could also be determined by measuring the protein content in the collected eluate fractions using a standard protein assay (e.g., Bradford, Lowry).

Table 1
Physicochemical Properties of PDE4D Constructs[a]

| Construct | $M_{theor.}$ (kDa) | Gel filtration | | Sucrose density gradient | | |
		Stokes radius (Å)	$M_{app.}$ (kDa)	Sedimentation coefficient ($s_{20,w}$) (S)	$M_{app.}$ (kDa)	$M_{calc.}$ (kDa)
PDE4D3	76	65 ± 2.3	388 ± 36	6.1 ± 0.3	108 ± 6.3	166
PDE4D2	58	47 ± 1.7	167 ± 16	3.7 ± 0.4	50 ± 9.0	73

[a]$M_{theor.}$, theoretical molecular weight for a monomer based on the amino acid sequence; $M_{app.}$, apparent molecular weight; $M_{calc.}$, molecular weight calculated according to Siegel and Monty (*24*). (Data reprinted from **ref. *15***).

7. When starting the pump, do not switch to a flow rate of 0.5 mL/min at once because a sudden change in pressure may damage the column. Instead, increase the flow rate gradually (e.g., 0.1 mL/min, 0.2 mL/min and so on) or select an automatic flow rate gradient from 0 to 0.5 mL/min.

8. To ensure that the previous sample is completely eluted from the column before the next sample is loaded, determine the total volume of the column using deuterium-labeled water.

9. The internal standard is applied as a control to ensure that all columns and gradients are performed under identical conditions. Ideally, the peak of the [^{14}C]-methylated BSA radioactivity elutes in identical fractions from all columns or from all density gradients of the same centrifugation run.

10. Although it is often recommended in the literature to plot K_{AV} values *see* subsequent equation instead of elution volumes against the molecular weight/Stokes radius and generate a standard curve by linear regression through the resulting points, we obtained similar results by plotting the elution volume directly against the molecular weight/Stokes radius and drawing a smooth curve through these points: K_{AV} values can be derived as follows: $K_{AV} = (V_e - V_o)/(V_t - V_o)$, in which V_e is the elution volume, V_o is the void volume, and V_t is the total volume.

11. A gradient of 10 mL cannot be obtained even if both chambers of the gradient maker are filled with 5 mL of solution because the magnetic stirrer and the tubing cause a certain dead volume. Alternatively, we filled each chamber with approx 6 mL of solution and labeled the centrifugation tubes in advance to indicate a filling level of 10 mL. The gradient formation was then stopped when the solution reached the 10-mL mark on the tube. As a consequence, we obtained linear gradients from 5 to 33% even though the gradient maker was filled with a 40% sucrose solution in the storage chamber.

12. Depending on the range of the sucrose concentrations used, each density gradient also has an optimal molecular weight range of proteins to be separated. A density gradient from 5 to 33% separates most proteins sufficiently well to determine their molecular weight. However, the anticipated molecular weight of the protein of interest should be considered when choosing the sucrose concentrations and the ultracentrifugation parameters. For proteins less than 50 kDa, e.g., sucrose concentrations from 5 to 25% or an increased centrifugation speed or duration will give a better separation.

13. To minimize the disturbance of the gradient when the tubing is removed from or inserted into the sucrose solution, we used a thin metallic HPLC tubing as the end piece.

14. Alternatively, store the gradient for at least 2 h at 4°C to allow the solution to cool down.

15. The SW41-TI rotor used in this study can hold six centrifugation tubes. One tube was loaded with a mixture of the molecular weight standards catalase and BSA and one with aldolase and ovalbumin, leaving four centrifugation tubes to separate the PDE samples in the same centrifugation run.

16. If the ultracentrifuge allows this option, use a gradual acceleration and deceleration for starting and stopping the centrifugation.
17. Alternative methods to elute the gradient from the centrifugation tube are described in the literature, such as punching a whole in the bottom of the tube from which the solution can drip out, or pipetting the solution starting from the top of the gradient. However, because the first method destroys the centrifugation tubes and the second is rather tedious and time-consuming, we prefer the use of a fraction collector and peristaltic pump.
18. As an additional control, the linearity of the gradient can be confirmed by measuring the sucrose concentration in all fractions using a refractometer.

Acknowledgments

We are indebted to Marco Conti for critical reading of the manuscript and Caren Spencer for editorial assistance. This work was supported by National Institutes of Health Grant HD20788.

References

1. Francis, S. H., Turko, I. V., and Corbin, J. D. (2001) Cyclic nucleotide phosphodiesterases: relating structure and function. *Prog. Nucleic Acid Res. Mol. Biol.* **65,** 1–52.
2. Kakkar, R., Raju, R. V., and Sharma, R. K. (1999) Calmodulin-dependent cyclic nucleotide phosphodiesterase (PDE1). *Cell. Mol. Life Sci.* **55,** 1164–1186.
3. Francis, S. H., Chu, D. M., Thomas, M. K., et al. (1998) Ligand-induced conformational changes in cyclic nucleotide phosphodiesterases and cyclic nucleotide-dependent protein kinases. *Methods* **14,** 81–92.
4. Grange, M., Sette, C., Cuomo, M., Conti, M., Lagarde, M., Prigent, A. F., and Nemoz, G. (2000) The cAMP-specific phosphodiesterase PDE4D3 is regulated by phosphatidic acid binding: consequences for cAMP signaling pathway and characterization of a phosphatidic acid binding site. *J. Biol. Chem.* **275,** 33,379–33,387.
5. Sharma, R. K. and Wang, J. H. (1986) Calmodulin and Ca2+-dependent phosphorylation and dephosphorylation of 63-kDa subunit-containing bovine brain calmodulin-stimulated cyclic nucleotide phosphodiesterase isozyme. *J. Biol. Chem.* **261,** 1322–1328.
6. MacKenzie, S. J., Baillie, G. S., McPhee, I., et al. (2002) Long PDE4 cAMP specific phosphodiesterases are activated by protein kinase A-mediated phosphorylation of a single serine residue in Upstream Conserved Region 1 (UCR1). *Br. J. Pharmacol.* **136,** 421–433.
7. Sette, C. and Conti, M. (1996) Phosphorylation and activation of a cAMP-specific phosphodiesterase by the cAMP-dependent protein kinase: involvement of serine 54 in the enzyme activation. *J. Biol. Chem.* **271,** 16,526–16,534.
8. Ahmad, F., Cong, L. N., Stenson Holst, L., Wang, L. M., Rahn Landstrom, T., Pierce, J. H., Quon, M. J., Degerman, E., and Manganiello, V. C. (2000) Cyclic

nucleotide phosphodiesterase 3B is a downstream target of protein kinase B and may be involved in regulation of effects of protein kinase B on thymidine incorporation in FDCP2 cells. *J. Immunol.* **164,** 4678–4688.

9. Thomas, M. K., Francis, S. H., and Corbin, J. D. (1990) Substrate- and kinase-directed regulation of phosphorylation of a cGMP-binding phosphodiesterase by cGMP. *J. Biol. Chem.* **265,** 14,971–14,978.
10. Stroop, S. D. and Beavo, J. A. (1991) Structure and function studies of the cGMP-stimulated phosphodiesterase. *J. Biol. Chem.* **266,** 23,802–23,809.
11. Martinez, S. E., Wu, A. Y., Glavas, N. A., Tang, X. B., Turley, S., Hol, W. G., and Beavo, J. A. (2002) The two GAF domains in phosphodiesterase 2A have distinct roles in dimerization and in cGMP binding. *Proc. Natl. Acad. Sci. USA* **99,** 13,260–13,265.
12. Fink, T. L., Francis, S. H., Beasley, A., Grimes, K. A., and Corbin, J. D. (1999) Expression of an active, monomeric catalytic domain of the cGMP-binding cGMP-specific phosphodiesterase (PDE5). *J. Biol. Chem.* **274,** 34,613–34,620.
13. Kameni Tcheudji, J. F., Lebeau, L., Virmaux, N., Maftei, C. G., Cote, R. H., Lugnier, C., and Schultz, P. (2001) Molecular organization of bovine rod cGMP-phosphodiesterase 6. *J. Mol. Biol.* **310,** 781–791.
14. Muradov, K. G., Boyd, K. K., Martinez, S. E., Beavo, J. A., and Artemyev, N. O. (2003) The GAFa domains of rod cGMP-phosphodiesterase 6 determine the selectivity of the enzyme dimerization. *J. Biol. Chem.* **278,** 10,594–10,601.
15. Richter, W. and Conti, M. (2002) Dimerization of the type 4 cAMP-specific phosphodiesterases is mediated by the upstream conserved regions (UCRs). *J. Biol. Chem.* **277,** 40,212–40,221.
16. Kovala, T., Sanwal, B. D., and Ball, E. H. (1997) Recombinant expression of a type IV, cAMP-specific phosphodiesterase: characterization and structure-function studies of deletion mutants. *Biochemistry* **36,** 2968–2976.
17. Rocque, W. J., Holmes, W. D., Patel, I. R., Dougherty, R. W., Ittoop, O., Overton, L., Hoffman, C. R., Wisely, G. B., Willard, D. H., and Luther, M. A. (1997) Detailed characterization of a purified type 4 phosphodiesterase, HSPDE4B2B: differentiation of high- and low-affinity (R)-rolipram binding. *Protein Expr. Purif.* **9,** 191–202.
18. Richter, W., Hermsdorf, T., Lilie, H., Egerland, U., Rudolph, R., Kronbach, T., and Dettmer, D. (2000) Refolding, purification, and characterization of human recombinant PDE4A constructs expressed in Escherichia coli. *Protein Expr. Purif.* **19,** 375–383.
19. Grange, M., Picq, M., Prigent, A. F., Lagarde, M., and Nemoz, G. (1998) Regulation of PDE-4 cAMP phosphodiesterases by phosphatidic acid. *Cell Biochem. Biophys.* **29,** 1–17.
20. Lario, P. I., Bobechko, B., Bateman, K., Kelly, J., Vrielink, A., and Huang, Z. (2001) Purification and characterization of the human PDE4A catalytic domain (PDE4A330-723) expressed in Sf9 cells. *Arch. Biochem. Biophys.* **394,** 54–60.
21. Saldou, N., Baecker, P. A., Li, B., Yuan, Z., Obernolte, R., Ratzliff, J., Osen, E., Jarnagin, K., and Shelton, E. R. (1998) Purification and physical characterization

of cloned human cAMP phosphodiesterases PDE-4D and -4C. *Cell Biochem. Biophys.* **28,** 187–217.

22. Xu, R. X., Hassell, A. M., Vanderwall, D., et al. (2000) Atomic structure of PDE4: insights into phosphodiesterase mechanism and specificity. *Science* **288,** 1822–1825.

23. Lee, M. E., Markowitz, J., Lee, J. O., and Lee, H. (2002) Crystal structure of phosphodiesterase 4D and inhibitor complex(1). *FEBS Lett.* **530,** 53–58.

24. Siegel, L. M. and Monty, K. J. (1966) Determination of molecular weights and frictional ratios of proteins in impure systems by use of gel filtration and density gradient centrifugation: application to crude preparations of sulfite and hydroxylamine reductases. *Biochim. Biophys. Acta* **112,** 346–362.

14

Crystallization of Cyclic Nucleotide Phosphodiesterases

Hengming Ke, Qing Huai, and Robert X. Xu

Summary

Selective inhibitors of cyclic nucleotide phosphodiesterases (PDEs) have been widely studied as therapeutic agents for the treatment of various human diseases. Three-dimensional structures are essential for the design of highly selective inhibitors, but their availability is limited by the speed of crystallization. We describe crystallization of the catalytic domains of the unligated PDE4B2B, rolipram-bound PDE4D2, and 3-isobutyl-1-methylxanthine-bound PDE5A1 using the methods of vapor diffusion and microdialysis. We also briefly describe general methods of protein crystallization to provide a background to readers outside of the crystallographic field. Finally, we discuss detailed procedures for and pitfalls of the crystallization of PDEs, which may be valuable for crystallization of other PDE members.

Key Words

Crystallization; phosphodiestases; vapor diffusion; microdialysis; seeding.

1. Introduction

Selective inhibitors against each of 11 families of cyclic nucleotide phosphodiesterases (PDEs) have been widely studied as therapeutic agents for the treatment of various diseases *(1–6)*. For example, the PDE5 inhibitor sildenafil (Viagra) is a drug for male erectile dysfunction *(6)*. It remains a mystery how the selective inhibitors with different structures bind to the conserved catalytic site of PDEs *(7–10)*. Three-dimensional structures are essential for the design of highly selective inhibitors of PDEs. However, the speed of structure solution is determined by protein crystallization, which is a difficult-to-predict event. The crystallization of PDE proteins is especially challenging owing to the difficulty of the preparation of a large quantity of soluble proteins and the unusual crystallization behaviors of PDE proteins. We describe in this chapter the crystallization of unligated PDE4B2B, rolipram-bound PDE4D2, and PDE5A1 in

From: *Methods in Molecular Biology, vol. 307: Phosphodiesterase Methods and Protocols*
Edited by: C. Lugnier © Humana Press Inc., Totowa, NJ

complex with 3-isobutyl-1-methylxanthine (IBMX). We briefly outline the principle and practice of protein crystallization for readers who have no background in protein crystallography. We also discuss the details of PDE crystallization and potential pitfalls. Because the success of protein crystallization depends on many unpredictable factors, we hope that the general background on and experiences with crystallization of the PDE proteins will enable readers to design themselves crystallization of other PDEs.

2. Materials

1. Crystal screen I kit (HR2-110, Hampton Research) (*see* **Note 1**).
2. Crystallization dishes (HR3-170, Hampton Research).
3. Cover slides (HR3-215, Hampton Research).
4. Microdialysis buttons (10 µL) (HR3-316, Hampton Research).
5. Crystallization dishes or beakers (50 mL).
6. Proteins:
   a. PDE4B2B: Concentrate two fragments of the PDE4B2B catalytic domain (amino acids 152–528 and 152–505) to 16–20 mg/mL in a storage buffer of 50 mM HEPES, pH 7.5, 100 mM NaCl, 5 mM dithiothreitol.
   b. PDE4D2: Store two fragments of the catalytic domain of PDE4D2 (amino acids 79–438 and 79–410) as 15 mg/mL in a buffer of 50 mM NaCl, 20 mM Tris-HCl, pH 7.5, 1 mM β-mercaptoethanol, and 1 mM EDTA.
   c. PDE5A1: Store the catalytic domain of human PDE5A1 (amino acids 535–860) as 10 mg/mL in a buffer of 50 mM NaCl, 20 mM Tris-HCl, pH 7.5, 1 mM β-mercaptoethanol, and 1 mM EDTA.
7. Crystallization buffers:
   a. Buffer 1: 11% polyethylene glycol 3000 (PEG3000), 15% glycerol; 50 mM Na cacodylate, pH 6.6, 100 mM Na acetate, 2% *t*-butanol, 10 mM CaCl$_2$.
   b. Buffer 2: 0.1 M HEPES, pH 7.5, 20% PEG3350, 30% ethylene glycol, 10% isopropanol, 5% glycerol.
   c. Buffer 3: 0.1 M Tris-HCl, pH 7.5, 17% PEG3350, 0.2 M MgSO$_4$.
   d. Buffer 4: 20 mM HEPES, 50 mM sodium cacodylate, 0.2 M Na acetate, 3% *t*-butanol, 5.5% PEG3350 at pH 7.0.
   e. Buffer 5: 10 mM HEPES, pH 7.5, 2% PEG3350, 5% glycerol, 5% dimethyl sulfoxide (DMSO), 4 mM CaCl$_2$.

3. Methods

3.1. Principle of Protein Crystallization

Crystallization of a protein is a physical process in which a protein in a solution state is transformed into a solid state. In principle, crystallization of proteins can be divided into three stages: (1) saturation or supersaturation of a protein solution, (2) nucleation, and (3) crystal growth. Supersaturation is a metastable solution state in which proteins will automatically precipitate out in a certain period of time. In the nucleation stage, the crystal nucleus is thermodynamically unstable and exists in equilibrium of formation and dissolving. Free energy of the nucleation is positive and thus disfavors nucleus formation until it reaches a critical size (Rc). After passing the Rc point or at the crystal growth stage, crystals grow steadily because the free energy of crystallization decreases. The principle of crystallization is described in many publications; for a review *see* **ref. *11***.

To grow crystals, a protein is dissolved in a buffer such as 20 m*M* Tris-HCL, and then a precipitant such as PEG4000 is added to bring the protein solution to saturation. Crystals will form if the precipitation rate is carefully controlled. Otherwise, amorphous precipitation will be obtained if the precipitation is too fast. Thus, the whole point of crystallization is to control the precipitation process that is impacted by many factors.

Factors affecting the growth of macromolecular crystals include purity and initial concentration of the proteins, precipitants, pH of the solution, buffer, organic solvents, salts or ions, detergents, volume of the crystallization sample, and temperature. Other factors such as gravity, vibration of the environment, and molecular size and shape also affect crystallization of proteins but are less controllable. In principle, proteins with any molecular weights are crystallizable because the formation of crystal lattice is determined by the molecular conformation but not the size of a protein. Binding of an inhibitor to an enzyme stabilizes the conformation of a protein and, thus, usually favors crystallization.

3.2. General Methods for Crystallization

Four methods have successfully produced crystals of macromolecules: batch crystallization, vapor diffusion, microdialysis, and seeding.

3.2.1. Batch Crystallization

A classic technique for early crystallization of enzymes, batch crystallization is still in use in modern protein crystallization. By laying a precipitant (salt, organic solvent, or PEG) over a protein solution, crystals may form in several days. The advantages of this method are its simplicity and ability to add precipitants step-by-step, but the disadvantage is that it often quickly brings a protein solution to saturation. Recently, a crystallization robot based on batch crystallization has been developed for automatically screening hundreds of conditions (*12*).

3.2.2. Vapor Diffusion

The most common method for crystallization of proteins is vapor diffusion. It uses a closed chamber in which a crystallization buffer with a precipitant is placed in a reservoir and a protein drop hangs over the reservoir (**Fig. 1**).

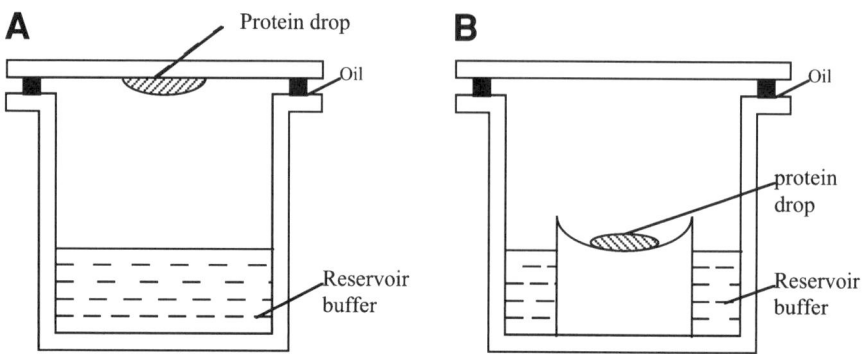

Fig. 1. Crystallization by vapor diffusion: (**A**) hanging drop and (**B**) sitting drop.

Vapors from the protein drop and the crystallization buffer exchanges in the chamber so as to gradually bring the protein solution to saturation for crystallization. Frequently used apparatuses for the vapor diffusion method include hanging drop and sitting drop (**Fig. 1**). The advantage of vapor diffusion is that it is a slow process allowing protein enough time to pack into a crystal lattice. Vapor diffusion is the major method used in the design of high-throughput crystallization robots for structural genomics *(12–16)*.

3.2.3. Microdialysis

Another method to crystallize proteins is microdialysis. A microdialysis button containing a protein solution is covered with dialysis membrane (**Fig. 2**) and immersed into a crystallization buffer that contains precipitants. The protein will crystallize after the precipitant gradually dialyzes into the button. Because all the chemical components inside the microdialysis button except for protein are identical to those of the crystallization buffer at equilibrium, the microdialysis method can precisely control the crystallization conditions and has good reproducibility. The disadvantage of the microdialysis method is that it requires a relatively large amount of materials, at least 5 μL of protein solution for each trial. However, large crystals are often obtained from microdialysis because a large volume of protein is used in crystallization.

3.2.4. Seeding

Seeding is a technique in which tiny crystal seeds are transferred to a protein solution so that proteins grow on the surface of the seeds to produce large crystals. Two seeding methods have been used: microseeding and macroseeding. Microseeding uses tiny crystal seeds to serve as the nuclei of crystallization. Macroseeding transfers one single crystal into a protein solution as the seed of

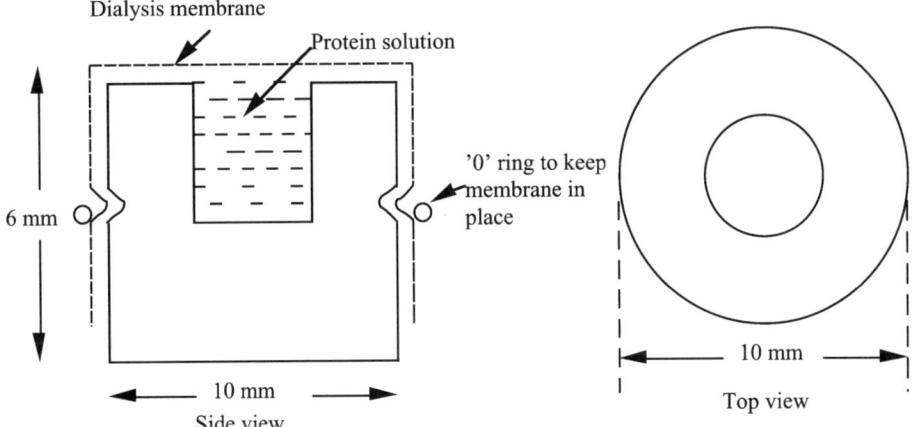

Fig. 2. Crystallization by microdialysis.

crystallization. In both seeding techniques, the starting protein solution for crystallization should be close to saturation and the seeds should not be completely dissolved after being transferred into the protein solution.

3.3. Practical Aspects of Crystallization

When a large amount of proteins (say 5 mg or more) is available for crystallization, three aspects must be tested to check if the protein is suitable for crystallization. First, the purity of a protein is essential for successful crystallization. If a single band is found in a mini sodium dodecyl sulfate-polyacrylamide gel electrophoresis (SDS-PAGE) with 20 µg of protein loaded, the protein is estimated to have a purity more than 95% and is suitable for crystallization. Second, isoforms and multiple conformations of a protein have different isoelectric points and will strongly impact crystallization. A native PAGE and isoelectrofocusing gel will show the conformational states. A protein sample having multiple bands in an isoelectrofocusing electrophoresis gel has a low probability of growing single crystals. Third, a protein that aggregates into a series of oligomers is unlikely to crystallize. Aggregation and a poorly folded protein from expression can be detected by dynamic light scattering and native PAGE (*see* **Note 2**).

There are three general features for crystallization of proteins: (1) crystallization is a trial-and-error process and no theory can guarantee the growth of protein crystals, (2) each protein has a characteristic profile of crystallization so that individual experiments have to be set up, and (3) hundreds or thousands of trials may be necessary to obtain quality crystals for X-ray diffraction. Therefore, protein crystallization is often very time-consuming and tedious. Several methods have been studied to reduce crystallization trials. The factorial method is the mathematic analysis of mutual interactions of multiple factors in crystallization *(17)*. However, its practical use is limited because crystallization behaviors of proteins do not always obey the statistical model. An improvement is a combination of the practical experiences with the factorial method, or "sparse matrix" *(18)*, listed as Screen I (*see* **Subheading 2.1.4.**) in "Crystallization Research Tools" (Hampton Research). Sparse matrix has been used as the first screen of crystallization of most proteins. The factorial method and sparse matrix have recently been modified for high-throughput crystallization robots *(12)*.

3.4. Crystallization of PDE4 and PDE5 by Vapor Diffusion

1. Concentrate the PDE proteins to 10–20 mg/mL with Amicon Stirred Cells and ultrafiltration membrane YM30 (Millipore, Bedford, MA) under the pressure of nitrogen gas.
2. Aliquot the concentrated protein into a 0.2-mL tube and store at –80°C for later use (*see* **Note 3**).
3. Run a mini SDS-PAGE, a native PAGE, or isoelectrofocusing gel with a load of 20 μg of protein, to check the purity and potential multiple conformations of the protein (*see* **Note 4**).
4. Because the crystallization conditions may vary for PDE proteins purified by different groups or from different expression systems, for the first trial set up a preliminary screen of crystallization such as with a crystal screen I kit from Hampton Research.
5. To grow cocrystals of PDE with an inhibitor, mix 1–5 m*M* inhibitor with the PDE protein for a couple of hours before setting up the crystallization screen (*see* **Note 5**).
6. Add 0.6–1 mL of the crystallization buffer into each well of a crystallization dish.
7. Place 2 μL of protein solution on the cover slide and add 2 μL of well buffer to the protein solution (*see* **Note 6**). Flip the cover slide and seal the crystallization well.
8. Store the crystallization dish at 4°C for crystal formation (*see* **Note 7**). If no significant precipitation is observed in most screen conditions, concentrate the protein to a higher concentration.
9. Observe the crystallization tray under a 40–100 magnifier microscope every day. Small crystals of PDE may grow under several conditions of the Hampton screen.
10. To optimize the crystallization conditions, make a series of dilutions of the well buffer with water, such as 90%, 80%, and so on, to let the crystals grow slower. The best precipitant concentration will produce crystals of most proteins in 2–7 d (*see* **Note 8**).
11. At the best concentration of the precipitant found in **step 10**, add 2–10% organic additives such as DMSO and *t*-butanol, or 0.01–1% detergents such as β-OG to improve further the growth of PDE crystals (*see* **Note 9**).
12. For crystallization of unligated PDE4B2B, crystallize the catalytic domain of PDE4B2B (amino acids 152–528) using hanging drop and macroseeding methods. Mix the protein solution at approx 20 mg/mL with an equal volume of crystallization buffer 1 at 4°C *(19)*. Improve the size of the crystals by macroseeding

(*see* **Note 10**) at 4°C using the same buffer. In our experiment, the crystal had the space group C222$_1$ with $a = 104.5$, $b = 159.6$, and $c = 109.0$ Å and diffracted to 1.8-Å resolution.

13. For crystallization of PDE4D2-rolipram, prepare the complex of the PDE4D2 catalytic domain (amino acids 79–438) with (R)-rolipram by mixing PDE4D2 with 1 mM (R)-rolipram predissolved in DMSO as a 20 mM stock solution. Grow the crystals of PDE4D2-(R)-rolipram by hanging drop against crystallization buffer 2 at 4°C *(20)*. In our experiment, the crystal had the space group P2$_1$2$_1$2$_1$ with unit cell dimensions of $a = 99.3$, $b = 112.5$, and $c = 160.9$ Å and diffracted to 2-Å resolution.

14. For crystallization of PDE5A1-IBMX, dissolve IBMX in DMSO as a 100 mM IBMX stock solution. Prepare the complex of PDE5A1-IBMX by mixing the catalytic domain of 10 mg/mL of human PDE5A1 (amino acids 535–860) with 5 mM IBMX and crystallize against crystallization buffer 3 at room temperature. In our experiment, the crystal of PDE5A1-IBMX had the space group P3$_1$21 with cell dimensions of $a = b = 74.5$ and $c = 129.0$ Å and diffracted to 2.5-Å resolution in a Rigaku imaging plate Raxis IV++.

3.5. Crystallization of PDE4 by Microdialysis

1. Follow **steps 1–3** in **Subheading 3.4.**
2. Prepare 20 mL of crystallization buffer in 50-mL crystallization dishes (*see* **Note 11**).
3. Fill 10 μL dialysis buttons with the catalytic domain of PDE4 and cover with a dialysis membrane presoaked in distilled water and with mol wt cutoff of 12–14 kDa.
4. Immerse the dialysis button into the crystallization buffer and store the crystallization dishes in a cold room.
5. Add 2% precipitant if the dialysis does not produce crystals or precipitation after 2 d. Increase the precipitant by 2% every 2 d until crystals or precipitation is observed.
6. If heavy precipitation is observed on the second day, reduce the starting concentration of the precipitant.
7. After the proper concentration of precipitant is found, add 2–10% organic solvents such as *t*-butanol into the crystallization buffer to improve the quality of the crystals.
8. For crystallization of the catalytic domain of unligated PDE4B2B (residues 152–505), dialyze 16 mg/mL of PDE4B2B against crystallization buffer 4 at 4°C. The crystals will have a diamond shape and reach a typical size of $0.4 \times 0.4 \times 0.6$ mm in 7–10 d (**Fig. 3**). In our experiment, they had the space group P6$_4$22 with cell dimensions of $a = b = 102.7$ and $c = 327.4$ Å and diffracted anisotropically to 2.5-Å resolution along the diamond axis and 3.2-Å resolution along the other two directions (*see* **Note 12**).
9. For crystallization of the catalytic domain of unligated PDE4D2 (amino acids 79–410), dialyze 17 mg/mL of PDE4D2 against crystallization buffer 5 at 4°C. Other organic additives such as 5% MPD, *t*-butanol, isopropanol, PEG400, and

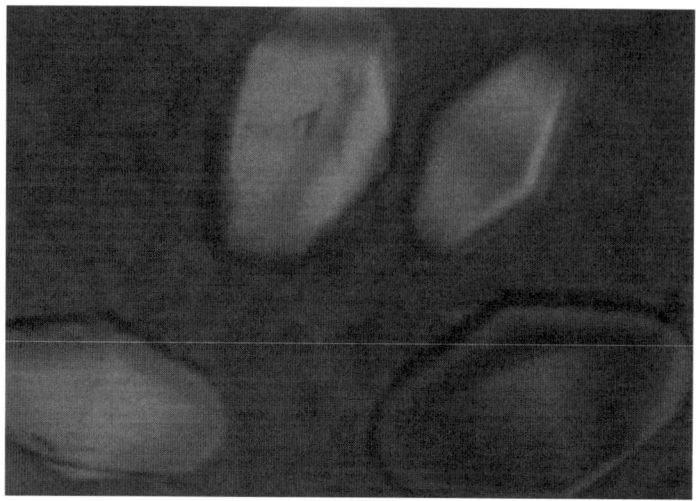

Fig. 3. Crystals of PDE4B2B grown by dialysis.

methanol will also produce similar crystals. In our experiments, the crystals had a space group of $I4_122$ with $a = 108.7$ and $c = 169.0$ Å and diffracted to 3-Å resolution.

4. Notes

1. The crystallization buffers may be prepared by oneself with chemically pure materials.
2. Because an aggregative protein can usually be concentrated to a maximum of a few milligrams per milliliter or will be lost at a high concentration with no visible precipitation, a good recovery yield during concentration is an indication of no significant aggregation of a protein. In some cases, high molecular weight aggregates may be separated by a molecular sieving column from corrected folded oligomers (monomer, dimer, tetramer, and so on). Protease cleavage is a common technique to remove fragments that cause aggregation.
3. The catalytic domains of PDE4B2B and PDE5A1 are not thermally stable. When these proteins are stored at 4°C for longer than 1 wk, they cannot reproduce crystals. PDE4D2 can be stored at 4°C for 1 mo without problems with crystal reproduction. No additives are necessary for storage of the PDE proteins at −80°C.
4. A protein showing multiple bands in a native PAGE or an isoelectrofocusing gel may be crystallizable, but it is much more difficult to obtain quality crystals for diffraction. Careful control of the crystallization conditions may produce crystals that diffract to medium or high resolution. For example, a catalytic domain of PDE4D2 that shows monomer to tetramer in a native PAGE and a major band in isoelectrofocusing gel requires organic solvents such as 25% ethylene glycol to obtain crystals diffracting to 2-Å resolution.

5. Most PDE inhibitors are only dissolvable in organic solvents such as DMSO. When an inhibitor in DMSO is mixed with PDE protein solution, solid precipitation may form. If this happens, let the mixture sit for several hours or overnight and then centrifuge to remove the precipitation before setting the crystallization.

6. In some cases, a variation in the ratio of protein solution vs well buffer in the protein drop may improve the quality of PDE crystals.

7. The crystals of PDE4B2B and PDE4D2 can be grown at either 4°C or room temperature under different conditions but have different space groups.

8. If the Hampton screen I does not produce tiny crystals, choose the crystallization conditions that produce sandlike or solid precipitation for further experiments. After finding an appropriate concentration of a precipitant by a series of dilutions, add the organic and detergent additives to help crystal formation.

9. In addition to optimization of the crystallization conditions, the quality of the PDE4 crystals is dependent on the length of a PDE fragment. For example, the PDE4B2B (residues 152–528) crystals diffracted to 1.8-Å resolution, but the PDE4B2B (residues 152–505) crystals diffracted to only 2.5-Å resolution. However, there are no good methods to predict the best length of a PDE fragment for crystallization.

10. To grow crystals with the macroseeding technique, transfer a single crystal into 20 µL of plain buffer (such as 20 mM Tris-HCl, pH 7.5; 50 mM NaCl) on the cover slide and let the crystal growing face dissolve slightly. Dry the buffer with a filter paper and then add protein solution and well buffer for crystallization.

11. The crystallization buffers should be designed for tests of various pH values and precipitants. A pH range of 4.5–8.5 is often screened in an interval of one pH unit. For example, 20 mM Tris-HCl will produce a stable pH at 8.5, HEPES for pH 7.5, MES at pH 6.5, maleic acid at pH 5.5, and sodium acetate at pH 4.5. Precipitant such as 6% PEG4000 or 10% ammonium sulfate is useful as the starting concentration.

12. The similar hexagonal crystals of PDE4B2B (amino acids 152–505) can also be grown by the hanging drop method but are often smaller and diffract to about 3.5-Å resolution.

References

1. Barnette, M. S. and Underwood, D. C. (2000) New phosphodiesterase inhibitors as therapeutics for the treatment of chronic lung disease. *Curr. Opin. Pulm. Med.* **6,** 164–169.

2. Huang, Z., Ducharme, Y., Macdonald, D., and Robichaud, A. (2001) The next generation of PDE4 inhibitors. *Curr. Opin. Chem. Biol.* **5,** 432–438.

3. Giembycz, M. A. (2002) Development status of second generation PDE 4 inhibitors for asthma and COPD: the story so far. *Monaldi Arch. Chest Dis.* **57,** 48–64.

4. Spina, D. (2003) Theophylline and PDE4 inhibitors in asthma. *Curr. Opin. Pulm. Med.* **9,** 57–64.

5. Reilly, M. P. and Mohler, E. R. III. (2001) Cilostazol: treatment of intermittent claudication. *Ann. Pharmacother.* **35,** 48–56.

6. Corbin, J. D. and Francis, S. H. (2002) Pharmacology of phosphodiesterase-5 inhibitors. *Int. J. Clin. Pract.* **56,** 453–459.
7. Torphy, T. J. (1998) Phosphodiesterase isozymes: molecular targets for novel anti-asthma agents. *Am. J. Respir. Crit. Care Med.* **157,** 351–370.
8. Conti, M. and Jin, S. L. (1999) The molecular biology of cyclic nucleotide phosphodiesterases. *Prog. Nucleic Acid Res. Mol. Biol.* **63,** 1–38.
9. Soderling, S. H. and Beavo, J. A. (2000) Regulation of cAMP and cGMP signaling: new phosphodiesterases and new functions. *Curr. Opin. Cell Biol.* **12,** 174–179.
10. Houslay, M. D. and Adams, D. R. (2003) PDE4 cAMP phosphodiesterases: modular enzymes that orchestrate signalling cross-talk, desensitization and compartmentalization. *Biochem. J.* **370,** 1–18.
11. Weber, P. C. (1997) Overview of protein crystallization, in *Methods in Enzymology* (Carter, C. W. Jr. and Sweet, R. M., eds.), Academic, New York, pp. 13–22.
12. Stevens, R. C. (2000) High-throughput protein crystallization. *Curr. Opin. Struct. Biol.* **10,** 558–563.
13. Juarez-Martinez, G., Steinmann, P., Roszak, A.W., Isaacs, N.W., and Cooper, J. M. (2002) High-throughput screens for postgenomics: studies of protein crystallization using microsystems technology. *Anal. Chem.* **74,** 3505–3510.
14. Chayen, N. E. and Saridakis, E. (2002) Protein crystallization for genomics: towards high-throughput optimization techniques. *Acta Crystallogr. D* **58,** 921–927.
15. Krupka, H. I., Rupp, B., Segelke, B. W., Lekin, T. P., Wright, D., Wu, H. C., Todd, P., and Azarani, A. (2002) The high-speed Hydra-Plus-One system for automated high-throughput protein crystallography. *Acta Crystallogr. D* **58,** 1523–1526.
16. Mueller, U., Nyarsik, L., Horn, M., Rauth, H., Przewieslik, T., Saenger, W., Lehrach, H., and Eickhoff, H. (2001) Development of a technology for automation and miniaturization of protein crystallization. *J. Biotechnol.* **85,** 7–14.
17. Carter, C. W. Jr. and Carter, C. W. (1979) Protein crystallization using incomplete factorial experiments. *J. Biol. Chem.* **254,** 12,219–12,223.
18. Jancarik, J. and Kim, S. H. (1991) Spare matrix sampling: a screening method for crystallization of proteins. *J. Appl. Crystallogr.* **24,** 409–411.
19. Xu, R. X., Hassell, A. M., Vanderwatt, D., et al. (2000) Atomic structure of PDE4: insight into phosphodiesterase mechanism and specificity. *Science* **288,** 1822–1825.
20. Huai, Q., Wang, H., Sun, Y., Kim, H. Y., Liu, Y., and Ke, H. (2003) Three dimensional structures of PDE4D in complex with rolipram and implication on inhibitor selectivity. *Structure* **11,** 865–873.

15

Generation of PDE4 Knockout Mice by Gene Targeting

S.-L. Catherine Jin, Anne M. Latour, and Marco Conti

Summary

The development of gene-targeting techniques has ushered in a new era in mouse genetics. Two discoveries have been instrumental: the finding that an exogenous DNA introduced in mammalian cells can recombine with homologous chromosomal sequences, a process known as gene targeting, and the revelation that cultured embryonic stem (ES) cells when injected into early stage mouse embryos can contribute to produce germ-line chimeras. On the basis of these seminal findings, gene targeting by homologous recombination in mouse ES cells in vitro has been established as a powerful means of altering specific loci in the mouse genome. As a result, gene function can be studied in vivo. By applying this technology, targeted disruption of PDE4 alleles is created in cultured ES cells and, subsequently, the mutant ES cells are injected into blastocysts and returned to pseudopregnant foster mothers to produce germ-line chimeric pups. In this chapter, we describe the basic protocols used to generate the PDE4 knockout mice.

Key Words

Cyclic nucleotide phosphodiesterase; cyclic adenosine monophosphate; homologous recombination; gene targeting; embryonic stem cells; knockout mice; signal transduction.

1. Introduction

The cyclic adenosine monophosphate (cAMP)-specific PDE4 family is composed of at least 20 isozymes, which are encoded by four distinct genes, named PDE4A, 4B, 4C, and 4D. Besides sequence homology, these enzymes share similar kinetic and regulatory properties and are all inhibited by rolipram and related compounds (1,2). PDE4s are expressed in almost all immune and inflammatory cells, and pharmacological inactivation of these enzymes attenuates several inflammatory responses (3). Thus, significant interest has emerged to develop therapeutic agents that selectively inhibit this enzyme family for the treatment of inflammatory diseases.

From: *Methods in Molecular Biology, vol. 307: Phosphodiesterase Methods and Protocols*
Edited by: C. Lugnier © Humana Press Inc., Totowa, NJ

To date, pharmacological studies using PDE4 inhibitors have provided valuable insights into the role of PDE4s in cAMP homeostasis. However, it is difficult to evaluate in vivo the functions of individual PDE4 isoforms because subtype-selective inhibitors are not available and more than one PDE4 is expressed in the same cell type or tissue *(1,2)*. To overcome this obstacle, a targeted gene inactivation by homologous recombination has been used. Briefly, with this technology one can alter a gene of interest initially in mouse embryonic stem (ES) cells in culture by recombination between an introduced DNA fragment and the endogenous homologous sequences *(4,5)*. Subsequently, the mutant ES cells are injected into recipient blastocysts and returned to pseudopregnant foster mothers to produce chimeric mice *(6,7)*. Because the ES cells potentially contribute to all tissues in the chimeras, including germ cells *(8)*, the germ-line transmission of the mutant allele can be achieved when a part or all of the germ cells are derived from the ES cells. Using this technique, we have produced PDE4D-, PDE4B-, and PDE4A-deficient (knockout [KO]) mice *(9,10)*. In general, this methodology provides investigators with a powerful tool to mutate virtually any gene in mice.

In this chapter, we describe the basic protocols used to generate PDE4 KO mice. First, we describe the methods in detail for generation of PDE4-targeted ES cells (*see* **Subheadings 3.1.–3.5.** and **3.7.**) and for karyotyping of ES cells (*see* **Subheading 3.6.**). Second, we briefly review the steps involving mouse and tissue/organ handling and surgery such as blastocyst preparation, blastocyst injection of ES cells, and transfer of injected blasotocysts into foster mothers (*see* **Subheading 3.8.**). Finally, we describe the process of KO mice production from chimera breeding (*see* **Subheading 3.9.**) and the polymerase chain reaction (PCR) for genotyping of PDE4 KO mice (*see* **Subheading 3.10.**). Because the general principles regarding gene targeting by homologous recombination, strategies involved in the design of targeting vectors, and considerations of ES cell quality for DNA delivery and ES cell culture conditions have been detailed in many articles *(11–15)*, these subjects are not discussed in this chapter.

1.1. Strategies for Generation of PDE4-Targeted ES Cells

The targeting vectors we used for inactivation of PDE4 alleles are replacement-type vectors *(12,14)*. Each contains a pBluescript plasmid backbone, two selection markers, and a large region of DNA with sequences homologous to those of the desired chromosomal integration site. The targeted disruption of PDE4B locus is used as an example in the text to illustrate the targeting strategy (**Fig. 1**).

To ensure that homologous recombinant generated by the replacement vector produces mice deficient in PDE4B enzymatic activity, a genomic region that encodes the catalytic domain of the enzyme is selected as a target. This region contains exons 8–10 of the PDE4B gene *(16)* (**Fig. 1**). The two selection

A

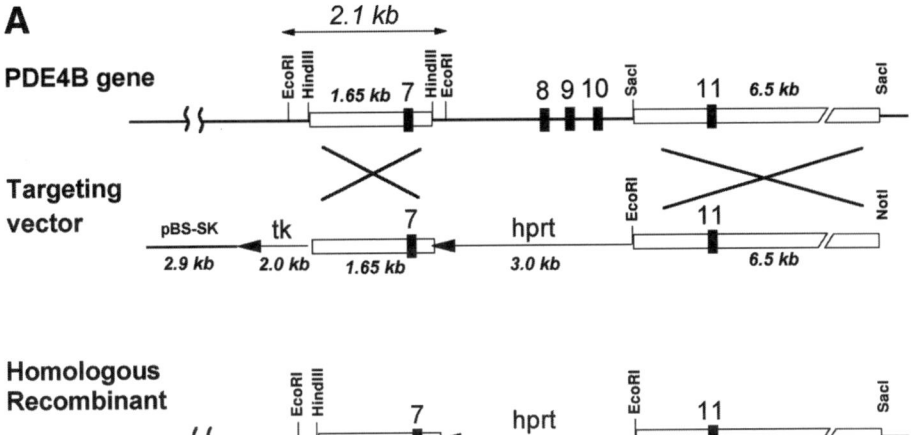

B

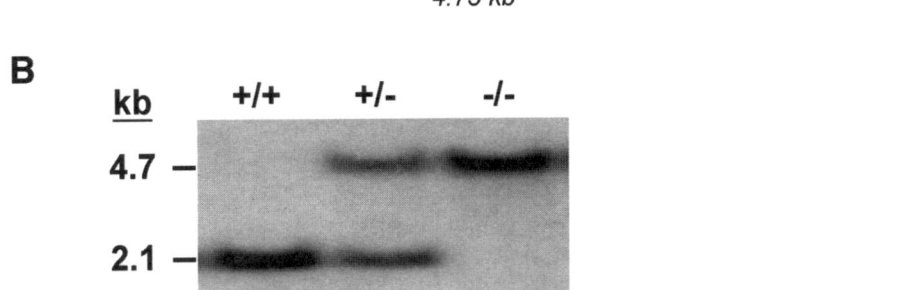

Fig. 1. **(A)** Strategy used for targeting PDE4B locus. The structures of the wild-type PDE4B gene in the region containing exons 7–11 (top), the targeting vector (middle), and the homologous recombinant (bottom) are shown. The targeting vector derived from pBluescript SK+ (pBS-SK) contains 5'- and 3'-flanking regions of homology (open boxes) and is designed to replace the *Hin*dIII-*Sac*I fragment that contains exons 8–10 with a PGK-hprt cassette. Both hprt and tk genes are inserted in the opposite transcriptional orientation as PDE4B. The probe used for Southern blot analysis of *Eco*RI-digested genomic DNA is indicated. Double arrows indicate the expected sizes of *Eco*RI fragments (2.1 and 4.75 kb). **(B)** Southern blot analysis of genomic DNA isolated from PDE4B$^{+/+}$, PDE4B$^{+/-}$, and PDE4B$^{-/-}$ mouse tails digested with *Eco*RI and hybridized to probe shown in (A).

markers, one positive and one negative, are included in the vector because transfection efficiency and targeting frequency of such vectors are usually very low and the selection is helpful to enrich the population of homologous recombinant clones (*17*). The selection markers that we used are the PGK-hprt cassette (positive) and the pMC1-tk cassette (negative). The PGK-hprt cassette is positioned within the PDE4B genomic sequences and the pMC1-tk cassette outside the homology regions (**Fig. 1**).

When the linearized vector is transfected into ES cells, it can integrate into the host genome either at the PDE4B target site as a consequence of two crossovers (**Fig. 1**) or into a random chromosomal location. When the targeted integration occurs, the region of exons 8–10 is replaced with the PGK-hprt cassette, and the pMC1-tk cassette is not recovered in the recombinant allele (**Fig. 1**). Subsequently, the targeted cells that carry the functional *hprt* gene can be isolated by their ability to survive in medium containing both HAT (hypoxanthine, aminopterin, and thymidine, for positive selection) and the drug ganciclovir (for negative selection). The cells that carry the herpes simplex thymidine kinase (*HSVtk*) gene as a result of random integration, however, are selectively killed by ganciclovir. Thus, this positive-negative selection process is useful to enrich the cells with targeted integration.

The ES cells that we use to generate PDE4 knockout mice are *hprt*-deficient E14TG2a cells, which are derived from strain 129/Ola mouse (*18*). The quality of ES cells is known to be critical for the successful creation of chimeric mice and of germ-line transmission (*19*). To maintain a pluripotent and undifferentiated state, the cells routinely are cultured on feeder layers of mitotically inactive fibroblasts, either primary mouse embryonic fibroblasts (MEFs) (*20*) or permanent STO cell lines (*21,22*). The feeder layers are used because these cells can release and present to ES cells a factor called leukemia inhibitory factor, which is capable of inhibiting ES cell differentiation (*23–26*). Moreover, as a general rule, the ES cells should be grown at high density and passed frequently to prevent cell differentiation. It is also important to dissociate clumps of cells at each passage (*11*).

1.2. Screening of ES Clones for Homologous Recombinants

Southern blot analysis is a common and reliable method to screen for ES clones targeted with gene replacement vectors. To achieve this, it is necessary to identify unique probes that are not contained in the homologous regions (external probe), and to select the restriction sites within the wild-type (WT) locus so that the WT and mutant alleles can be distinguished. As shown in **Fig. 1**, *Eco*RI is the restriction site used to distinguish the PDE4B WT locus (2.1-kb fragment) from the corresponding mutant (4.75-kb fragment). The probe used is the 105-bp sequence of the *Eco*RI-*Hind*III fragment, which flanks the 1.65-kb homologous sequences. Using this probe, the targeted ES cells can be identified by detection of the hybridization signals of both the 2.1- and 4.75-kb bands in Southern blot whereas the nontargeted cells yield only the WT 2.1-kb band signal.

Mutant PDE4B locus

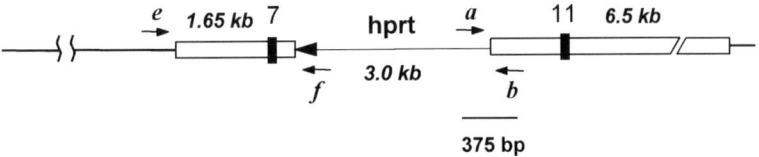

WT PDE4B locus

Fig. 2. Schematic diagram of PCR strategy used to genotype PDE4B mutant mice. The forward primers *a* and *c* are complementary to the sequences in the hprt cassette and in the PDE4B-deleted region, respectively. A reverse primer, *b*, hybridizes to a sequence within the 6.5-kb homologous region. The mutant and WT alleles are amplified with primers *a* and *b* and primers *c* and *b*, respectively, to produce 375- and 700-bp fragments as indicated. The primers *e* and *f* are presented to illustrate the PCR strategy used to screen homologous recombinants in ES cells (*see* **Note 3** for details).

An alternative screening method is the PCR. PCR is rapid and less costly. However, it has some drawbacks. For example, because the method is extremely sensitive, false positives could be detected owing to contamination of reagents with exogenous nucleic acids. Moreover, amplification efficiency decreases when the distance between the primers is more than 2 kb, and, consequently, the number of false negatives may increase. For detection of a gene-targeting event by PCR, it is necessary to design two primers with one specific for the mutation region and the other for the chromosomal DNA just outside the homologous sequences (*see* **Fig. 2**, primers *e* and *f*). The strategies and considerations for the PCR method are detailed in **Note 1**.

2. Materials
2.1. Generation of PDE4-Targeted ES Cells

1. 129svj lambda FIXII genomic library (Stratagene, La Jolla, CA).
2. pBluescript II SK containing PGK-hprt and pMC1-tk cassettes (a gift from Dr. D. Repaske, University of Cincinnati, Cincinnati, OH).
3. Restriction digestion enzymes and T4 DNA ligase.
4. Phenol (buffer saturated), chloroform, 3*M* sodium acetate, pH 5.2, and absolute alcohol.

5. Sterile TE buffer: 10 mM Tris-HCl, pH 8.0, 1 mM EDTA, pH 8.0.
6. Phosphate-buffered saline (PBS), pH 7.5: 2.7 mM KCl, 1.1 mM KH$_2$PO$_4$, 138 mM NaCl, 8.1 mM Na$_2$HPO$_4$•7H$_2$O. Filter sterilize and store at room temperature.
7. Trypsin solution: Mix 100 mL of 0.25% trypsin in 500 mL of buffer solution for trypsin (*see* **item 9**). Store at 4°C. Warm the solution to 37°C before use.
8. 0.25% Trypsin (Gibco-BRL, Gaithersburg, MD).
9. Buffer solution for trypsin: For 1 L, mix 8.0 g of NaCl, 0.4 g of KCl, 1.0 g of D-glucose, 0.35 g of NaHCO$_3$, and 0.2 g of EDTA (not disodium form) in distilled, deionized water. Adjust the pH to 7.0 with 0.5 M NaOH. Filter sterilize and store at 4°C.
10. MEF medium: Dulbecco's modified Eagle's medium (DMEM), 10% fetal bovine serum (FBS), 2 mM L-glutamine, penicillin (100 U/mL)/streptomycin (100 μg/mL).
11. DMEM, FBS (ES cell tested), L-glutamine (100X stock), penicillin/streptomycin (100X stock) (Gibco-BRL).
12. γ irradiator (University of North Carolina Hospital, Chapel Hill, NC).
13. ES cell medium: DMEM-high glucose (DMEM-H), 15% FBS, 2 mM L-glutamine, penicillin (100 U/mL)/streptomycin (100 μg/mL), 0.1 mM β-mercaptoethanol.
14. DMEM-H (Gibco-BRL).
15. 14.3 M β-Mercaptoethanol, cell culture tested (Sigma, St. Louis, MO).
16. BTX Electro Cell Manipulator and electroporation cuvet (BTX, San Diego, CA).
17. 50X HAT medium (Sigma).
18. Ganciclovir (Sigma).
19. Freezing medium: DMEM-H, 10% FBS, 10% dimethyl sulfoxide; store at 4°C.

2.2. Extraction of DNA From ES Cells and Mouse Tail Tips and Southern Blot Analysis

1. Tail buffer: 0.05 M Tris-HCl, pH 8.0, 0.1 M NaCl, 0.1 M EDTA, pH 8.0, 1% sodium dodecyl sulfate (SDS).
2. Proteinase K: Make a 10 mg/mL solution in water.
3. Saturated NaCl: 175 g of NaCl brought up to 500 mL with water. Mix well with stirring overnight. Let undissolved NaCl settle.
4. Pasteur pipet (9-in).
5. TE buffer: 10 mM Tris-HCl, pH 7.5, 1 mM EDTA, pH 8.0.
6. *Eco*RI, agarose, and absolute alcohol.
7. Denaturation solution: For 1 L, mix 87.75 g of NaCl and 20 g of NaOH in water. Store at room temperature.
8. Neutralization solution: For 1 L, mix 87.75 g of NaCl and 60.55 g of Tris-HCl in water. Adjust the pH to 8.0 with concentrated HCl. Store at room temperature.
9. Biotrans nylon membrane (ICN, Irvine, CA).
10. Ultraviolet (UV) Stratalinker 1800 (Stratagene).
11. Prehybridization and hybridization solution: 50% formamide, 5X Denhardt's, 5X SSPE, 0.1% SDS, 100 μg/mL of boiled sperm DNA.
12. Random primers DNA labeling system (Gibco-BRL).

13. 50X Denhardt's solution: For 500 mL, mix 5 g of Ficoll 400, 5 g of polyvinyl-pyrrolidone, and 5 g of bovine serum albumin (Fraction V) in water. Filter and store at −20°C in 50-mL aliquots.
14. 20X SSPE: For 1 L, mix 175 g of NaCl, 27.6 g of $NaH_2PO_4 \cdot H_2O$, and 7.4 g of EDTA in water. Adjust the pH to 7.4 with 10 N NaOH. Store at room temperature.
15. 20X Saline sodium citrate (SSC): For 1 L, mix 175 g of NaCl and 88 g of $Na_3citrate \cdot 2H_2O$ in water. Adjust the pH to 7.0 with 1 M HCl. Store at room temperature.
16. Herring sperm DNA (Roche, Indianapolis, IN).
17. X-ray film (Eastman Kodak, Rochester, NY).

2.3. ES Cell Karyotyping

1. Colcemid (10 µg/mL) (Gibco-BRL).
2. Ethidium bromide (EtBr) (1 mg/mL).
3. KCl solution, 0.56% in water.
4. Fixation solution: 3 parts methanol to 1 part glacial acetic acid.
5. Microscope slides.
6. Giemsa stain, 3% in PBS (Sigma).

2.4. Microinjection and Production of Chimeras

1. CO_2-independent medium (Gibco-BRL).
2. Mineral oil, animal tested.
3. ES cell microinjection equipment, injection and holding pipets, and blastocyst transfer pipet.
4. Female mice: strains C57BL/6 (B6) and C57BL/6 × DBA/2 (B6D2) F_1 hybrids.
5. Male mice: fertile B6 and vasectomized sterile B6D2 mice.
6. Avertin: Dissolve 0.5 g of 2,2,2-tribromoethanol in 0.63 mL of *tert*-amylalcohol by vortexing. Mix 0.5 mL of this solution with 19.5 mL of prewarmed 0.9% saline and shake to dissolve the anesthetic. Inject 0.012 mL/g of body weight.
7. Dissecting tools.
8. Sterile 1-mL syringes and 25- and 26-gage needles.
9. Iodine solution.

3. Methods

3.1. Generation of PDE4B Targeting Vector (see Note 2)

1. Subclone the PGK-hprt and the pMC1-tk cassette by blunt-end ligation in pBluescript II SK at the *Hin*dIII and *Kpn*I site, respectively.
2. Screen a mouse 129svj λ FIXII genomic library to isolate clones that contain sequences flanking the region of exons 8–10 of the PDE4B locus (*see* **Note 3**).
3. Isolate two restriction fragments from the PDE4B genomic clones: the 1.65-kb *Hin*dIII fragment and the 6.5-kb *Sac*I fragment, which are 5′ and 3′ to exons 8–10, respectively (*see* **Fig. 1** and **Note 1**).

4. Insert the two homology fragments in either side of the hprt cassette by blunt-end ligation. The 1.65-kb *Hin*dIII fragment is subcloned at the *Cla*I site and subsequently the 6.5-kb *Sac*I fragment at the *Bam*HI site. All sites are blunt-ended prior to ligation reactions. Orientation of the inserted DNA is confirmed by restriction mapping and DNA sequencing (*see* **Note 4**).
5. Linearize the vector at the unique *Not*I site, which is located 3′ to the 6.5-kb homology region (**Fig. 1**).
6. Clean the *Not*I-digested DNA with phenol/chloroform (1 : 1) and then chloroform. Precipitate DNA with sodium acetate, pH 5.2, and ethanol. Wash the DNA pellet with 70% ethanol, and then remove the ethanol in a tissue culture hood to prevent microbial contamination of the DNA.
7. In the tissue culture hood, air-dry the DNA and resuspend in sterile TE buffer at a concentration of 0.5–1.0 mg/mL.

3.2. Transfection of ES Cells

3.2.1. Preparation of Primary MEF Feeder Cells

1. Sacrifice 15 d postcoitus (dpc) pregnant mice by cervical dislocation, open the abdominal cavity, and dissect out the uterus.
2. Transfer the uterus to a 10-cm Petri dish containing PBS, open the uterine wall, and dissect the embryos away from the uterus and associated membranes.
3. Transfer 8–12 embryos into a new 10-cm Petri dish containing PBS. Remove all internal organs.
4. Wash the carcasses twice in 50 mL of PBS to remove blood.
5. Treat the carcasses individually as follows. Transfer a carcass into a fresh dish containing 2 mL of trypsin solution and chop into fine pieces. Incubate at 37°C for 5 min. Inactivate trypsin by adding 5 mL of MEF medium, and transfer the contents of a dish to a conical tube. Let the large pieces of debris settle out and then transfer the supernatant to a new conical tube.
6. Combine the supernatant prepared from all the carcasses. Plate the cell suspension onto ten 10-cm tissue culture dishes and incubate in a 37°C, 5% CO_2 incubator.
7. Change the medium the next day to remove the cellular debris, and grow the cells to confluence.
8. Aspirate the medium and wash the cells twice with PBS.
9. Trypsinize the cells with trypsin solution and pool all the cells.
10. Centrifuge at 200g for 5 min, aspirate the supernatant, and resuspend the cells in 20 mL of freezing medium. Prepare 1-mL aliquots in cryogenic vials. Store the vials at –80°C overnight and then transfer to liquid nitrogen.

3.2.2. Preparation of MEF Feeder Layers

1. Thaw one frozen vial of MEF feeder cells and plate onto three 10-cm dishes each containing 10 mL of MEF medium. Change the medium the next day, and grow the cells until they form a confluent monolayer (~3 d).

2. Wash the cells twice with PBS, detach the cells with trypsin solution, and pass at a 1:5 ratio. Grow the cells until they form a confluent monolayer (3 d). Repeat this step two more times.
3. Trypsinize and pool all the cells, centrifuge at 200g for 5 min, aspirate the supernatant, and resuspend the cells in 50 mL of MEF medium.
4. Inactivate the cells with 3000 rad of γ irradiation.
5. Centrifuge the cells at 200g for 5 min. Aspirate the supernatant and resuspend the pellet in MEF medium to give a density of approx 1.8×10^6 cells/mL.
6. Transfer 1 mL of irradiated feeders onto each 10-cm dish containing 10 mL of MEF medium. Incubate dishes overnight to allow the cells to attach.
7. Change the medium to ES cell medium before adding ES cells.

3.2.3. Preparation of ES Cells and Electroporation

1. Thaw one frozen vial of E14TG2a cells *(18)*, and plate onto a 10-cm dish with a feeder layer in ES cell medium. Change the medium the next day and grow the cells until confluent.
2. Feed the cells with fresh ES cell medium 3 h before harvesting for electroporation.
3. Wash the cells twice with PBS and trypsinize with trypsin solution. Gently pipet up and down to create a single-cell suspension.
4. Centrifuge the cells at 200g for 5 min.
5. Resuspend the cells in 5 mL of ES cell medium and centrifuge again.
6. Resuspend the cells in ES cell medium to give a total volume of 400 μL (*see* **Note 5**).
7. Add 13 μg of the linearized PDE4B targeting vector to the ES cells (*see* **Note 6**).
8. Transfer the DNA and ES cell mixture to a sterile electroporation cuvet.
9. Electroporate the cells in BTX Electro Cell Manipulator at 270 V (675 V/cm), 50 μF, 360 Ω (the time constant is 0.5 to 0.6 s).
10. Transfer the electroporated cells from the cuvet to 10 mL of ES cell medium. Plate 2 mL of the suspended cells on 10-cm dishes (total of five dishes) containing feeder layers in ES cell medium.
11. Supplement HAT (1X) in ES cell medium (for positive selection) 24 h after electroporation. Add ganciclovir (2 mM, for negative selection) to the medium 48 h after electroporation.
12. Change the medium with selection daily until colonies form. This usually takes 7–10 d. Widespread cell death should be apparent after 2 or 3 d of drug selection. Ten days after electroporation, the drug-resistant colonies reach the size of 1.5–2 mm and are ready to be picked.

3.3. Picking and Expansion of Colonies Following Electroporation

1. Wash the colony containing dishes twice with PBS and add PBS to cover the colonies.
2. Prepare 24-well dishes with feeder layers in 1 mL of ES cell medium/well.

3. Pick individual colonies in 5–10 µL of PBS with a micropipettor and sterile tips under a dissecting microscope. Add each colony to a separate well in the 24-well dishes prepared in **step 2**, and break up the colonies by pipetting up and down. About 60 colonies are picked for each PDE4 targeting construct.

4. Grow the cells in a 37°C, 5% CO_2 incubator for 3 d.

5. To further grow the dish to confluence, trypsinize the cells as follows:
   a. Wash the cells twice with PBS.
   b. Add 0.3 mL of trypsin solution to each well and incubate at 37°C until the cells detach.
   c. Add 1.5 mL of ES cell medium/well and pipet up and down to break the cell clumps.
   d. Leave the cell suspension in the wells to reattach.
   e. Change the medium 12–24 h later and grow the cells to confluence (2 d).

6. Expand the cells in the 24-well dishes into 6-well dishes by washing and trypsinizing the cells in the 24-well dishes and then plating to the 6-well dishes with feeder layers. Grow the cells to confluence.

7. Wash the cells twice with PBS and add 0.5 mL of trypsin solution to each well. Once the cells are detached, add 1 mL of ES cell medium/well and pipet up and down to break up the cell clumps.

8. Transfer half (0.75 mL) of the cells to individual Eppendorf tubes, and then centrifuge at 200g for 5 min. Wash each pellet with 1 mL of PBS and centrifuge again. Remove the supernatant and store the pellets at –80°C until DNA extraction.

9. Centrifuge the remaining half of the cells from each clone at 200g for 5 min, resuspend the pellet in 0.5 mL of freezing medium, and store at –80°C.

3.4. Screening of ES Clones for Homologous Recombinants

3.4.1. Extraction of DNA From ES Cells

The following DNA extraction method is adapted from Miller et al. (*27*).

1. Retrieve the ES cell pellets (prepared in **Subheading 3.3., step 8**) from the –80°C freezer.

2. Resuspend each pellet in 200 µL of tail buffer containing 150 µg of proteinase K.

3. Incubate the tubes overnight at 56°C.

4. Add 90 µL of saturated NaCl/tube and shake the tubes vigorously for 15 s.

5. Centrifuge the tubes at 16,000g for 10 min at room temperature.

6. Transfer 260 µL of the supernatant in each tube to new tubes.

7. To precipitate DNA, add 520 µL of 100% ethanol and mix by inverting the tubes three to four times.

8. Spool the precipitated DNA with heat-sealed Pasteur pipets.

9. Immediately wash the spooled DNA by dipping the pipet five times into a tube of 70% ethanol. Repeat the wash in a second tube of 70% ethanol, followed by dipping five times into a tube of 95 or 100% ethanol.

10. Allow the DNA to air-dry, and then release the DNA from the pipet tip by soaking the tip in an Eppendorf tube containing 60 µL of TE buffer.

11. Allow the DNA to resuspend overnight at 4°C.
12. Flick the tubes to break up the DNA clumps and incubate at 56°C for 1 h. The DNA samples are now ready for Southern blot analysis.

3.4.2. Southern Blot Analysis

1. Digest 20 µL of DNA prepared in **Subheading 3.4.1.** in a 50-µL reaction volume with 3 µL of *Eco*RI (10 U/µL). The enzyme is added in two steps: First add 2 µL and incubate at 37°C for 2 h, and then add an additional 1 µL and incubate overnight.
2. Load the digested DNA onto a well of a 0.8% agarose gel and separate by electrophoresis for 4 to 5 h.
3. After electrophoresis, immerse the gel in denaturation solution, rock gently for 20–30 min, and then decant the solution. Repeat the DNA denaturation step.
4. To neutralize the DNA, immerse the gel in neutralization solution, rock gently for 20–30 min, and then decant the solution. Repeat the neutralization step.
5. Transfer the DNA in gel to a Biotrans nylon membrane overnight using the blotting method of Southern *(28)*.
6. After the transfer is completed, crosslink the DNA to the nylon membrane by a UV light using UV Stratalinker 1800.
7. Prehybridize the blot in prehybridization solution at 40–42°C for 4 h to overnight.
8. Prepare the hybridization probe by labeling the *Eco*RI-*Hin*dIII fragment (*see* **Fig. 1**) with [α-^{32}P]dCTP using a random primers DNA labeling kit.
9. Hybridize the blot in hybridization solution containing the labeled probe at 40–42°C overnight.
10. Wash the blot with shaking twice in 0.4X SSC/0.1% SDS at room temperature for 15–20 min, followed by a wash in the same solution at 45–48°C for 30–60 min.
11. Air-dry the blot and then expose to X-ray film at –70°C for 2 to 3 d.

3.5. Expansion and Freezing of Targeted ES Cell Clones

The ES clones positive for homologous recombination are expanded for karyotyping and blastocyst injection. For each PDE4 mutation, two clones are selected for these procedures. Routinely, the expanded cells are subject to Southern blot analysis again to confirm the heterozygous genotype. In addition, these cells at this early passage are frozen in aliquots for future needs.

1. Identify two positive ES clones and retrieve them from the –80°C freezer (*see* **Subheading 3.3., step 9**).
2. Quickly thaw the vials and add the cells to a six-well dish with feeder layers in ES cell medium. Change the medium 6–8 h later.
3. Grow the cells to confluence (2 d) and then pass the cells of each clone to a 10-cm dish with feeder layers in ES cell medium.

4. When the cells reach confluence (3 to 4 d), expand each clone to three 10-cm dishes. Grow the dishes to confluence again.

5. Wash the cells twice with PBS. Add 2 mL of trypsin solution/dish and incubate at 37°C until the cells detach. Then add 8 mL of ES cell medium to each dish and pipet up and down to create a single-cell suspension. Pool the cells from the three dishes.

6. Prepare two aliquots of cells per clone for DNA extraction and Southern blot analysis as described in **Subheadings 3.4.1.** and **3.4.2.**

7. Take the necessary cells for the karyotyping procedures (*see* **Subheading 3.6.**).

8. Pellet the rest of the cells by centrifuging at $200g$ for 5 min. Resuspend the cells in 4 mL of freezing medium/clone, and freeze 1-mL aliquots in cryogenic vials. Store the vials at −80°C overnight and then transfer to liquid nitrogen.

3.6. Karyotyping of ES Cells

The mouse has 40 chromosomes, 19 pairs of autosomes and 2 sex chromosomes. After gene targeting, some ES cell lines do not contain such a normal diploid (euploid) karyotype. Therefore, it is recommended that the ES clones be karyotyped before blastocyst injection.

1. Seed the ES cells prepared in **Subheading 3.5.**, **step 5** onto a six-well dish at a 1:10 ratio. Grow for 1.5 d to approx 50% confluence.

2. Aspirate the medium. To arrest the cells in mitosis, add 4 mL/well of medium containing colcemid (20 ng/mL) and EtBr (375 ng/mL). Incubate the cells at 37°C for 1.5–2 h.

3. Trypsinize the cells, add the medium, and break the cell clumps as mentioned in **Subheading 3.3.**, **step 7**.

4. Centrifuge the cells at $200g$ for 5 min.

5. Aspirate the supernatant. Resuspend the cells in 5 mL of 0.56% KCl hypotonic solution, and incubate at room temperature for 20 min to swell the cells and nuclei.

6. Centrifuge the cells at $50g$ for 5 min and brake slowly.

7. Remove the supernatant except for 1 mL. Resuspend the cells in the remaining 1 mL of supernatant.

8. Add 5 mL of ice-cold fixation solution dropwise with flicking or vortexing of the tubes to prevent cell clumping. Incubate the tubes at room temperature for 5 min.

9. Repeat **steps 6–8** twice.

10. Repeat **steps 6** and **7**. The resultant cell suspension is ready to be transferred to microscope slides as follows.

11. Clean the microscope slides with 70% ethanol.

12. Transfer the cells with a Pasteur pipet and hold the pipet approx 10 cm above each slide. Squeeze out one drop of the fixed cells onto each slide with the slide tilted to a 35–45° angle so that the fixative and the cells spread out over the slide.

13. After the slides are dried, stain the slides in 3% giemsa stain for 20 min.

14. Rinse the slides twice in distilled water.
15. Air-dry the slides. Count the chromosomes using a microscope with ×200 magnification.

3.7. Preparation of ES Cells for Microinjection

After the positive ES clones are confirmed by Southern blot and karyo-typing, the cells are prepared for blastocyst injection.

1. Thaw the ES cells prepared in **Subheading 3.5.** and plate onto 10-cm dishes with feeder layers in ES cell medium 4 to 5 d prior to injection. Grow the cells to confluence (2 d).
2. Pass the cells at a 1:5 to 1:10 ratio. Grow for 2 to 3 d to the exponential phase.
3. Change the medium the day before harvesting for injection.
4. Wash the cells twice with PBS. Add 2 mL of trypsin solution/plate, and incubate at 37°C until the cells are detached.
5. Add 10 mL of ES cell medium and pipet up and down to create a single-cell suspension.
6. Centrifuge the cells at 200g for 5 min.
7. Wash the cells with 10 mL of ES cell medium and centrifuge again.
8. Resuspend the cells in 1 mL of CO_2-independent medium.
9. Dilute the cells by transferring 100 µL of the cell suspension to a sterile 1.5-mL Eppendorf tube containing 900 µL of CO_2-independent medium.
10. Prepare a microinjection droplet by adding 100 µL of CO_2-independent medium to the bottom of the lid of a 10-cm dish. Cover the droplet with a layer of mineral oil.
11. Transfer 20 µL of the diluted cells and 10–20 of the 3.5-d blastocysts (*see* **Subheading 3.8.1.** for preparation of blastocysts) to the microinjection droplet. Proceed to the injection procedures (*see* **Subheading 3.8.2.**).

3.8. Production of Chimeras

To produce chimeras successfully, sophisticated micromanipulatory equipment is essential. The operators must be familiar with the apparatus and setup for microinjection. In addition, the skills involved in the preparation and handling of injection and holding pipets as well as in animal surgery to retrieve and transfer blastocysts also are necessary. Because these techniques have been described in detail elsewhere (*29–31*), in this section we only present the steps involved in the generation of PDE4 mutant chimeras.

The mouse strains that we used to obtain the blastocysts and to receive the injected blastocysts are C57BL/6 (B6) and C57BL/6 × DBA/2 (B6D2) F$_1$ hybrids, respectively (*see* **Note 7**).

3.8.1. Preparation of Recipient Blastocysts

1. Four days before ES cell injection, mate the female B6 mice to fertile B6 males (*see* **Note 8**).

2. The next morning, determine whether mating has taken place by checking the vaginal plug for each female. Separate the plugged females in individual cages.
3. Three days later (3.5 dpc), sacrifice the females by cervical dislocation.
4. Open the abdominal cavity and locate the uterus.
5. Cut across the cervix to lift the uterine horns with blunt forceps.
6. Remove the attached mesentery and cut at the uterotubal junction to free the entire uterus.
7. Place the uterus in a Petri dish. Make a longitudinal clip at the end of each horn.
8. Insert a 25-gage needle attached to a 1-mL syringe into the uterus at the cervical end, and then flush the blastocysts out of each horn with CO_2-independent medium (0.5 mL/horn).
9. When finished with all the animals, collect all the blastocysts in a culture dish containing CO_2-independent medium. Store in a 37°C incubator before transferring to a microinjection droplet (*see* **Subheading 3.7., step 11**).

3.8.2. Microinjection of ES Cells Into Blastocysts

1. Bring the microinjection droplet-containing dish to the injection apparatus. The injection is performed at room temperature under a fixed-stage microscope.
2. Prepare the microinjection setup so that a successful injection can be achieved. This includes preparing the injection and holding pipets so that they work properly and appropriately connecting the pipets to the two micromanipulators (the holding manipulator and the injection manipulator) to ensure that no bubbles are trapped in the system. Orient and position the pipets such that picking and injection of ES cells can be accomplished (*see* **refs. *29–31*** for details).
3. In the microinjection droplet, select viable individual ES cells and draw up slowly into the injection pipet. Subsequently, pick up one blastocyst with the holding pipet and immobilize the embryo at the tip of the pipet by suction.
4. Insert the injection pipet into the blastocoel cavity by penetrating the pipet tip between the cells. Release 10–20 ES cells slowly and smoothly into the cavity. When finished, slowly withdraw the injection pipet.
5. Repeat the injection procedure for each blastocyst in the droplet.
6. Transfer the injected blastocysts to pseudopregnant recipients (*see* **Subheading 3.8.4.**).

3.8.3. Preparation of Pseudopregnant Recipients

Mice in estrus can be pseudopregnant by mating with sterile males. After mating, these females undergo hormonal changes associated with pregnancy even though no fertilization has occurred. The uteri of these pseudopregnant mice are suitable for embryos to develop to term when the embryos are transferred at the appropriate time.

1. Three days before the ES cell injection (*see* **Note 9**), mate the B6D2 F_1 hybrid females to vasectomized sterile B6D2 males (*see* **Note 10**).
2. The next morning, put the females with vaginal plugs in separate cages.

3. Two days later (2.5 dpc), use these mice as hosts for blastocyst transfer (*see* **Subheading 3.8.4.**).

3.8.4. Transfer of Injected Blastocysts to Pseudopregnant Recipients

1. Anesthetize the pseudopregnant females by ip injection of Avertin.
2. Pick up six to seven blastocysts to be transferred in a transfer pipet. Set aside until the mouse is ready.
3. Remove the fur and swab the back of the mouse with iodine solution. Across the midline on the back, make a 1-cm incision in the skin at the level slightly posterior to the thoracic cavity. Progress the skin incision laterally until the fat pad and ovary are seen under the peritoneal wall.
4. Make an incision no more than 5 mm in the peritoneum. Grasp the ovarian fat with blunt forceps and pull out the ovary, oviduct, and a part of the uterine horn.
5. Insert a 26-gage needle into the uterus to make a hole about 2 mm from the uterotubal junction.
6. Through the hole insert the tip of the transfer pipet in the uterine horn and release the blastocysts into the lumen. When finished, remove the pipet and gently push the organs back.
7. Repeat the procedures for the other uterine horn.
8. To finish, close the peritoneal wall with a suture and apply a metal clip to close the skin. The chimeras should be born in 16–18 d.

3.9. Generation of PDE4 Knockout Mice From Chimeras

After transfer of the injected blastocysts to pseudopregnant females, the ES cells in each embryo migrate to form a part of the inner cell mass of the blastocyst and potentially contribute to all tissues of the developing chimeric mouse (*8*). Because the PDE4 mutant ES cells are injected into blastocysts of C57BL/6 mice, the resultant chimeras are composed of 129/Ola cells and C57BL/6 cells. If the 129/Ola ES cells differentiate into germ cells in the developing embryos, the resulting chimeras can be bred to produce mice heterozygous for the PDE4 allele. The germ-line transmission can be monitored by mating the chimeras to C57BL/6 mice. The presence of the agouti coat color of the offspring (F_1 generation) is evidence of transmission of the ES genome through the chimeras because agouti (ES derived) is dominant to black (C57BL/6 derived). Half of the agouti offspring in F_1 should carry the targeted mutation. These heterozygous mutant animals subsequently are bred to produce animals homozygous for the disrupted allele (F_2 generation). Here, we outline the steps to produce the PDE4 homozygous null mice from the chimeras.

1. Mate the chimeras to C57BL/6 mice to produce F_1 generation mice (*see* **Note 11**).

2. Approximately 1 wk after birth, check the coat color of the pups to identify the agouti mice. Perform genotyping (*see* **steps 3–5**) for these mice at the age of 3 to 4 wk by Southern blot analysis.

3. Clip a piece of tail tip (~1 cm) from each agouti mouse.

4. Extract DNA from the tail tips by following the protocol described in **Subheading 3.4.1.** except that the quantities of all reagents used for ES cells are doubled here.

5. Digest the DNA and perform Southern blot analysis as described in **Subheading 3.4.2.** The F_1 offspring can be WT or heterozygous for the targeted allele.

6. When the F_1 heterozygous animals are identified, mate these mice to produce F_2 generation mice.

7. Perform genotyping for all F_2 mice at the age of 3 to 4 wk as in **step 2**. Following the Mendelian inheritance, the offspring will have the genotype of 25% WT, 50% heterozygous, and 25% homozygous mutant. Our genotyping results of the PDE4B-targeted mice are shown in **Fig. 1B** (*see* **Note 12**).

3.10. Genotyping of PDE4 Knockout Mice by PCR

After PDE4 mutant alleles are characterized by Southern blot, PCR is used as an alternative method to genotype the mouse colonies because it is much faster and less laborious than Southern blot. In comparison with the design of PCR primers used for screening the mutant ES cells (*see* **Note 1**), it is easier to design primers for maintenance of mouse colonies. Given that the disruption of the PDE4 locus has been confirmed by Southern blot, the primers therefore can be selected within the PDE4 homologous regions in the targeting vector to amplify fragments with desired lengths, ideally not more than 1 kb. For example, to identify the PDE4B mutant allele, we use one primer (primer *a*) complementary to the sequence in the hprt cassette and a second primer (primer *b*) complementary to the sequence within the 6.5-kb homologous region to amplify a 375-bp product (*see* **Fig. 2**). Identification of the PDE4B WT allele is accomplished by using the same reverse primer (primer *b*) and a forward primer (primer *c*) with sequence located in the PDE4B-deleted region. The two primers amplify a 700-bp product (**Fig. 2**). It is necessary to design the primers such that the sizes of the two PCR products are different and can be distinguished in the agarose gel.

4. Notes

1. As a general rule, targeting frequency increases by increasing the length of homologous sequences in the vector (*5,32*). However, if PCR is to be used to screen ES cells for a targeting event, the positive selection marker should be positioned near one end of the homologous sequence. As a result, the vector contains a long arm and a short arm of homologous sequences (e.g., the 6.5-kb long arm and 1.65-kb short arm shown in **Figs. 1** and **2**). The short arm region can be used for the PCR design. This is accomplished by designing one primer complementary to the

sequences of the positive selection marker and a second primer complementary to the chromosomal sequences just outside the short homologous sequences (primers *e* and *f* in **Fig. 2**). It is recommended that the distance between the two primers be less than 2 kb to achieve efficient PCR amplification. However, it is desirable to make the short arm long enough (not less than 0.5 kb) for efficient DNA pairing and crossover formation.

2. The details of protocols associated with molecular cloning are beyond the scope of this chapter and are not described. These include genomic DNA library screening, identification and isolation of PDE4 genomic clones, restriction enzyme digestion, DNA fragment ligation, transformation of plasmid DNA into *Escherichia coli* competent cells, and agarose gel electrophoresis.

3. Because the existence of sequence mismatches between the homology in the vector and the target locus can reduce the targeting frequency, it is recommended that the DNA used to construct the vector be isolated from genomic libraries where the sequences are isogenic with the ES cells used for homologous recombination *(33,34)*.

4. The targeting vector is constructed such that the hprt cassette is in the reverse orientation with respect to the reading frame of the PDE4B gene. This is to avoid a possible complication of expression of a translatable PDE4B sequence 3′ to the hprt cassette by the PGK promoter.

5. The ES cell volume of 400 μL is for the BTX electroporator cuvet (2-mm gap).

6. The DNA concentration required for the ES cell transfection is 3 n*M*. This is equivalent to 1.2 pmol of DNA in a volume of 400 μL. For every kilobase of DNA, 1.2 pmol weighs approx 0.8 μg, so 16.3 kb of PDE4B targeting DNA needs 13 μg to give the quantity of 1.2 pmol.

7. F_1 hybrid females are used as recipients for microinjected blastocysts because they usually take good care of their litters.

8. The B6 females that we use are treated with exogenous hormones to induce superovulation. The mice are first injected with pregnant mare serum gonadotropin, a surrogate for follicle-stimulating hormone. After 48 h, the mice are injected with human chorionic gonadotropin (hCG), which mimics the effects of luteinizing hormone. Ovulation occurs approx 12 h after administration of hCG. Routinely, we place these mice for mating immediately after injection of hCG.

9. The pseudopregnant recipients are mated 1 d after the females that produce blastocysts so that the microinjected blastocysts at the stage of 3.5 dpc are transferred to the uterus of 2.5-dpc pseudopregnant recipients. This allows the blastocysts to recover in vivo from the in vitro manipulation.

10. The procedures to generate vasectomized mice are detailed in **refs. *29–31***.

11. It is crucial to breed chimeras efficiently so that identification of germ-line transmittants and production of F1 heterozygous mice from each germ-line chimera can be achieved in a timely manner.

12. After confirming the disruption of the PDE4 allele by Southern and/or PCR analysis, it is necessary to verify that the full-length PDE4 message and protein are not expressed in the knockout mice. Because the PDE4 targeting constructs

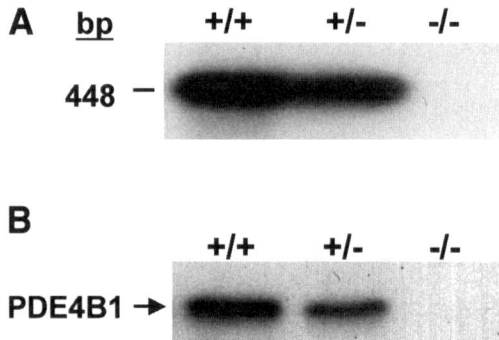

Fig. 3. (**A**) Expression of PDE4B mRNA in mouse brain. RT-PCR is carried out for poly(A)$^+$ RNA extracted from pooled cortex and cerebellum of PDE4B$^{+/+}$, PDE4B$^{+/-}$, and PDE4B$^{-/-}$ mice. The 5′ and 3′ primers used in the PCR are specific to the exon 8 and exon 10 sequence of PDE4B, respectively. The 448-bp PCR fragment is detected by Southern blot analysis using an exon 9-specific oligonucleotide probe. (**B**) Expression of PDE4B protein in mouse brain. Cortexes and cerebella dissected from PDE4B$^{+/+}$, PDE4B$^{+/-}$, and PDE4B$^{-/-}$ mice are homogenized, followed by centrifugation at 16,000g for 20 min. The supernatant is subjected to Western blot analysis using a PDE4B-specific polyclonal antibody, K118.

are designed to delete the 3′ region of the gene (e.g., the region of exons 8–10 of the PDE4B locus), it is conceivable that truncated message and protein can be produced. By reverse transcriptase-PCR (RT-PCR) followed by Southern blot analysis, as expected, no mRNA containing the exon 8–10 sequence is detected in the PDE4B knockout mice (**Fig. 3A**). However, expression of truncated message with the sequence 5′ to the deleted region is observed in the PDE4B knockout mice (data not shown). In spite of this expression, the PDE4B protein is not detected in the knockout mice by Western blot analysis using antibodies that recognize epitopes at the carboxyl terminus (**Fig. 3B**). Conversely, PDE4B mRNA and protein are readily detected in the brain of WT and PDE4B$^{+/-}$ mice although their levels are decreased in the PDE4B$^{+/-}$ mice (**Fig. 3**).

Acknowledgments

We are indebted to Dr. Beverly Koller for assistance with the technical details described herein. This work was supported by National Institutes of Health grants HD20788, HL67674, and HD5344.

References

1. Conti, M. and Jin, S.-L. C. (1999) The molecular biology of cyclic nucleotide phosphodiesterases. *Prog. Nucleic Acid Res. Mol. Biol.* **63,** 1–38.
2. Houslay, M. D. (2001) PDE4 cAMP-specific phosphodiesterases. *Prog. Nucleic Acid Res. Mol. Biol.* **69,** 249–315.

3. Torphy, T. J. (1998) Phosphodiesterase isozymes: molecular targets for novel anti-asthma agents. *Am. J. Respir. Crit. Care Med.* **157,** 351–370.
4. Doetschman, T., Gregg, R. G., Maeda, N., Hooper, M. L., Melton, D. W., Thompson, S., and Smithies, O. (1987) Targeted correction of a mutant HPRT gene in mouse embryonic stem cells. *Nature* **330,** 576–578.
5. Thomas, K. R. and Capecchi, M. R. (1987) Site-directed mutagenesis by gene targeting in mouse embryo–derived stem cells. *Cell* **51,** 503–512.
6. Robertson, E., Bradley, A., Kuehn, M., and Evans, M. (1986) Germ-line transmission of genes introduced into cultured pluripotential cells by retroviral vector. *Nature* **323,** 445–448.
7. Gossler, A., Doetschman, T., Korn, R., Serfling, E., and Kemler, R. (1986) Transgenesis by means of blastocyst-derived embryonic stem cell lines. *Proc. Natl. Acad. Sci. USA* **83,** 9065–9069.
8. Bradley, A., Evans, M., Kaufman, M. H., and Robertson, E. (1984) Formation of germ-line chimaeras from embryo-derived teratocarcinoma cell lines. *Nature* **309,** 255–256.
9. Jin, S. L. C. and Conti, M. (2002) Induction of the cyclic nucleotide phosphodiesterase PDE4B is essential for LPS-activated TNF-alpha responses. *Proc. Natl. Acad. Sci. USA* **99,** 7628–7633.
10. Jin, S.-L. C., Richard, F. J., Kuo, W. P., D'Ercole, A. J., and Conti, M. (1999) Impaired growth and fertility of cAMP-specific phosphodiesterase PDE4D-deficient mice. *Proc. Natl. Acad. Sci. USA* **96,** 11,998–12,003.
11. Ramirez-Solis, R., Davis, A. C., and Bradley, A. (1993) Gene targeting in embryonic stem cells. *Methods Enzymol.* **225,** 855–878.
12. Hasty, P., Abuin, A., and Bradley, A. (2000) Gene targeting, principles, and practice in mammalian cells, in *Gene Targeting: A Practical Approach*, 2nd ed. (Joyner, A. L., ed.), Oxford Unversity Press, New York, pp. 1–35.
13. Matise, M. P., Auerbach, W., and Joyner, A. L. (2000) Production of targeted embryonic stem cell clones, in *Gene Targeting: A Practical Approach*, 2nd ed. (Joyner, A. L., ed.), Oxford University Press, New York, pp. 101–132.
14. Koller, B. H. and Smithies, O. (1992) Altering genes in animals by gene targeting. *Annu. Rev. Immunol.* **10,** 705–730.
15. Bradley, A., Zheng, B., and Liu, P. (1998) Thirteen years of manipulating the mouse genome: a personal history. *Int. J. Dev. Biol.* **42,** 943–950.
16. Monaco, L., Vicini, E., and Conti, M. (1994) Structure of two rat genes coding for closely related rolipram-sensitive cAMP phosphodiesterases: multiple mRNA variants originate from alternative splicing and multiple start sites. *J. Biol. Chem.* **269,** 347–357.
17. Mansour, S. L., Thomas, K. R., and Capecchi, M. R. (1988) Disruption of the proto-oncogene int-2 in mouse embryo-derived stem cells: a general strategy for targeting mutations to non-selectable genes. *Nature* **336,** 348–352.
18. Hooper, M., Hardy, K., Handyside, A., Hunter, S., and Monk, M. (1987) HPRT-deficient (Lesch-Nyhan) mouse embryos derived from germline colonization by cultured cells. *Nature* **326,** 292–295.

19. Bradley, A. (1990) Embryonic stem cells: proliferation and differentiation. *Curr. Opin. Cell. Biol.* **2,** 1013–1017.
20. Doetschman, T. C., Eistetter, H., Katz, M., Schmidt, W., and Kemler, R. (1985) The in vitro development of blastocyst-derived embryonic stem cell lines: formation of visceral yolk sac, blood islands and myocardium. *J. Embryol. Exp. Morphol.* **87,** 27–45.
21. Martin, G. R. (1981) Isolation of a pluripotent cell line from early mouse embryos cultured in medium conditioned by teratocarcinoma stem cells. *Proc. Natl. Acad. Sci. USA* **78,** 7634–7638.
22. Evans, M. J. and Kaufman, M. H. (1981) Establishment in culture of pluripotential cells from mouse embryos. *Nature* **292,** 154–156.
23. Nichols, J., Evans, E. P., and Smith, A. G. (1990) Establishment of germ-line-competent embryonic stem (ES) cells using differentiation inhibiting activity. *Development* **110,** 1341–1348.
24. Moreau, J. F., Donaldson, D. D., Bennett, F., Witek-Giannotti, J., Clark, S. C., and Wong, G. G. (1988) Leukaemia inhibitory factor is identical to the myeloid growth factor human interleukin for DA cells. *Nature* **336,** 690–692.
25. Williams, R. L., Hilton, D. J., Pease, S., Willson, T. A., Stewart, C. L., Gearing, D. P., Wagner, E. F., Metcalf, D., Nicola, N. A., and Gough, N. M. (1988) Myeloid leukaemia inhibitory factor maintains the developmental potential of embryonic stem cells. *Nature* **336,** 684–687.
26. Smith, A. G., Heath, J. K., Donaldson, D. D., Wong, G. G., Moreau, J., Stahl, M., and Rogers, D. (1988) Inhibition of pluripotential embryonic stem cell differentiation by purified polypeptides. *Nature* **336,** 688–690.
27. Miller, S. A., Dykes, D. D., and Polesky, H. F. (1988) A simple salting out procedure for extracting DNA from human nucleated cells. *Nucleic Acids Res.* **16,** 1215.
28. Southern, E. M. (1975) Detection of specific sequences among DNA fragments separated by gel electrophoresis. *J. Mol. Biol.* **98,** 503–517.
29. Bradley, A. (1987) Production and analysis of chimaeric mice, in *Teratocarcinomas and Embryonic Stem Cells: A Practical Approach* (Robertson, E. J., ed.), IRL, Oxford, UK, pp. 113–151.
30. Papaioannou, V. and Johnson, R. (2000) Production of chimeras by blastocyst and morula injection of targeted ES cells, in *Gene Targeting: A Practical Approach*, 2nd ed. (Joyner, A. L., ed.), Oxford Unversity Press, New York, pp. 133–176.
31. Stewart, C. L. (1993) Production of chimeras between embryonic stem cells and embryos. *Methods Enzymol.* **225,** 823–855.
32. Hasty, P., Rivera-Perez, J., and Bradley, A. (1991) The length of homology required for gene targeting in embryonic stem cells. *Mol. Cell Biol.* **11,** 5586–5591.
33. te Riele, H., Maandag, E. R., and Berns, A. (1992) Highly efficient gene targeting in embryonic stem cells through homologous recombination with isogenic DNA constructs. *Proc. Natl. Acad. Sci. USA* **89,** 5128–5132.
34. Deng, C. and Capecchi, M. R. (1992) Reexamination of gene targeting frequency as a function of the extent of homology between the targeting vector and the target locus. *Mol. Cell Biol.* **12,** 3365–3371.

16

Immunoprecipitation of PDE2 Phosphorylated and Inactivated by an Associated Protein Kinase

J. Kelley Bentley

Summary

A PDE2A2-associated protein kinase phosphorylates PDE2A2 in vivo and in vitro to inhibit its catalytic activity. Rat brain PDE2A2 may be solubilized using nona (ethylene glycol) mono dodecyl ether (Lubrol 12A9). PDE2A2 exists in a complex with a protein kinase regulating its activity in an adenosine triphosphate-dependent manner. When native or recombinant PDE2 is immunoprecipitated from PC12 cells using an antibody to the amino terminus in a buffer containing Lubrol 12A9, protease inhibitors, and phosphatase inhibitors, a coimmunoprecipitating nerve growth factor-stimulated protein kinase acts to phosphorylate it. PDE2A2 phosphorylation occurs optimally at pH 6.5 in a sodium 2-(4-morpholino)-ethane sulfonate buffer with 5 mM MgCl$_2$ and 1 mM Na$_3$VO$_4$. I describe protocols for producing an antibody to an amino-terminal bacterial fusion protein encoding amino acids 1–251 of PDE2A2 as well as the use of this antibody in immunoprecipitating a PDE2:tyrosine protein–kinase complex from rat brain or PC12 cells.

Key Words

Cyclic adenosine monophosphate; cyclic guanosine 5′-monophosphate; PDE2; tyrosine phosphorylation; nerve growth factor; brain.

1. Introduction

The cyclic guanosine 5′-monophosphate (cGMP)-stimulated cyclic adenosine monophosphate and cGMP phosphodiesterase PDE2 is abundant in neuronal and adrenal tissues by immunochemical criteria *(1–5)*. PDE2 is a route for cyclic nucleotide breakdown in neurons of the brain *(6)*. In neurons, cyclic nucleotides stimulate many of the same pathways stimulated through the nerve growth factor (NGF) receptor tyrosine kinases *(7)*. PDE2 is inactivated in response to NGF *(8)*. The NGF-mediated inactivation of PDE2 is associated with the formation of a detergent-soluble complex PDE2 and multiple

From: *Methods in Molecular Biology, vol. 307: Phosphodiesterase Methods and Protocols*
Edited by: C. Lugnier © Humana Press Inc., Totowa, NJ

phosphoproteins. To study PDE2 regulation by NGF, an antibody recognizing the amino-terminal 251 amino acids of PDE2A2 was developed. This antibody immunoprecipitates PDE2A2 from rat brain or PC12 cells.

As a primary step to understanding PDE2 regulation by phosphorylation, reaction conditions were found that allow for its phosphorylation by coimmunoprecipitated protein kinase(s).

2. Materials

1. Common chemicals: These can be purchased from Sigma-Aldrich (St. Louis, MO), Fisher (Hanover Park, IL), New England Biolabs (Beverley, MA), or Invitrogen (Carlsbad, CA).
2. Constitutive expression, G418 resistance vector such as pcDNA3 (without epitope tag) or pcDNA3.1/His[6] (Invitrogen) (with both epitope tags) or pFLAG-CMV-3 (Sigma-Aldrich) (only enterokinase epitope tag). A 5′ sense oligonucleotide, GAA TTC AGC ATG GAC TAC AAG GAC GAC GAT GAC AAG CAT CAC CAT CAC CAT CAC CAT CAC CAT CAC GTC CTG GTG TTG CAC CAC ATC C, is useful for introduction of the N-terminal amino acid sequence MDYKDDDDKHHHHHHHHHH epitope tag into a recombinant PDE2A2 protein.
3. PC12 cells: These can be obtained from American Type Culture Collection.
4. Polymerase chain reaction (PCR) components: 20 mM Tris HCl, pH 8.8, 10 mM KCl, 10 mM $(NH_4)_2SO_4$, 2 mM $MgSO_4$, 0.1% (v/v) Triton X-100, 10 ng of PC12 cells or rat brain random-primed first-strand cDNA, 1 μM of each sense and antisense oligonucleotide primer, 400 μM of each deoxynucleotide triphosphate, and 2 U of Vent polymerase. A PCR fragment-cloning vector such as pGEM-T easy (Promega, Madison, WI) or pCR 2.1 (Invitrogen) proves useful. Alternatively, the full-length open reading frame of PDE2A2 can be cloned in frame with an amino-terminal epitope tag vector such as pcDNA3.1/His[6].
5. Agarose flat-bed gel systems for DNA analysis and purification.
6. DNA sequencing gels or access to a sequencing facility.
7. Corning Spin-X 0.45-μ cellulose acetate centrifugal filtering apparatus and microcentrifuge.
8. pET 28a vector or comparable bacterial fusion protein expression vector with a defined epitope tag along with a poly histidine (H_6 or greater) tag (Calbiochem-Novagen, San Diego, CA).
9. BL21 (DE3) bacterial cells for expression of a pET vector (Calbiochem-Novagen).
10. Nickel nitriloacetic acid (Ni-NTA) agarose (Qiagen, Valencia, CA).
11. Polyprep minicolumns with Econo-column 250-mL reservoir (Bio-Rad, Hercules, CA).
12. Minigels for sodium dodecyl sulfate-polyacrylamide gel electrophoresis (SDS-PAGE) such as precast 4–12% acrylamide Bis-Tris gels (Invitrogen).
13. Antibody production facility services such as those of Research Genetics (Invitrogen) for the production of polyclonal or monoclonal immunoprecipitating antibodies to an amino-terminal epitope of PDE2A.

14. Preimmune serum or nonimmune serum (Sigma).
15. Buffer A: 25% (w/v) sucrose, 1% (v/v) Lubrol 12A9, 20 mM KF, 20 mM benzamidine HCl, 20 mM imidazole HCl, pH 7.0, 10 mM Na$_2$EDTA, 1 mM phenyl-methyl-sulfonyl fluoride, two "Complete" protease inhibitor cocktail tablets (Roche, Indianapolis, IN)/50 mL (2X manufacturer-recommended concentration). This is used to conduct homogenizations of mammalian cells.
16. Buffer B: 20 mM N-morpholinoethane sulfonic acid (MES), pH 6.5, 20 mM benzamidine HCl, 2X Complete protease inhibitor cocktail (EDTA-free), 1% Lubrol 12A9. This is used to wash immunoprecipitants prior to phosphorylation reactions.
17. PDE assay buffer: 20 mM Tris-HCl; 20 mM imidazole HCl, pH 7.0, 20 mM MgCl$_2$, 1 mM EGTA, 1 mM Na$_3$VO$_4$.
18. 5X Sample buffer for SDS-PAGE: 0.5% bromophenol blue, 5% (w/v) lithium dodecyl sulfate, 25% glycerol, 100 mM NaF, 300 mM EDTA (acid form), and 1 M Tris-HCl. This is used to help retain protein phosphorylation during electrophoresis. The final pH should be 6.5 without adding any additional acid or base.
19. Solution of 20% methanol and 0.1% Coomassie Blue R-250, for staining Western blots.
20. 50% Methanol, for destaining Western blots.
21. Antiphosphotyrosine, e.g., antiphosphotyrosine PY20 (BD Biosciences, Palo Alto, CA) or antiphosphotyrosine PY100 (Cell Signaling, Beverly, MA).
22. Secondary antirabbit or antimouse antibodies conjugated with either horseradish peroxidase (HRP) or alkaline phosphatase (Cell Signaling).
23. Long-term storage buffer for the active complex: 50% (v/v) glycerol, 20 mM imidazole HCl, pH 7.0, 20 mM benzamidine HCl, 20 mM KF, 10 mM EDTA, 2X Complete protease inhibitors: this buffer is used to wash the immunoprecipitate.

3. Methods

The following methods outline the production of a FLAG-epitope-tagged PDE2A2 cDNA and its transfection to produce a PC12 cell line containing both endogenous and epitope-tagged PDE2A2; the production of an amino-terminal antibody to PDE2A2 enabling immunoprecipitation; the solubilization and immunoprecipitation of PDE2A2 under conditions that maintain cGMP stimulation with phosphorylation-state-dependent activity differences; the in vitro phosphorylation of PDE2A2; and the PDE assay, gel electrophoresis, and Western blotting conditions that maintain PDE2 in the phosphorylated state.

3.1. Production of a Stable PC12 Cell Line Containing a FLAG Epitope-Tagged PDE2A2 and Cell Transfections

1. Produce the cDNA encoding an epitope-tagged PDE2A2 by reverse transcriptase-PCR protocols *(8)* (*see* **Note 1**). To do this, use a 3′ oligonucleotide containing the unique *Hin*dIII site in rat PDE2A2 or downstream from this site *(9)* with Vent polymerase and the oligonucleotide above to obtain a 5′ cDNA fragment.

2. Denature the polymerase mixture at 94°C for 4 min. Process for 30 cycles of 92°C for 30 s, followed by 50°C for 30 s, then 72°C for 2 min. Use a final extension reaction of 10 min at 72°C. This fragment can be used to replace the native sequence in a PDE2A2 cDNA.

3. Purify the cDNA by agarose gel electrophoresis and excision of the visualized band. Place the agarose plug in a cellulose acetate centrifugal filtering apparatus and freeze. Then centrifuge the frozen plug for 5 min at 10,000*g* to extract the DNA from the agarose plug. After a phenol-chloroform extraction, precipitate the DNA with isopropanol, wash with ethanol, and dry.

4. Tail the cDNA for T/A subcloning with 5 U of Taq DNA polymerase under reaction conditions identical to those described in **Subheading 2.4.** without any added primers or template other than the purified cDNA (~100 ng to 1 μg). Instead of thermal cycling, carry out the deoxyadenosine tailing reaction for 10 min at 72°C. Afterward remove the Taq again by phenol-chloroform extraction, and precipitate and process the cDNA for ligation into a T/A cloning vector.

5. After sequencing of the cDNA insert in the vector, excise the recombinant epitope-tagged PDE2A2 with *Eco*RI (or any other unique site that is introduced) and *Hin*dIII for subcloning by ligating into the PDE2 sequence 3′ of the *Hin*dIII site.

6. Produce a cell line with epitope-tagged PDE2A2 by transfection of PC12 cells with a G418 resistance expression vector encoding PDE2A2 with the amino-terminal FLAG-enterokinase substrate epitope (DYKDDDDK) followed by a 10 amino acid polyhistidine sequence. This recombinant construct is designated FH-PDE2A2 *(8)*. Transfect PC12 cells with Lipofectamine 2000 following the protocol described by the manufacturer.

7. Initially select stable PC12 transfectants using 400 μg of G418/mL for 2 wk until >99.9% of the transfected cells die off (*see* **Note 2**). Lower the G418 concentration to 200 μg/mL for 2 wk, and allow colonies of 10–100 cells to form. Isolate 12 colonies and serially dilute to 1 cell/well of a 96-well plate. Pick wells growing single isolated colonies for expansion.

3.2. Production of Antibody to Amino Terminus of PDE2A2

Using the PCR with full-length recombinant PDE2A2 cDNA as a template, a PCR product from –3 to +753 of PDE2A2 is synthesized with 5′ *Eco*RI and 3′ *Not*I sites. This reaction uses 5′- GAA TTC AGC ATG GTC CTG TTG CAC C as the forward primer and GCG GCC GCT GTC CTC TGA CAC CAG CAG GAG -3′ as the reverse primer.

After subcloning into pGEM T for propagation and sequencing *(8)*, the cDNA is excised with an *Eco*RI and *Not*I digest for subcloning into a pET 28a vector to create pET 28a PDE 2A2 : 1-251. The pET 28a PDE 2A2 : 1-251 vector is used to produce a recombinant fusion protein in bacteria. The fusion protein contains N-terminal His$_6$ and T7 epitopes as well as an extra C-terminal His$_6$ epitope.

Isolation of the fusion protein requires solubilization in urea with protease inhibitors and 1% (v/v) Triton X-100 (*see* **Note 3**). This produces a 296 amino acid protein migrating at approx 38,000 Daltons on SDS-PAGE as a fusion protein. The protein is purified by Ni-NTA agarose chromatography *(8)* for use as an antigen in rabbits.

The polyclonal antibody is produced commercially using this antigen and is used to immunoprecipitate native PDE2A2. A PDE2 monoclonal antibody (MAb) has been obtained from ICOS as a secondary screen for PDE2A2 in Western blots *(10)*.

3.3. Homogenization, Solubilization, and Immunoprecipitation

1. In assays using PC12 cells, suspend cells (10^5–10^6) in 1 mL of buffer A. Scrape the cells on ice and sonicate with 30 times at 50% duty cycle maximum-intensity bursts of a Branson W-350 sonifier (*see* **Notes 4** and **5**).
2. In assays using rat brain, homogenize 25 rat brains (about 40 g wet wt) in buffer A at 4°C in 250 mL by dispersal for 30 s at maximum intensity in a Polytron homogenizer apparatus.
3. Centrifuge the homogenate initially at 4°C for 30 min at 10,000g, and filter the supernatant fluid through five layers of cheesecloth to adsorb the Lubrol-insoluble floating fraction (*see* **Note 6**). Although this fraction has significant amounts of PDE2, lipid rafts also have many proteins not associated with PDE2.
4. Centrifuge the supernatant fluid at 200,000g for 1 h, and carefully remove.
5. Centrifuge homogenates of the cultured cells for 30 min at 10,000g. Always preclear soluble fractions by incubating with either 10 μL of nonimmune serum (Sigma Chemical Corporation) plus 40 μL of protein A (Sigma Chemical Corporation) or protein G (Upstate, Lake Placid, NY) agarose/mL of soluble fraction for 1 h on a rotary mixer at 4°C. After incubation, centrifuge the samples for 5 min at 10,000g to remove any insoluble material.
6. Perform immunoprecipitations as described previously *(8)*. Use approx 10 μg of antiserum with 10 μL of protein A or protein G agarose (50% [v/v])/mL of soluble fraction immunoprecipitated. Perform anti-FLAG epitope immunoprecipitations using an anti-FLAG M2 agarose conjugate (Sigma). Immunoprecipitations are usually carried out for at least 1 h and at most 12 h at 4°C on a rotary mixer (*see* **Note 7**).
7. Wash the immunoprecipitates three times in more than 10 vol of a buffer composed of buffer A. For long-term storage, *see* **Note 8**.

3.4. In Vitro Phosphorylation of PDE2

1. Bring the washed immunoprecipitates to 4°C and wash with more than 10 vol of buffer B.
2. Suspend the immunoprecipitates to 50% (v/v) slurry in buffer B.
3. Conduct a phosphorylation reaction in buffer B additionally containing 5 mM MgCl$_2$, 10–1000 μM adenosine triphosphate (ATP), and an approx 2–10% (v/v) slurry of the immunoprecipitate. Proceed with the reactions for 30–60 min at 30°C.

4. Stop the reactions directly by adding a 5X sample buffer for SDS-PAGE. The results of electrophoresis and Western style immunoblotting of such samples are shown in **Fig. 1A**. Alternatively, dilute the immunoprecipitates more than 10-fold in buffer A at 4°C for further processing for PDE assays as described in the next steps.

5. After dilution in buffer A, centrifuge the resin conjugate for 10 s at 10,000g, aspirate the supernatant fluid, and wash three more times with at least 10 vol of buffer A.

6. Resuspend the pellet in an equal volume of buffer A, and process for assay of PDE activity as discussed in **Subheading 3.5.**

3.5. PDE Assays

Enzymatic assays of in vitro phosphorylated PDE2 are performed after three more washes of the phosphorylated immunoprecipitate in buffer A. Protein and PDE assays are performed as described previously *(8)* with the modification that the PDE in all assays contains 1 mM Na$_3$VO$_4$. This phosphatase inhibitor has no direct effect on PDE2 activity at 1 mM concentrations (**Fig. 1**) (*see* **Note 9**).

PDE2 activity immunoprecipitated from rat brain or PC12 cells with the polyclonal antibody raised to PDE2A2 amino acids 1–251 also was lowered after ATP washout (*see* **Note 10**) (**Fig. 2**). This effect is most pronounced for enzyme immunoprecipitated from NGF-treated PC12 cells compared with that from untreated cells incubated without ATP or Na$_3$VO$_4$ (*see* **Note 11**). As with the brain enzyme, ATP treatment does not decrease the recovery of PDE2 after washing. The persistence of inhibition after ATP washout suggests that it is a result of the phosphorylation of PDE2 (*see* **Note 11**).

PDE2 mass is assessed following PDE assay by SDS-PAGE and Western analysis. All PAGE described here is performed using precast 4–12% acrylamide Bis-Tris gels with an MES-based running buffer (Invitrogen). Samples are made with 1% LiDS, 20 mM NaF, 200 mM Tris-HCl, 50 mM EDTA (pH 6.5), and 5% 2-mercaptoethanol. They are boiled for 5 min prior to electrophoresis. Silver staining for protein and autoradiography are performed as described previously *(8)*. Western blotting is performed against polyvinyl difluoride (PVDF) membranes. Transfer is assessed by staining the blot for 15 min. Destaining is performed to visualize distinct bands. Autoradiography may be performed directly on the PVDF membrane when covered with a thin transparent plastic food wrap or sleeve cover. An autoradiograph is shown in **Fig. 3A** for an anti-FLAG immunoprecipitation of an extract from a stable FH-PDE2-expressing PC12 cell line (*see* **Note 11**).

Anti-FLAG M2 immunoprecipitates many phosphoproteins along with phospho-FH-PDE2 from the PC12B line in an NGF-sensitive manner *(8)*. When 10^5 cells in a 100-mm dish are incubated with either NGF or orthovanadate, and then solubilized, immunoprecipitated, and incubated in Mg-ATP,

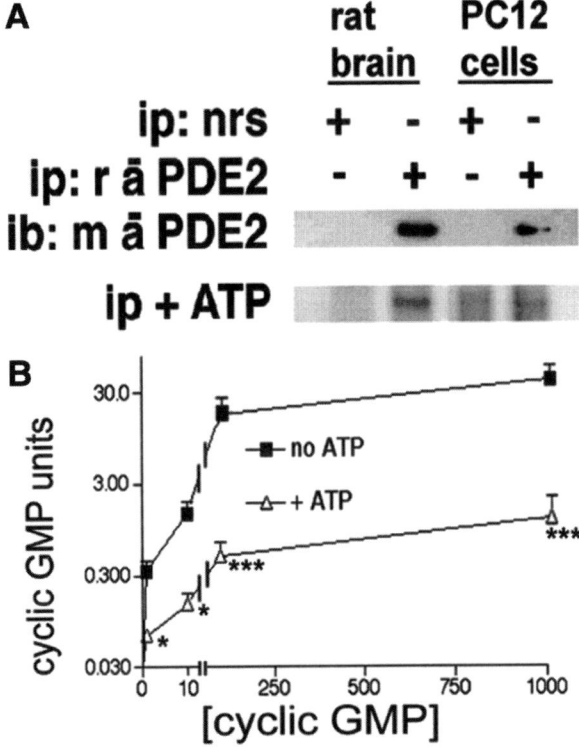

Fig. 1. A polyclonal antibody to PDE2 precipitates a PDE2–protein kinase complex from rat brain or PC12 cells that is ATP inhibited. (A) A 100,000g brain (50 mL at 3 mg/mL) or PC12 cell (50 mL at 100 µg/mL) soluble fraction was immunoprecipitated (ip) with 10 µg of polyclonal (rā) anti-PDE2A2 1–251/mL and 5% vol protein A agarose/vol reaction for 1 h. The immunoprecipitate was washed three times; incubated with 5 mM MgCl$_2$, 10 µM ATP, and 10 µCi of [γ-^{32}P] ATP for 1 h at 30°C; and processed for Western transfer, autoradiography, and PDE2 immunoreactivity. The immunoreactive PDE2 was visualized on the membrane using MAb Mā recognizing the epitope AHPLFYRG at amino acids 471–478 of PDE2A2. (B) PDE2 from a rat brain 100,000g soluble fraction (50 mL at 3 mg/mL) in homogenization buffer with 1% Lubrol 12A9, 20 mM KF, 10 mM EDTA, and a 4X Complete cocktail of protease inhibitors was immunoprecipitated with 10 µg of polyclonal anti-PDE2A2 1–251/mL and 5% vol protein A agarose/vol reaction for 12 h. The immunoprecipitate was washed three times in buffer A and once in buffer B and was incubated with 5 mM MgCl$_2$ and without (■) or with (△) 10 µM ATP + 1 mM Na$_3$VO$_4$ for 30 min at 30°C. It was processed for cGMP PDE assay in 1 mM Na$_3$VO$_4$ (both samples) in the presence (△) or absence (■) of ATP. For each sample ($n = 3$), the data shown are the SEM. *Sample with significant difference of $p < 0.05$; ***sample with significant difference of $p < 0.001$. When first incubated with the immunoprecipitate in the presence of sodium orthovanadate, ATP significantly lowered PDE2 activity.

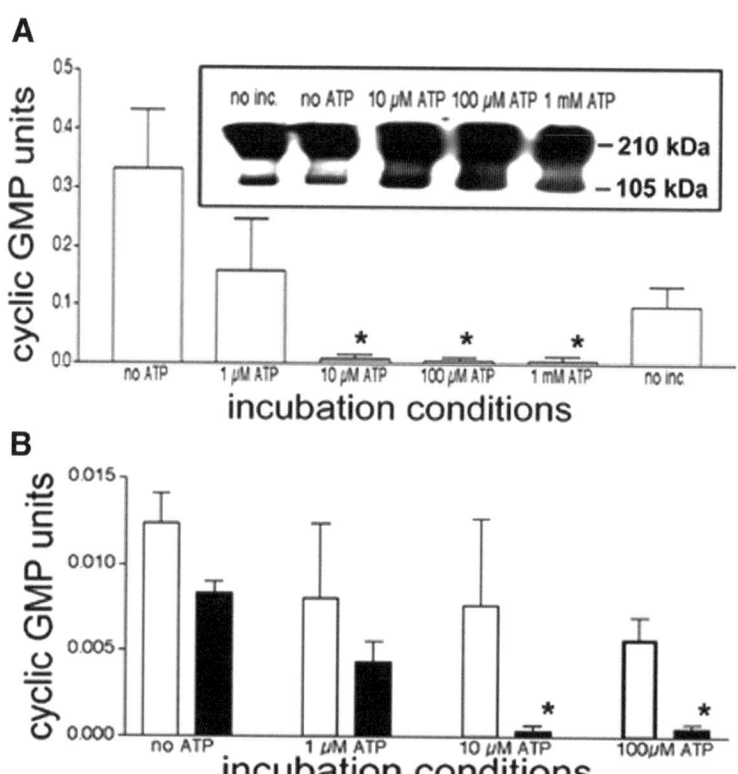

Fig. 2. ATP inhibition of PDE2 complex from rat brain or NGF-treated PC12 cells. (A) A rat brain soluble fraction was immunoprecipitated with anti-PDE2A2 1–251, and the washed immunoprecipitate in buffer B was incubated without ATP (no ATP), or in the presence of 1 mM Na$_3$VO$_4$ and varying (1 μM to 1 mM) concentrations of ATP in the primary incubation for 1 h at 30°C, or maintained in buffer A at 4°C for this period of time (no inc.). Immunoprecipitates were subsequently washed four times in buffer A and processed for PDE activity at 1 μM cGMP in the presence of 1 mM Na$_3$VO$_4$ without added ATP for any of the samples. The inset shows the immunoreactivity with monoclonal 107B after giving 7.5 μL of immunoprecipitates with no primary incubation; or incubation without ATP; or 10 μM, 100 μM, or 1 mM ATP and washing. Data are the mean for three experiments and show the SEMs for each. *Samples with significant differences from immunoprecipitates primarily incubated without ATP or orthovanadate ($p < 0.05$). (B) PC12 cells were either untreated (□) or treated (■) with 100 ng of NGF/mL for 24 h and homogenized. The soluble extracts were immunoprecipitated and washed as described in (A). Immunoprecipitates were incubated without ATP or orthovanadate; or with 1, 10, or 100 μM ATP and 1 mM Na$_3$VO$_4$. After 1 h at 30°C, samples were washed and assayed for cGMP PDE as just described. Data are the mean for three experiments and show the SEMs for each. *Samples with significant differences from untreated cell immunoprecipitates primarily incubated without ATP or orthovanadate ($p < 0.05$).

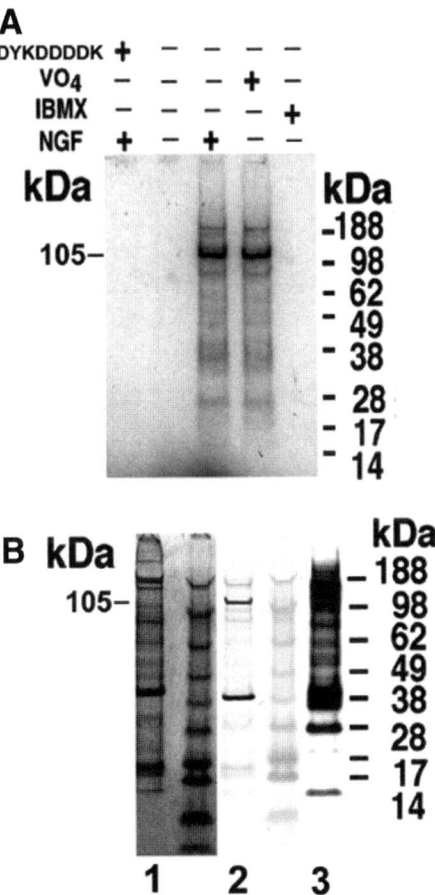

Fig. 3. Proteins in anti-FLAG immunoprecipitates from PC12B cells are phosphorylated in vitro by a PDE2-associated kinase. PC12B cells are a stable cell line expressing a FLAGged PDE2. (**A**) Cells were treated with 100 μ*M* 3-isobutyl-1-methylxanthine (IBMX), 100 ng of NGF/mL, or 30 μ*M* sodium orthovanadate for 24 h. Soluble precleared extracts of the cells were processed for anti-FLAG M2 immunoprecipitation for 1 h at 4°C. The pellet (5% vol resin/vol reaction) was washed three times in buffer A and once in buffer B. The immunoprecipitate was incubated with 5 m*M* MgCl$_2$ and 10 μ*M* [γ-^{32}P] ATP (50 μCi) for 1 h. When present, the competing FLAG peptide (DYKDDDDK) was 100 μ*M*. Samples shown in (A) are immunoprecipitates from 1 mg of soluble protein/lane. (**B**) All cells were treated for 24 h with 100 ng of NGF/mL, and samples 1–3 are anti-FLAG immunoprecipitates from 50 mg of soluble protein washed as described for (A). Lane 1, a Coomassie blue stain of the blot; lane 2, the same sample destained, immunoblotted with a biotinylated anti-FLAG M2, and stained with an avidin–alkaline phosphatase development; lane 3, an autoradiograph of the proteins phosphorylated by PDE2-associated kinases in this sample. Shown between lanes 1 to 2 and 2 to 3 are shown the prestained protein standards for alignment.

most of the radiolabel is incorporated into PDE2A2 (**Fig. 3**). Cells cultured at low density were found to have the most consistent PDE2 response to NGF *(8)*. Very little protein-staining material is seen in these immunoprecipitates (*see* **Note 13**). The radiolabel is found in the protein-staining band corresponding to FH-PDE2 (**Fig. 3B**).

Immunoreactive bands can be developed using either secondary antirabbit or antimouse antibodies conjugated with either HRP or alkaline phosphatase (**Fig. 3B**). HRP development uses chemiluminescence reagents with protocols from the manufacturer (Pierce, Rockford, IL). Alkaline phosphatase development uses colorimetric nitroblue tetrazolium and bromo chloro indolyl phosphate development with protocols available from the manufacturer (Promega).

4. Notes

1. The DYKDDDDK epitope contains the enterokinase recognition site *(11)*, and vectors with the "FLAG" epitope (Sigma) and the "Xpress" epitope of pcDNA3.1/ His^6 (Invitrogen) produce fusion proteins that crossreact with each company's antibodies. Many Novagen vectors also have the enterokinase epitope. If these are used to generate a fusion protein for antigen production, it will result in antibodies that crossreact with this epitope.
2. It is important to maintain PC12 cells at a density of no greater than 10^5/mL to keep them responsive to NGF. These cells produce NGF as well as respond to it. Because NGF differentiation halts down the process of cell division, cells grown to high density will tend to lose their NGF receptors *(8)*. At high density, NGF-nonresponsive cells will overgrow NGF-responsive cells to take over the culture.
3. The PDE2A2 amino-terminal 1–251 amino acid fusion protein with the His$_6$-T7 epitope is soluble from BL21 (DE3) expression systems, but its solubility is greatly increased by 8 *M* urea. The initial homogenization and fractionation without urea (but with lysozyme) removes cellular contaminants that would otherwise bind to the Ni-NTA agarose in later steps. Although 8 *M* urea is a potent denaturant, and the BL21 cells are a protease-deficient cell line, substantial proteolysis of the fusion protein will occur if protease inhibitors are not added and purification is not carried out in the cold.
4. A higher than normal concentration of a broad spectrum of protease inhibitors, such as 2X or greater of the recommended Complete protease inhibitor cocktail, is required for working with PDE2A2. Even under these conditions, significant proteolysis occurs at room temperature. It is essential to keep the enzyme complex as cool as possible unless it is being phosphorylated or enzymatic activity is being assayed. Broad-spectrum phosphatase inhibitors are also essential. Fluoride or orthovanadate at high concentrations work well. Fluoride is useful in the presence of EDTA but tends to precipitate out with magnesium ion. Orthovanadate is useful under these conditions, but it becomes reduced and inactivated easily. Whenever possible, adjust the pH with HCl and KOH instead of NaOH.

5. The nonionic detergent Lubrol 12A9 (also known as Brij 9 lauryl ether or nona oxyethylene lauryl ether or Polidocanol) is used in these studies exclusively to solubilize PDE2A2 from PC12 cells or rat brain. Nonionic detergents of the Triton X-100 class should not be used because solubilized particulate PDE2 only shows cGMP stimulation with the Lubrol class of compounds *(1,2)*.

6. Using a filtered, high-speed (>100,000g) supernatant fraction significantly decreases the nonspecific background from rat brain. Unlike the case with brain tissues, no observable floating Lubrol-insoluble fraction is found after a 10,000g centrifugation of PC12 cell homogenate. Samples from cultured cells gave the same results on immunoprecipitation whether or not the initial soluble supernatant fluid fraction was from the 10,000 or 200,000g soluble fractions.

7. Shorter (1 h) immunoprecipitations were less efficient in precipitating immunoreactive PDE2A2, but longer immunoprecipitation incubations resulted in more proteolysis, especially when anti-FLAG antibodies were used. Increasing the antibody conjugate concentration increased the amount of background material immunoprecipitated. When 12-h incubations were used for immunoprecipitation, complete protease inhibitors were added to a final concentration of 4X.

8. Higher salt washes and storage resulted in a lower background in Western blots but lower recoveries of enzyme as assessed by Western blots or PDE assay. Immunoprecipitates were typically stored at –20°C indefinitely in the glycerol storage buffer detailed in **Subheading 2., item 23**. Under these conditions there was little breakdown of PDE2 protein or loss of catalytic activity after 4 wk of storage.

9. Although sodium fluoride and sodium orthovanadate are both effective protein phosphatase inhibitors at nanomolar concentrations, NaF tends to precipitate with divalent cations, and orthovanadate tends to reduce and become ineffective. Fluoride is used in the homogenization buffers with EDTA because it is more stable for long-term storage of the enzyme complex. Final concentrations of fluoride more than 10 mM will precipitate the magnesium in the PDE assay. Any vanadate-containing buffer that is yellow contains vanadate in the reduced state. Impure water can contain trace reductant that will act to reduce vanadate to vanadyl with a color change. Discard any yellow solutions.

10. Under these conditions of the assay, the phosphorylated PDE2 had a higher apparent K_m for the hydrolysis of cGMP and a lower maximum velocity (**Fig. 1B**) than the nonphosphorylated enzyme. A primary incubation with 10 μM ATP and 5 mM MgCl$_2$ under phosphorylating conditions significantly lowered PDE2 activity from rat brain even when saturated with cGMP (**Fig. 1B**). All values of cGMP PDE were significantly higher when ATP and orthovanadate were absent from the primary (phosphorylation) incubation. At 1–10 μM cGMP, ATP-treated samples were lower in activity using a nondirectional two-tailed t-test for independent means, with $p < 0.05$. The ATP-treated samples had significantly ($p < 0.001$) lower catalytic activity at 100 μM to 1 mM cGMP. Inhibition was evident at 1 μM cGMP even after the immunoprecipitate had been washed free of ATP (**Fig. 2A**). Under these conditions, there was a significant difference ($p < 0.05$) between samples

incubated with between 10 and 1000 µ*M* ATP and the sample incubated for 1 h at 30°C without ATP. The samples incubated without ATP or Na$_3$VO$_4$ had a higher mean activity and a greater variance in activity than the nonincubated samples (cf. no ATP and no inc. in **Fig. 2A**). The reaction conditions did not affect the amount of PDE2 present, nor did phosphorylation conditions influence the recovery of PDE2 in the washes postphosphorylation (**Fig. 2A**, inset).

11. The amount of phosphate incorporated was related to the NGF treatment of the cells (**Fig. 3A**). Presumably rat neurons already exist in a neurotrophin-rich environment.
12. The nonhydrolyzable ATP analog 5′-adenylylimidodiphosphate does not substitute for ATP in phosphorylation but does lower the activity of PDE2 when added directly to enzyme assays. This may be owing to a competitive interaction with the catalytic domain since it can be washed out (data not shown).
13. The reaction conditions do not affect the amount or solubility of PDE2 present, nor do phosphorylation conditions influence the recovery of PDE2 in the washes postphosphorylation. As with the brain enzyme, ATP treatment did not decrease the recovery of PDE2 after washing (data not shown).

Acknowledgments

Thanks go to Sharon Wolda, Vince Florio, and ICOS for the anti-PDE2 MAbs. This work was supported by grants 5 R29 NS35802 from the National Institute of Neurological Disease and Stroke (NINDS) and 5P60 DK20572-24 from the National Institute of Diabetes and Digestive and Kidney Diseases (NIDDK), and by a McKay-Cardiovascular Center Award from the University of Michigan.

References

1. Whalin, M. E., Strada, S. J., and Thompson, W. J. (1988) Purification and partial characterization of membrane-associated type II (cyclic GMP–activatable) cyclic nucleotide phosphodiesterase from rabbit brain. *Biochim. Biophys. Acta* **972,** 79–94.
2. Murashima, S., Tanaka, T., Hockman, S., and Manganiello, V. (1990) Characterization of particulate cyclic nucleotide phosphodiesterases from bovine brain: purification of a distinct cyclic GMP–stimulated isoenzyme. *Biochemistry* **29,** 5285–5292.
3. MacFarland, R. T., Zelus, B. D., and Beavo, J. A. (1991) High concentrations of a cGMP-stimulated phosphodiesterase mediate ANP-induced decreases in cAMP and steroidogenesis in adrenal glomerulosa cells. *J. Biol. Chem.* **266,** 136–142.
4. Whalin, M. E., Scammell, J. G., Strada, S. J., and Thompson, W. J. (1991) Phosphodiesterase II, the cyclic GMP–activatable cyclic nucleotide phosphodiesterase, regulates cyclic AMP metabolism in PC12 cells. *Mol. Pharmacol.* **39,** 711–717.
5. Repaske, D. R., Corbin, J. G., Conti, M., and Goy, M. F. (1993) A cyclic GMP stimulated cyclic nucleotide phosphodiesterase gene is highly expressed in the limbic system of rat brain. *Neuroscience* **56,** 673–686.

6. Wykes, V., Bellamy, T. C., and Garthwaite, J. (2002) Kinetics of nitric oxide–cyclic GMP signalling in CNS cells and its possible regulation by cyclic GMP. *J. Neurochem.* **83,** 37–47.
7. Frodin, M., Peraldi, P., and Van Obberghen, E. (1994) Cyclic AMP activates the mitogen-activated protein kinase cascade in PC-12 cells. *J. Biol. Chem.* **269,** 6207–6214.
8. Bentley, J. K., Juilfs, D. M., and Uhler, M. D. (2001) Nerve growth factor inhibits PC12 cell PDE2 phosphodiesterase activity and increases PDE 2 binding to phosphoproteins. *J. Neurochem.* **76,** 1252–1263.
9. Yang, Q., Paskind, M., Bolger, G., Thompson, W. J., Repaske, D. R., Cutler, L. S., and Epstein, P. M. (1994) A novel cyclic GMP stimulated phosphodiesterase from rat brain. *Biochem. Biophys. Res. Commun.* **205,** 1850–1858.
10. Sadhu, K., Hensley, K., Florio, V., and Wolda, S. (1999) Differential expression of the cyclic GMP–stimulated phosphodiesterase PDE2A in human venous and capillary endothelial cells. *J. Histochem. Cytochem.* **47,** 895–905.
11. Einhauer, A. and Jungbauer, A. (2001) The FLAG peptide, a versatile fusion tag for the purification of recombinant proteins. *J. Biochem. Biophys. Methods* **49,** 455–465.

17

Investigation of Extracellular Signal-Regulated Kinase 2 Mitogen-Activated Protein Kinase Phosphorylation and Regulation of Activity of PDE4 Cyclic Adenosine Monophosphate-Specific Phosphodiesterases

Elaine V. Hill, Miles D. Houslay, and George S. Baillie

Summary

Recently, it has been shown that enzymes of the cyclic adenosine monophosphate (cAMP)-specific phosphodiesterase (PDE) family 4 can be directly phosphorylated by extracellular signal-regulated kinase 2 (ERK2). Phosphorylation of PDE4s by ERK2 is dependent on two docking domains on either side of the target serine that allow specificity and high-fidelity binding of the kinase. The functional consequence of PDE4 phosphorylation by ERK is either an increase or a decrease in PDE activity, depending on whether the PDE4 contains only one of the upstream conserved regions (UCR1) that are typical of PDE4s or both (UCR1 and UCR2). We detail some of the methods that have been crucial in elucidating these important discoveries that represent a novel point of cross talk between the cAMP signaling system and the ERK mitogen-activated protein kinase cascade.

Key Words

Phosphodiesterase; mitogen-activated protein kinase; cyclic adenosine monophosphate; phosphorylation; KIM domain; PDE4.

1. Introduction

Cyclic adenosine monophosphate (cAMP) is an intracellular second messenger that is critical in the determination of many aspects of cellular function *(1,2)*. Its hydrolysis provides an important route whereby cAMP levels are regulated and provides the temporal and spatial gradients of cAMP that are required for many cell-signaling responses. Hydrolysis of cAMP is achieved by members of the large multigene family of cyclic nucleotide phosphodiesterases (PDEs) *(3,4)*. Members of the PDE4 enzyme family specifically hydrolyze

From: *Methods in Molecular Biology, vol. 307: Phosphodiesterase Methods and Protocols*
Edited by: C. Lugnier © Humana Press Inc., Totowa, NJ

cAMP and have recently been shown to have important functions in immune cells and in the central nervous system *(3,5,6)*. PDE4 enzymes are encoded by four genes (A, B, C, D), each of which generates a number of distinct isoenzymes by alternative mRNA splicing *(7)*. All PDE4 enzymes contain a common catalytic unit but are distinguished by a unique N-terminal region that is thought to confer specific targeting of the enzyme within a cell. "Long" PDE4 isoforms are characterized by two PDE4-specific homology regions called upstream conserved regions 1 and 2 (UCR1 and UCR2), which are situated between the N-terminal region and catalytic unit. The "short" PDE4 isoforms lack UCR2.

The extracellular signal-regulated kinase mitogen-activated protein kinase (ERK MAP) cascade provides a pivotal route where various growth factors and hormones can exert action on important transcriptional events and other cellular processes *(8)*. In the past, the ERK MAPK system has been shown to cross talk with the cAMP signaling pathway at the level of the protein kinase Raf. Such regulation, however, must be regarded as cell-type specific, being dependent on the expression of a selection of the three raf isoforms. Recently, though, a novel point of cross talk between the MAPK/cAMP system was identified when it was discovered that a long PDE4 isoform (PDE4D3) could be directly phosphorylated by ERK2, causing a profound, transient decrease in the PDE's activity *(9)*. The drop in activity was associated with a concomitant increase in intracellular cAMP levels. On closer inspection, it was shown that all PDE4s contained both ERK2 docking regions (FQF and KIM domains) essential for specific, high-fidelity ERK2 binding and phosphorylation *(10)*. Indeed, the existence of both functional docking regions allowed coimmunoprecipitation of ERK2 with overexpressed PDE4 enzymes. Moreover, it was discovered that the functional outcome of PDE4 phosphorylation by ERK2 was determined by the regulatory signal integration model provided by UCR1 and UCR2 regions within each PDE4 isoform. In effect, long isoforms are inhibited and short isoforms are activated by ERK2 phosphorylation, and supershort forms are weakly inhibited. This is true except for the multiple products of the PDE4A gene, which, although they contain both ERK docking sites, they lack an ideal consensus site for ERK phosphorylation *(11)*.

We describe the methods used in making the aforementioned discoveries. In particular, we highlight the direct phosphorylation of PDE4s by ERK2, the coimmunoprecipitation of ERK2 with PDE4s, and the activity changes measured as a direct consequence of PDE4 phosphorylation by ERK2.

2. Materials

1. DEAE-dextran (Sigma Chemical Corporation, St. Louis, MO).
2. Chloroquine (Sigma Chemical Corporation).

3. Phosphate-free Dulbecco's modified Eagle's medium (DMEM) (Sigma Chemical Corporation).
4. [^{33}P]-Orthophosphate (Amersham Biosciences).
5. [γ-^{32}P]-Adenosine triphosphate (ATP) (Amersham Biosciences).
6. Okadaic acid (Calbiochem).
7. Monoclonal anti-VSV antibody (Sigma Chemical Corporation).
8. Active GST-ERK2 (Upstate).
9. Epidermal growth factor (EGF) (Sigma Chemical Corporation).
10. Cos-1 cells.
11. KHEM buffer: 50 mM KCl, 50 mM HEPES-KOH, pH 7.2, 10 mM EGTA, 1.92 mM MgCl$_2$.
12. Sodium dodecyl sulfate-polyacrylamide gel electrophoresis (SDS-PAGE) and transfer equipment (Novex NuPAGE, Invitrogen, Carlsbad, CA): 4–12% Bis-Tris gels, 3-(n-morpholino)propane sulfonic acid (MOPS) SDS-PAGE running buffer, transfer buffer.
13. *Escherichia coli* expressing tagged PDE isoforms (such as VSVPDE4D3, MBP-PDE4D3).
14. Isopropyl-β-D-thiogalactopyranoside (IPTG).
15. EDTA-free protease inhibitor cocktail tablets (Roche Molecular Biochemicals, Indianapolis, IN).
16. Amylose resin (New England Biolabs).
17. Bradford protein assay reagent (Bio-Rad, Hercules, CA).
18. 3T3 lysis buffer: 25 mM HEPES, pH 7.5, 2.5 mM EDTA, 50 mM NaCl, 50 mM NaF, 30 mM sodium pyrophosphate, 10% glycerol, 1% Triton X-100.
19. 2X Phosphorylation assay buffer: 100 mM Tris-HCl, pH 7.4, 30 mM MgCl$_2$, 30 mM β-mercaptoethanol, 20% glycerol.
20. TE buffer: 10 mM Tris-HCl, pH 8.0, 0.1 mM EDTA.
21. Phosphorimager.
22. Monoclonal anti-ERK2 antibody 3A7 (Cell Signaling).
23. MEK inhibitor UO126 (Promega, Madison, WI).
24. PDE assay Tris/Mg^{2+} buffer: 20 mmol/L of Tris-HCl, pH 7.4, 10 mmol/L of MgCl$_2$.
25. PDE assay Tris buffer: 20 mmol/L of Tris-HCl, pH 7.4.
26. PDE assay substrate solution: 2 μmol/L of cAMP containing 3 μCi of [^{3}H]cAMP (Amersham)/mL in Tris/Mg^{2+} buffer (100,000 cpm/assay tube).
27. Snake venom (V-0376; Sigma Chemical Corporation): 1 mg/mL in water.
28. Anion-exchange resin: Supelco MTO-Dowex 1 × 8 200–400 mesh CL form. Use as a 1:1:1 slurry with Dowex:water:ethanol.

3. Methods

3.1. DEAE-Dextran Transfection of Cos-1 Cells

This method efficiently and reproducibly transfects Cos-1 cells (*see* **Note 1**). Cells are optimally transfected when the dish is about 50–60% confluent so that the cells are still actively growing. It is best to seed the cells onto a new dish the day before transfection. All amounts given here are for a 10-cm dish of cells, but quantities may be adjusted for different sized dishes or flasks.

1. Use 10 μg of DNA/dish. Dilute the DNA to 250 μL in TE buffer.
2. Add 200 μL of DEAE-dextran (10 mg/mL in phosphate-buffered saline [PBS]; filter sterilized). Mix and leave at room temperature for 15 min.
3. To 5 mL of DMEM containing 5% newborn calf serum, add 5 μL of chloroquine (100 mM; filter sterilized).
4. Aspirate the medium from a 10-cm dish of Cos-1 cells and add the DMEM/chloroquine mix. Add the DNA/DEAE-dextran to the medium dropwise, mix by swirling, and then incubate at 37°C for 3 to 4 h.
5. Shock the cells with 10% dimethyl sulfoxide (DMSO). Aspirate the medium (containing DNA/DEAE-dextran) from the cells, add 5 mL of 10% DMSO in PBS, and leave for 2 min at room temperature. Aspirate the DMSO and wash once with PBS; then add DMEM containing 10% fetal calf serum (FCS) (normal growth medium) to the cells and incubate at 37°C. The following day, the cells may be trypsinized and seeded onto six-well dishes if required. The maximum expression of the transfected protein is obtained 48–72 h posttransfection.

3.2. Determination of Transfected PDE4 Activity

When Cos-1 cells are subjected to EGF treatment, PDE isoforms are transiently phosphorylated by ERK. Long-form PDE4 isoforms are inhibited by this phosphorylation, and here we describe a method for the EGF treatment of cells and analysis of transfected PDE4D3 (a long isoform) activity. This method may be adapted for different agonists or PDE isoforms. Transfection of Cos-1 cells with VSV-tagged PDE4D3 (VSVPDE4D3) leads to the VSVPDE4D3 being responsible for more than 97% of PDE activity in the cells (*9*). Therefore, the PDE activity of VSVPDE4D3 may be analyzed directly from the cell lysates, and it is not necessary to immunoprecipitate the VSVPDE4D3. However, because transfection efficiency may vary, it is necessary to equalize different cell lysates for expression of PDE4D3, and this is carried out by Western blotting.

3.2.1. Equalization of Transfected VSVPDE4D3

1. Transfect Cos-1 cells with VSVPDE4D3 using the DEAE-dextran method (*see* **Subheading 3.1.**), and 24 h posttransfection seed the cells onto six-well culture dishes. After a further 24 h, incubate the cells with medium containing 50 ng/mL of EGF for incubation times between 2 and 30 min. Discard the medium and wash the cells three times with ice-cold KHEM before disrupting the cells with KHEM buffer containing protease inhibitors. Freeze the cells in liquid nitrogen or on dry ice, and thaw on ice before homogenization by passing through a 26-gage

needle 10 times. Pellet the cell debris by centrifuging at 16,000g for 10 min in a refrigerated microcentrifuge and retain the supernatant.

2. Determine the relative amounts of VSVPDE4D3 in the cell lysates by subjecting 20 µg of lysate to SDS-PAGE and analyzing by Western blot using a monoclonal anti-VSV antibody (Sigma Chemical Corporation) and enhanced chemiluminescence (ECL) detection (Amersham Biosciences).

3. Quantitate the relative amounts of VSVPDE4D3 by scanning the Western blot and analyzing the bands using software such as Quantity One (Bio-Rad) or Kodak 1D. It is important that the blot not be overexposed to the film because the bands quantitated must be in the linear range of the ECL detection. It may be necessary to subject a titration of cell lysate to SDS-PAGE and Western blotting in order to ensure that the amount quantitated is in the linear range of detection.

4. Normalize the amounts of cell lysate used to determine PDE activity with respect to the amount of VSVPDE4D3 expressed.

3.2.2. Assay of PDE Activity

The cAMP-specific PDE activity assay is adapted from the procedure of Marchmont and Houslay *(12)*. In this two-step procedure, the proteins to be assayed are first incubated with 1 µM 8-[^{3}H]-labeled cAMP. In the second step, the [^{3}H]-labeled cAMP product of cAMP hydrolysis, 5′AMP, is dephosphorylated to adenosine by incubation with snake venom. The negatively charged, nonhydrolyzed cAMP is then separated from the uncharged adenosine by incubating with Dowex ion-exchange resin. The amount of unbound [^{3}H]-adenosine in the supernatant is determined by scintillation counting, and the rate of cAMP hydrolysis is calculated. A detailed protocol follows:

1. Dilute the sample to be assayed in PDE assay Tris buffer. The total sample volume should be 50 µL in Eppendorf tubes.
2. Add 50 µL of PDE assay substrate solution and mix.
3. Incubate at 30°C for 10 min. Remember to include a negative control/background, i.e., 50 µL of PDE assay Tris buffer plus 50 µL of substrate.
4. Boil for 2 min to stop the reaction.
5. Place on ice for 15 min to cool.
6. Add 25 µL of snake venom solution to each tube and mix.
7. Incubate for 10 min at 30°C.
8. Remove the tubes, place on ice, and add 400 µL of Dowex slurry, and mix.
9. Incubate on ice for 15 min, mixing every 5 min.
10. Centrifuge on a microfuge for 3 min (14,000g).
11. Carefully remove 150 µL and add to a new Eppendorf tube.
12. Add 1 mL of scintillation fluid and count on a scintillation counter that can detect tritium. Remember to count 50 µL of the PDE assay substrate (also known as "total") to enable final calculation of PDE activity (*see* **Note 2**).

13. Calculate PDE activity using the following equation:

$$\text{Activity} = 2.61 \times (\text{value} - \text{background/average total}) \times 10^{-11} \times 10^{-12} \times (1000/\mu g \text{ protein})$$

Units are expressed as picomoles of cAMP per minute per milligrams of protein.

3.3. Immunoprecipitation of PDE4

The following protocol details the immunoprecipitation of VSVPDE4D3 from transfected Cos-1 cells, but it may be adapted for different isoforms or endogenous PDEs (*see* **Note 3**).

3.3.1. Immunoprecipitation

1. Use 5×10^5 cells (approx 1×10 cm dish) per immunoprecipitation. Wash three times in ice-cold PBS, and scrape the cells into 500 μL of 3T3 lysis buffer containing protease inhibitors.
2. Incubate the lysates for 30 min at 4°C while gently inverting. Remove insoluble matter from the lysates by centrifuging at 14,000*g* for 10 min at 4°C. Discard the insoluble pellet and retain the supernatant for immunoprecipitation.
3. Add 3–5 μL of monoclonal anti-VSV antibody (Sigma Chemical Corporation) and gently invert for 1 h at 4°C.
4. Wash protein G–coupled Sepharose (Amersham) (30-μL bead volume per immunoprecipitation) three times in 3T3 lysis buffer. Add 30 μL to each lysate and gently invert for 1 h at 4°C.
5. Pellet the protein G–coupled Sepharose by centrifuging at 14,000*g* for 1 min at 4°C. Wash the pellet three times in 3T3 lysis buffer. The immunoprecipitated PDE may now be used in a PDE activity assay or analyzed by Western blot. If the PDE activity is to be measured, perform a final wash of the pellet in PDE assay buffer.

3.3.2. Analysis of Immunoprecipitate by Western Blot

1. Add 30 μL of SDS-PAGE sample buffer to the immunoprecipitate pellet. Heat at 95°C for 5 min and centrifuge at 14,000*g* for 1 min in a microcentrifuge.
2. Subject the supernatant to SDS-PAGE and transfer to nitrocellulose. Western blot the membrane using an antibody to either the PDE or the tag.

ERK has been shown to interact with PDE4D and can be coimmunoprecipitated with transfected VSVPDE4D3 *(10)*. To detect ERK in the immunoprecipitation, the membrane may be probed with an anti-ERK antibody.

3.4. In Vitro Phosphorylation of MBP-PDE4D3

Purified recombinant MBP-PDE4D3 may be phosphorylated in an in vitro kinase assay in order to assess phosphorylation by purified kinases or cell lysates. This approach may be used to investigate the phosphorylation of PDE4D3 at various sites using mutants of MBP-PDE4D3.

3.4.1. Purification of MBP-PDE4D3

1. Set up overnight 35-mL cultures from glycerol stocks of *E. coli* expressing MBP-PDE4D3 in Luria-Bertani (LB) containing ampicillin at 37°C with shaking.
2. Dilute the overnight cultures into 400 mL of LB containing ampicillin in a 2-L flask. Continue to grow at 37°C with shaking until the OD_{600} is more than 0.6 but less than 1.0 (about 3 h). Induce with 0.1 mM IPTG for 4 h.
3. Pellet the bacteria by centrifuging at 5000g for 10 min at 4°C. The pellets may be stored overnight at –80°C.
4. On the next day, thaw the cell pellets and resuspend thoroughly in 10 mL of PBS containing protease inhibitors and 1 mM dithiothreitol (DTT)/400 mL of original culture. Sonicate the resuspended pellets on ice for 6 min, sonicating 1 min and resting 1 min alternately. Clear the bacterial lysates of insoluble matter by centrifuging at 9000g for 30 min at 4°C.
5. Wash amylose resin (0.5-mL bead volume/400 mL of original culture) three times in PBS. Add the cleared bacterial lysates to the washed amylose resin and incubate for 1 h at 4°C, gently inverting.
6. Wash the amylose resin three times with 20 mL of PBS containing protease inhibitors and 1 mM DTT. Add 1 mL of PBS, transfer the amylose resin to a 1.5-mL Eppendorf tube, and centrifuge at 14,000g in a refrigerated microcentrifuge for 1 min. Remove the supernatant, and add 500 μL of 10 mM maltose/50 mM Tris, pH 8.0, to elute the MBP-PDE4D3 from the amylose resin. Incubate for 30 min at 4°C, gently inverting. Repeat and pool the elutions. The protein concentration of the eluate may be determined by the Bradford assay *(13)*. The purity of the MBP-4D3 may be determined by subjecting 20 μL to SDS-PAGE and Coomassie staining the gel. The molecular mass of MBP-PDE4D3 is approx 145 kD (*see* **Fig. 1**). Aliquot and store at –80°C.

For use in in vitro kinase assays, it is more convenient to use MBP-PDE4D3 coupled to amylose resin rather than the eluted protein in solution. It is therefore not necessary to perform the elutions in this case. Instead, proceed to **step 6** and wash the amylose resin as directed. Resuspend the amylose resin in 0.5 mL of PBS containing protease inhibitors and 1 mM DTT in order to make a 1 : 1 suspension. To assess the quantity and purity of the MBP-PDE4D3, subject 20 μL to SDS-PAGE and compare with a protein standard to estimate the quantity bound to the amylose resin.

3.4.2. In Vitro Phosphorylation of MBP-PDE4D3 by Purified Kinase

The purified MBP-PDE4D3 may be used in a kinase assay with purified kinase or cell lysate (*see* **Note 4**).

1. To each 1.5-mL Eppendorf assay tube, add the following:
   a. 1 μg of MBP-PDE4D3.
   b. 50 ng of GST-ERK2 (about 10 mU).

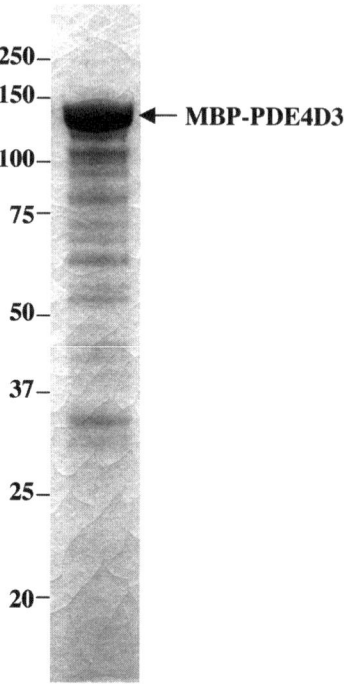

Fig. 1. Purification of MBP-PDE4D3. Twenty microliters of purified MBP-PDE4D3 bound to amylose resin was subjected to SDS-PAGE and Coomassie stained. The position of the 145-kDa MBP-PDE4D3 band is indicated.

    c. 20 μL of kinase assay buffer mix (for 1 mL: 250 μL 0.5 *M* Tris, 250 μL of 1 m*M* EGTA, 25 μL of 1.0 *M* MgCl$_2$, 250 μL of 1% β-mercaptoethanol, 62.5 μL of 20 μ*M* microcystine, 162.5 μL of distilled H$_2$O).

Make up the volume of the reaction to 50 μL using the kinase assay buffer mix.

2. Vortex to mix and start the kinase reaction by adding 5 μL of [^{32}P]-ATP mix (9 μL of [γ-^{32}P]-ATP + 241 μL of 50 mM Tris, pH 7.5, containing 100 μ*M* ATP). Incubate each tube at 30°C for 10 min, and stop the assay by adding 12.5 μL of 5X sample buffer. Heat the samples at 70°C for 10 min in a heat block.

3. Pellet the amylose resin by centrifuging at 14,000*g* for 1 min in a microcentrifuge, and subject the supernatants to SDS-PAGE on a 4–12% Bis-Tris Novex gel (Invitrogen) with MOPS SDS-PAGE running buffer (Invitrogen). Run 20 μL of each sample.

4. Transfer the proteins to nitrocellulose. Dry the mitrocellulose membrane and expose to a phosphorimager plate. The phosphorylated MBP-PDE4D3 will be a band at about 145 kDa (*see* **Fig. 2**), and the extent of phosphorylation may be quantitated using phosphorimager software such as Quantity One.

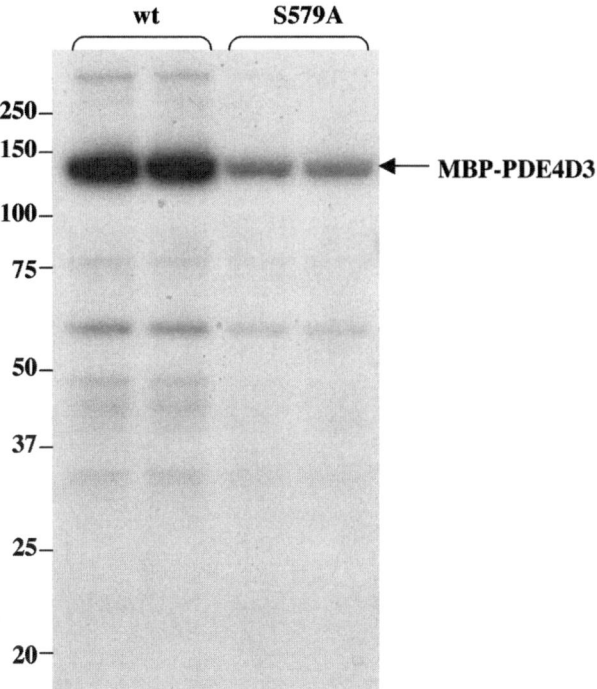

Fig. 2. In vitro phosphorylation of MBP-PDE4D3 by ERK2. One microgram of purified MBP-PDE4D3 (wild type [WT] or S579A mutant) was incubated with recombinant active ERK2 and [^{32}P]-ATP as described in **Subheading 3.4.2.** Radioactively labeled MBP-PDE4D3 was visualized and quantitated using a phosphorimager. The MBP-PDE4D3 is the labeled band at approx 145 kDa.

3.5. In Vivo Labeling of PDE4D Using [^{33}P]-Orthophosphate

To establish that PDE4D3 is phosphorylated by ERK in vivo, metabolic labeling of Cos-1 cells transfected with VSV-tagged human PDE4D3 may be carried out as described in the following steps:

1. Transfect 5×10^5 (10-cm dish) Cos-1 cells using the DEAE-dextran method (*see* **Subheading 3.1.**). Remove the culture medium 48-h posttransfection; wash the monolayer with phosphate-free DMEM; and incubate the cells overnight with 5 mL of phosphate-free DMEM containing 2% FCS, 20 m*M* HEPES (pH 7.4), and 500 µCi of [^{33}P]-orthophosphate. This will lead to an isotopic equilibrium of the ATP pool.
2. Treat the cells with 50 ng/mL of EGF for 5 min. As a control to determine the extent of phosphorylation owing to ERK, also treat the cells with the MEK inhibitor UO126 (10 µ*M*, 30 min) before EGF treatment. Then harvest the cells on

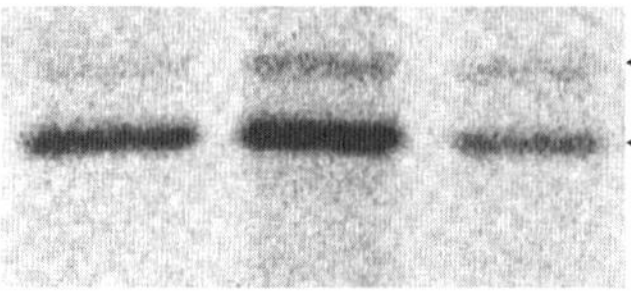

untreated   OA   UO126

←— PDE4D5

←— PDE4D3

Fig. 3. In vivo labeling of PDE4D using [^{33}P]-orthophosphate. Cos-1 cells were labeled overnight with [^{33}P]-orthophosphate and treated with either 10 nM okadaic acid (OA) (a phosphatase inhibitor) or 10 μM UO126 (an MEK/ERK inhibitor). Endogenous PDE4D was immunoprecipitated using an anti-PDE4D antibody, and the [^{33}P]-labeled proteins was visualized using a phosphorimager. The positions of phosphorylated PDE4D3 and PDE4D5 (as confirmed by Western blotting) are indicated.

    ice as follows: Remove the labeling medium, wash the monolayer twice with ice-cold PBS, and scrape into 500 μL of 3T3 lysis buffer containing protease inhibitors and 10 μM okadaic acid.

3. Immunoprecipitate transfected VSVPDE4D3 using a monoclonal anti-VSV antibody (Sigma Chemical Corporation), subject to SDS-PAGE, and transfer to nitrocellulose. [^{33}P]-labeled proteins may be visualized and quantitated using a phosphorimager, and a labeled band at about 100 kDa indicates phosphorylated PDE4D3.

This method may also be used to analyze in vivo phosphorylation of endogenous PDE4D. Untransfected Cos-1 cells are labeled overnight with [^{33}P]-orthophosphate as described in **step 1**, and the PDE4D is immunoprecipitated using an anti-PDE4D antibody. Phosphorylated bands corresponding to PDE4D3 and PDE4D5 (as determined by Western blotting the same membrane) are shown in **Fig. 3**.

3.6. Analysis of In Vivo Phosphorylation of PDE4D by ERK

Described next is a method of determining the extent of phosphorylation of PDE4D in vivo using a "back-phosphorylation" technique. Cells are stimulated and the PDE4D is isolated. The PDE4D undergoes an incubation with active ERK2 and [γ-^{32}P]-ATP, which radioactively labels any "empty" ERK phosphorylation sites. The extent of labeling thereby gives an inverse value for the amount of ERK phosphorylation in vivo.

3.6.1. In Vivo Phosphorylation and Immunoprecipitation

1. Stimulate approx 5 × 10^6 cells with EGF or other agonist. Cells stably or transiently transfected with PDE4D give clearer results than endogenous PDEs owing to the low expression level of PDEs.

2. Transfer the cells to ice, wash three times with ice-cold PBS, and lyse each dish of cells in 500 µL of 3T3 lysis buffer containing protease inhibitors and okadaic acid. Divide each lysate into two equal volumes, and use 3T3 lysis buffer to increase the volume of each to 700 µL. Immunoprecipitate the PDE4D from all lysates (as in **Subheading 3.3.**).

3. Wash the immunoprecipitates in 1 mL of 3T3 lysis buffer; 1 mL of PBS; and, finally, 1 mL of 1X phosphorylation assay buffer.

3.6.2. [γ-^{32}P]-ATP ERK Labeling of Immunoprecipitates

1. Use 50 µL of reaction mix per immunoprecipitate. For 1 mL: 500 µL of 2X phosphorylation assay buffer, 1 µL of 1 mM okadaic acid, 2 µL of 100 mM ATP, 100 µCi of [γ-^{32}P]-ATP. Make up to 1 mL with distilled H_2O.

2. Add 0.5 µL of active GST-ERK (about 10 mU) to appropriate samples. As a control, retain one immunoprecipitate with no added GST-ERK.

3. Add 50 µL of reaction mix to all samples, mix, and incubate at 30°C for 10 min.

4. Pellet the immunoprecipitates by centrifuging at 14,000g for 1 min in a microcentrifuge and remove the supernatant. Add sample buffer to each immunoprecipitate, heat at 70°C for 10 min, and repellet the beads by centrifuging at 14,000g for 1 min. Subject the supernatants to SDS-PAGE on a 4–12% Bis-Tris Novex gel with MOPS SDS-PAGE running buffer and transfer to nitrocellulose. Dry the nitrocellulose membrane and analyze the ^{32}P-labeled PDE4D using a phosphorimager (*see* **Fig. 4**). To quantitate the ^{32}P-labeled PDE4D, it is necessary to Western blot the membrane using an anti-PDE4D antibody (or antitag antibody if a tagged PDE4D such as VSVPDE4D3 has been transfected) to normalize the amount of PDE4D in each immunoprecipitate.

4. Notes

1. PDE isoforms are expressed at low levels endogenously, so it is often necessary to transfect PDE isoforms for efficient analysis. Cell lines other than Cos-1 may be used, and other methods of transfection are also available, such as DOTAP (Roche Molecular Biochemicals) or lipofectamine (Invitrogen); the method of transfection should be optimized for the cell line used.

2. When measuring PDE activity using this method, it is important to remember that the assay has a linear range into which the measured number of counts must fall for the results to be meaningful. In this case, once the background counts have been removed (negative control value) from each result, the maximum detected counts should be no more than 18,000. If the counts are greater than this, the assay should be repeated with less protein.

3. Endogenous PDE4 isoforms may be immunoprecipitated from cell lysates using subfamily-specific antibodies such as anti-PDE4D or isoform specific antibodies such as anti-PDE4D5. Transfected PDE4 isoforms may also be immunoprecipitated using anti-PDE antibodies or, alternatively, if the transfected isoform is tagged, an antibody against the tag such as anti-VSV.

A

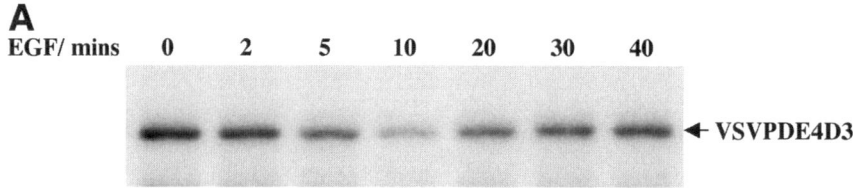

B

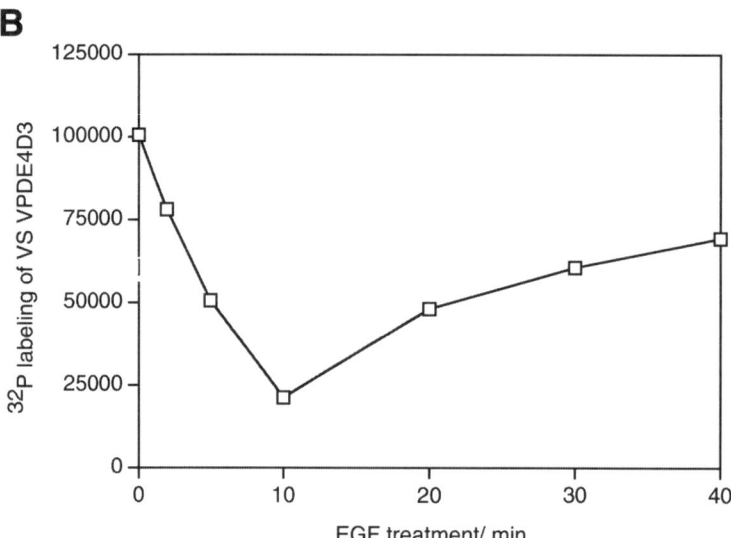

Fig. 4. Analysis of in vivo phosphorylation of VSVPDE4D3 by ERK using "back-phosphorylation" method. Cos-1 were transfected with VSV-tagged PDE4D3 and treated as described in **Subheading 3.6. (A)** Immunoprecipitated VSVPDE4D3 was subjected to SDS-PAGE and [^{32}P]-labeled VSVPDE4D3 was visualized using a phosphorimager. **(B)** The transient decrease in phosphorylation correlates with a transient increase in ERK labeling in vivo.

4. The protocol described here details the method for phosphorylation by recombinant activated ERK2, but other kinases may alternatively be used. If using cell lysate rather than a purified kinase, in **step 2** after the 30°C incubation, pellet the amylose resin by centrifuging for 1 min at 14,000g and carefully remove all of the supernatant. Wash the resin with 1X phosphorylation assay buffer. Add 20 µL of sample buffer to the beads, heat at 70°C for 10 min, and continue with **step 3**.

References

1. Houslay, M. D. and Milligan, G. (1997) Tailoring cAMP signaling responses through isoform multiplicity. *Trends Biochem. Sci.* **22,** 217–224.

2. Beavo, J. A. and Brunton, L. L. (2002) Cyclic nucleotide research: still expanding after half a century. *Nat. Rev. Mol. Cell Biol.* **3,** 710–718.
3. Houslay, M. D., Sullivan, M., and Bolger, G. B. (1998) The multi-enzyme PDE4 cAMP specific phosphodiesterase family: intracellular targeting, regulation and selective inhibition by compounds exerting anti-inflammatory and anti-depressant actions. *Adv. Pharmacol.* **44,** 225–342.
4. Conti, M. and Jin, S. L. C. (1999) The molecular biology of cyclic nucleotide phosphodiesterases. *Prog. Nucleic Acids Res.* **63,** 1–38.
5. Torphy, T. J. (1998) Phosphodiesterase isozymes: molecular targets for novel anti-asthma agents. *Am. J. Respir. Crit. Care Med.* **157,** 351–370.
6. Giembycz, M. A. (2000) Phosphodiesterase 4 inhibitors and the treatment of asthma: where are we now and where do we go from here? *Drugs* **59,** 193–212.
7. Houslay, M. D. and Adams, D. R. (2003) PDE4 cAMP phosphodiesterases: modular enzymes that orchestrate signaling cross-talk, desensitization and compartmentalization. *Biochem J.* **370,** 1–18.
8. Blumer, K. J. and Johnson, G. L. (1994) Diversity in function and regulation of MAP kinase pathways. *Trends Biochem. Sci.* **19,** 236–240.
9. Hoffman, R., Baillie, G. S., MacKenzie, S. J., Yarwood, S. J., and Houslay, M. D. (1999) The MAP kinase ERK2 inhibits the cyclic AMP-specific phosphodiesterase, HSPDE4D3 by phosphorylating it at Ser579. *EMBO J.* **18,** 893–903.
10. MacKenzie, S. J., Baillie, G. S., McPhee, I., Bolger, G. B., and Houslay, M. D. (2000) ERK2 MAP kinase binding, phosphorylation and regulation of PDE4D cAMP specific phosphodiesterases: the involvement of C-terminal docking sites and N-terminal UCR regions. *J. Biol. Chem.* **275,** 16,609–16,617.
11. Baillie, G. S., MacKenzie, S. J., Mc Phee, I., and Houslay, M. D. (2000) Sub-family selective actions in the ability of ERK2 MAP kinase to phosphorylate and regulate the activity of PDE4 cyclic AMP-specific phosphodiesterases. *Br. J. Pharmacol.* **131,** 811–819.
12. Marchmont, R. J. and Houslay, M. D. (1980) A peripheral and an intrinsic enzyme constitute the cyclic AMP phosphodiesterase activity of rat liver plasma membranes. *Biochem. J.* **187,** 381–392.
13. Bradford, M. M. (1976) A rapid and sensitive method for the quantitation of microgram quantities of protein utilizing the principle of protein-dye binding. *Anal. Biochem.* **72,** 248–254.

18

Radiolabeled Ligand Binding to the Catalytic or Allosteric Sites of PDE5 and PDE11

James L. Weeks II, Mitsi A. Blount, Alfreda Beasley, Roya Zoraghi, Melissa K. Thomas, Konjeti Raja Sekhar, Jackie D. Corbin, and Sharron H. Francis

Summary

Cyclic nucleotide phosphodiesterases (PDEs) have been investigated for years as targets for therapeutic intervention in a number of pathophysiological processes. Phosphodiesterase-5 (PDE5), which is highly specific for guanosine 3′-5′-cyclic-monophosphate (cGMP) at both its catalytic site and its allosteric sites, has generated particular interest because it is potently and specifically inhibited by three drugs: sildenafil (Viagra™, Pfizer), tadalafil (Cialis™, Lilly-ICOS), and vardenafil (Levitra™, Bayer GSK). Previously, we have used [³H]cGMP to directly study the interaction of cGMP with the allosteric sites of PDE5, but because cGMP binds with relatively low affinity to the catalytic site, it has been difficult to devise a binding assay for this particular binding reaction. This approach using measurement of radiolabeled ligand binding continues to allow us to more precisely define functional features of the enzyme. We now use a similar approach to study the characteristics of high-affinity [³H]inhibitor binding to the PDE5 catalytic domain. For these studies, we have prepared [³H]sildenafil and [³H]tadalafil, two structurally different competitive inhibitors of PDE5. The results demonstrate that radiolabeled ligands can be used as probes for both catalytic site and allosteric site functions of PDE5. We describe herein the methods that we have established for studying the binding of radiolabeled ligands to both types of sites on PDE5. These techniques have also been successfully applied to the study of binding of radiolabeled PDE5 inhibitors to PDE11, suggesting that these methods are applicable to the study of other PDEs, and perhaps other enzyme families.

Key Words

Phosphodiesterase-5; sildenafil; tadalafil; GAF domains; catalytic site; regulatory domain; allosteric sites; phosphodiesterase-11.

1. Introduction

Cyclic nucleotides participate in the regulation of most physiological processes, and modulation of their levels is of utmost importance to the cell.

From: *Methods in Molecular Biology, vol. 307: Phosphodiesterase Methods and Protocols*
Edited by: C. Lugnier © Humana Press Inc., Totowa, NJ

Mammalian tissues contain multiple intracellular binding sites for cyclic guanosine 3′-5′-cyclic-monophosphate (cGMP) and adenosine 3′-5′-cyclic-monophosphate (cAMP) including allosteric cyclic nucleotide-binding sites on cyclic nucleotide-dependent protein kinases, cyclic nucleotide-gated cation channels, cyclic nucleotide-regulated guanine nucleotide exchange factors (GEFs) and both allosteric and catalytic sites of cyclic nucleotide phosphodiesterases (PDEs). The cyclic nucleotide-binding sites of the kinases, cation channels, and GEFs are evolutionarily related to the bacterial catabolite gene activator protein (CAP) family, but the catalytic and allosteric sites of mammalian PDEs are biochemically and evolutionarily distinct from each other and from the CAP-related family (1–5).

PDEs provide the major cellular mechanism for termination of cyclic nucleotide signaling by hydrolyzing cAMP and cGMP. It is therefore important to identify precisely the biochemical determinants that provide for interactions of the PDEs with cyclic nucleotides and related ligands. All known mammalian PDEs are class I PDEs and contain a well-conserved catalytic domain (C domain) that interacts with the cyclic nucleotide substrate or with numerous competitive inhibitors such as sildenafil and tadalafil. The regulatory domains (R domain) of certain PDEs (PDEs 2, 5, 6, and perhaps 10 and 11) also contain allosteric cGMP-binding sites that are provided by one or more GAF domains that occur in a broad range of proteins (cGMP-Anabaena adenylyl cyclase–FhlA transcription factor) (6). Insights into the interaction of a particular PDE with specific ligands are likely to be useful in designing and screening for pharmacological interventions in these systems. One way to approach this is to determine directly the properties of radiolabeled ligand (radioligand) binding to the cyclic nucleotide-binding sites of PDEs (7–9).

Allosteric sites of PDE5 have been studied for many years using this approach. Like all known class I PDEs, PDE5 is a chimeric protein (**Fig. 1**); PDE5 comprises an N-terminal R domain and a more C-terminal C domain that contains the catalytic site (2). cGMP binds with high affinity and specificity to the GAF a domain in PDE5 and has potential to bind to the GAF b domain (11,12). Many studies of [³H]cGMP binding to the PDE5 allosteric sites have been conducted, and we have recently gained new insights into factors that impact the measurement of this cyclic nucleotide binding. We have recently extended this approach to the measurement of binding of radiolabeled PDE5 inhibitors (**Fig. 2**) to the PDE5 catalytic site (9).

This chapter describes two approaches that we are using to study the interaction of ligands with PDE5: [³H]cGMP binding to the allosteric sites of PDE5 holoenzyme and the isolated R domain, as well as binding of the catalytic site inhibitors, [³H]sildenafil and [³H]tadalafil, to the PDE catalytic site holoenzyme. We are also utilizing these techniques to study PDE11. Although both types of

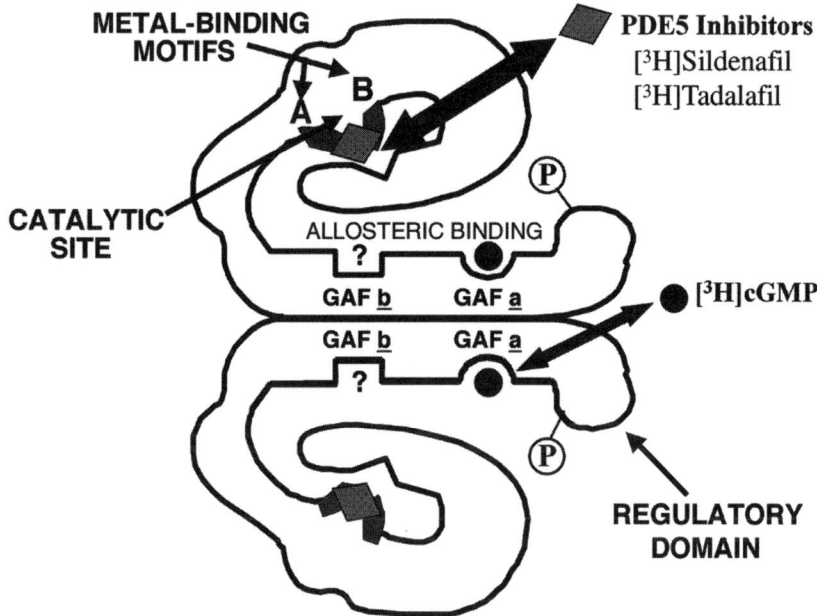

Fig. 1. Working model of PDE5. PDE5 is a chimeric protein that contains a catalytic domain and a regulatory domain. The catalytic domain contains metal-binding motifs; it hydrolyzes cGMP specifically and binds cGMP analogs and PDE5 inhibitors such as sildenafil (Viagra™; Pfizer) and tadalafil (Cialis™; Lilly-ICOS). The regulatory domain binds cGMP at allosteric cGMP-binding sites and contains a serine that is phosphorylated by protein kinase G (PKG) or the C-subunit of PKA.

binding studies utilize the same protein preparations of PDE5 and involve ligands with related structures, the conditions that foster maximum binding of the respective ligands differ quite substantially. The biochemical basis for some of these differences is not well understood. However, the need for close attention to detail when developing such an assay cannot be overemphasized.

1.1. Advantages of Using Radioligand Binding to Study Proteins

Studies that directly measure the binding of specific radioligands to a protein have several important advantages and allow a number of parameters to be determined. These include stoichiometry of ligand binding (mol ligand/mol protein); association and dissociation kinetics of the ligand; effects of other molecules (salts, divalent cations, chelators, analogs, proteins), temperature, and other conditions on the binding properties of each ligand (e.g., the K_D [affinity], B_{MAX} [total binding], EC_{50} [competitor affinity]); overall topography and spatial features of the respective sites; modulation of interactions between

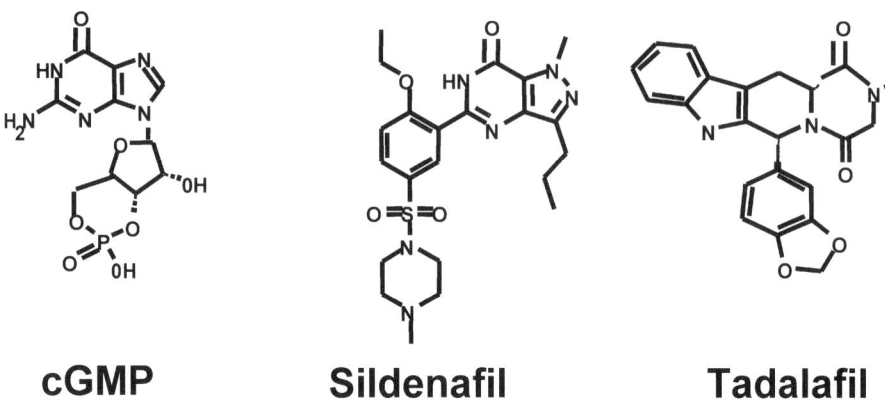

cGMP   **Sildenafil**   **Tadalafil**

Fig. 2. Molecular structures of cGMP, sildenafil, and tadalafil. cGMP, sildenafil, and tadalafil share several molecular features. This property is likely to provide for interaction of these inhibitors with the catalytic site to inhibit cGMP hydrolysis competitively.

functional domains within a protein; and estimation of enzyme concentration in tissue or cell extracts. Tritiated radioligands are advantageous in that they possess long half-lives and offer low health risk to the investigator. Many compounds can be easily prepared by tritiation of a preexisting, purified unlabeled compound as we have done here.

The use of radiolabeled ligands that are specific for either the allosteric cyclic nucleotide–binding sites or the catalytic site of PDEs is particularly valuable for proteins such as PDE5 and PDE6 in which both types of sites are highly specific for the same ligand, i.e., cGMP *(5)*. Studies of the cGMP interaction with the catalytic site of PDE5 are typically complicated by the ongoing hydrolysis of the ligand and the necessity of maintaining the enzyme in a catalytically competent state. This is especially important for PDE5, because in the absence of occupation of the catalytic site, essentially no cGMP binding can be detected at the allosteric sites *(7,13)*. The availability of a high-affinity, highly specific, nonhydrolyzable catalytic-site radioligand, such as [³H]sildenafil or [³H]tadalafil, provides a powerful tool for probing the properties of the C domain of PDE5, and other PDEs such as PDE11. To achieve this objective, we have developed a method for purifying the commercially available PDE5 inhibitor sildenafil from Viagra™ tablets, and we have synthesized tadalafil (Cialis™; Lilly-ICOS) according to US patent no. 6,140,329 (inventor: Alain Claude-Marie Daugan, ICOS, Bothell, WA). Both compounds were then tritiated commercially by Amersham Biosciences. Following radiolabeling, we have shown that the IC_{50} for each [³H]inhibitor of PDE5 agrees closely with that of the unlabeled compound, and the structural integrity of each of the radiolabeled compounds was further verified using both high-performance liquid chromatography and mass spectroscopy.

2. Materials

1. Purified his-tagged PDE5 *(9)*.
2. 3-Isobutyl-1-methylxanthine (IBMX) (Sigma Chemical Corporation, St. Louis, MO).
3. [^{3}H]cGMP, specific activity of 7–18 Ci/mmol (Amersham Biosciences); this is stable for several years when stored in 50% EtOH.
4. [^{3}H]sildenafil *(10)*.
5. Millipore HAWP nitrocellulose filter membranes, 0.45 μm (Millipore, Bedford, MA).
6. Whatman GF/B glass fiber filters, 1.0 μm (Whatman, UK).
7. Brandel M24T cell harvester (24 well) with Teflon-coated lines (Brandel, Gaithersburg, MD).
8. EDTA (Sigma Chemical Corporation).
9. Bovine serum albumin (BSA) (alcohol precipitated; Sigma Chemical Corporation).
10. Histone IIA-S (Sigma Chemical Corporation).
11. β-Mercaptoethanol (Sigma Chemical Corporation); add fresh to buffers each time.
12. Triton X-100 (Sigma Chemical Corporation).
13. KP buffer: 10 mM potassium phosphate, pH = 6.8; this is stable at 4°C for several months.
14. KPT: KP with 0.1% Triton X-100; store up to 6 mo.
15. KPM: KP with 15 mM β-mercaptoethanol; make fresh before experiments.
16. KPM-His: KPM with 0.2 mg/mL of histone IIA-S; this is stable for 4 to 5 d.
17. KPM-BSA: KPM with 1 mg/mL of BSA; this is stable for 4 to 5 d.
18. Millipore glass microanalysis filter holders with fritted glass filter (Millipore).
19. Poly-L-lysine (M_r ~7000) (Sigma Chemical Corporation).
20. [^{3}H]Tadalafil (synthesized as described herein and radiolabeled by Amersham Biosciences).
21. Semiautomatic, self-refilling, spring-loaded adjustable syringe pipet (VWR, Bristol, CT).
22. Theophylline (no. T-1633; Sigma Chemical Corporation).
23. cAMP (no. A-6885; Sigma Chemical Corporation).
24. cGMP (no. G-6129; Sigma Chemical Corporation).

3. Methods

This chapter describes the general protocols and emphasizes the variables that should be considered when attempting to establish a radioligand-binding assay. Two filter-based assays have been developed for detection of radioligand binding to PDE5. Method A is a single-well assay that has been used extensively for measuring [^{3}H]cGMP binding to PDE5 allosteric sites as well as binding of [^{3}H]sildenafil and [^{3}H]tadalafil to the PDE5 catalytic site *(2,7–9,13)*.

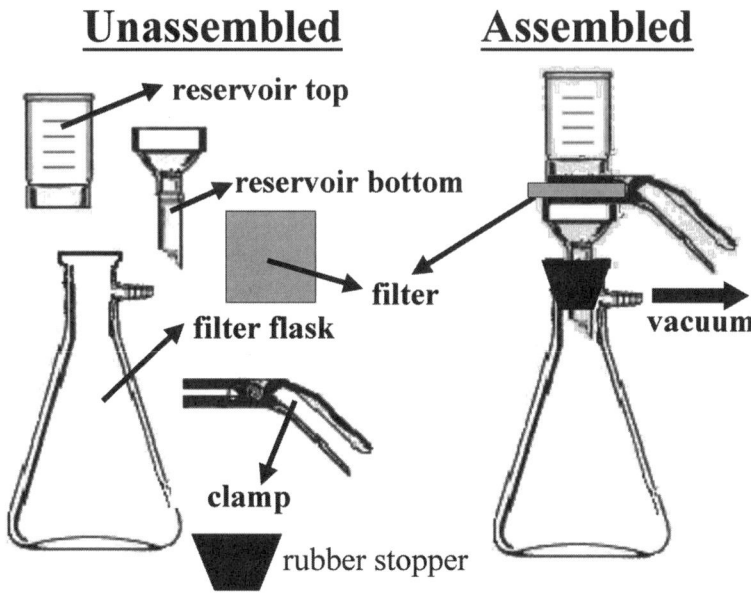

Fig. 3. Single-well Millipore filtration apparatus used in method A.

Method B utilizes the Brandel M24T cell harvester, which allows the simultaneous filtration of up to 24 samples; this method has been used for studying the binding of [³H]sildenafil to PDE5 *(9)* as well as binding of radioligands to PDE11. Because Method B has not been used to study [³H]cGMP binding to the PDE5 allosteric sites, it is further described in a **Subheading 3.3.3.** later section dealing with binding studies for [³H]sildenafil and [³H]tadalafil.

3.1. Method A

This method utilizes a single-well vacuum-filtration apparatus that is available from Millipore (**Fig. 3**; *see* **Note 1**). Various filter papers are commercially available, but HAWP nitrocellulose filter paper (pore size of 0.45 µm) has proven to be satisfactory for measuring [³H]cyclic nucleotide binding to allosteric cGMP-binding sites on PDE5 *(2,7–9)*, cGMP-dependent protein kinase (PKG) *(14)*, and the regulatory subunit of cAMP-dependent protein kinase (PKA) *(15,16)*. The general protocol is as follows:

1. Assemble the Millipore apparatus, connect it to a vacuum source, and then clamp the dry filter (Millipore HAWP nitrocellulose filter, pore size of 0.45 µm, 2.5 × 2.5 cm) in place while the vacuum is open. The vacuum force should be strong and consistent to afford a rapid and comparable filtration for all samples within an experimental set (*see* **Note 2**).

2. After clamping the membrane filter into place, prewet the membrane by allowing 2 mL of an appropriate buffer to pass through. We typically use 10 mM potassium phosphate, pH 6.8 (KP), but other buffers may also be acceptable. The contents of the sample tube are immediately applied to the membrane, and the fluid is pulled through the filter by the vacuum.

3. Rapidly wash the sample tube and the membrane with the same buffer; the volume of the sample as well as the number of washes must be determined for each ligand and protein to be studied. Fewer washes may enhance the recovery of bound ligand, but will invariably yield higher blanks. The buffer washes of the tube and membrane are rapidly delivered using a semiautomatic, self-refilling, spring-loaded adjustable syringe pipet (VWR) that is mounted above the filtration apparatus.

4. Air-dry the filter, place in a scintillation vial with either nonaqueous or aqueous scintillation fluid, and count in a scintillation counter. To correct for potential quench of the radioligand by the membrane, spot an aliquot of the radioligand directly onto a Millipore nitrocellulose filter, and then dry and count the filter. This general protocol can be altered in numerous ways to achieve particular experimental goals.

3.2. Typical Binding Assay for Studying the Interaction of [³H]cGMP With PDE5 Allosteric cGMP-Binding Sites

The [³H]cGMP-binding assay is typically conducted in glass tubes with a total volume of 100 µL, but volumes ranging from 50 to 250 µL are commonly used. The assay is routinely conducted at 0–4°C in a reaction mixture comprising 10–20 mM sodium or potassium phosphate, pH 6.8; 12 mM β-mercaptoethanol; 1 mM EDTA; 0.03–6 µM [8-³H]cGMP; 0.5–2 mg/mL of histone IIA-S; and an inhibitor of the PDE5 catalytic site, such as 200 µM IBMX or 0.1 µM sildenafil citrate. When using impure preparations of PDEs, a nonspecific PDE inhibitor or inhibitors such as IBMX should be included in the reaction mix in order to block hydrolysis of the cyclic nucleotide, in this case, [³H]cGMP. Unlabeled cGMP can be added in order to achieve the desired final concentration of cGMP. The binding reaction is initiated by the addition of enzyme (0.05–5 µg of purified PDE5/tube) followed by incubation for 30–60 min on ice. The reaction is terminated by the addition of 1 mL of ice-cold KP to the reaction tube, and the combined mixture is rapidly filtered onto a premoistened Millipore nitrocellulose HAWP filter. The tube is then rinsed with 1 mL of the same buffer, and this is applied to the membrane. Finally, the filter is rapidly washed with 3 × 2 mL of the same buffer.

3.2.1. Reagents

Other buffers that have been used in the reaction mixture include 50 mM Tris-HCl, pH 7.5; 1 mM EDTA or 10 mM MOPS, pH 7.5. The [³H]cGMP stock is typically provided in 50% ethanol. If it is necessary to use undiluted [³H]cGMP, it is advisable first to evaporate the ethanol using a speed vacuum followed by resuspension of the [³H]cGMP in water to avoid excessive carry-over of ethanol into the reaction. Concentrations of ethanol as low as 1% can have deleterious effects on some binding assays. However, prolonged storage of the [³H]cGMP in water results in significant exchange of tritium with water and a decrease in the specific radioactivity.

3.2.2. Effect of Temperature

Approximately 20–30 min is typically required for PDE5 allosteric cGMP binding to reach equilibrium, and the time required to reach equilibrium is similar at 4° and 30°C. This applies to both the holoenzyme and isolated R domain. Temperature has little effect on either the stoichiometry or affinity of allosteric cGMP binding by PDE5. This differs significantly from allosteric cGMP binding by PKG, in which we have found that cGMP-binding affinity increases approx 10-fold by lowering the temperature from 30° to 0°C *(17)*.

3.2.3. Protein Concentration and Preparation

Depending on the conditions used for studies of PDE5 and the preparation protocol used for purifying PDE5, the K_D for cGMP can range from 0.02 to 2 μM, so cGMP concentrations from 0 to 6 μM in the assay are typical. The inclusion of sulfhydryl reagents such as β-mercaptoethanol (12–25 mM) or dithiothreitol (DTT) (10–30 mM DTT), a chelator such as EDTA, or a divalent cation such as Mg^{2+} is optional, depending on the goals of the experiment. We have found that the addition of some of these agents (e.g., the reductants and/or $MgCl_2$) can significantly increase cGMP binding by PDE5. The concentration of PDE5 chosen for a particular experiment also varies, depending on the needs of the particular experiment. For straightforward determination of the K_D for radioligand binding, the concentration of PDE5 should be significantly lower than the K_D value. If the radiolabel signal is weak at lower concentrations of the ligand, increasing the reaction mixture volume may be helpful, although this may also increase the blank.

3.2.4. Specificity of Binding

To verify the specificity of ligand binding, a number of unlabeled related compounds (e.g., a collection of nucleotides) should be tested separately in the binding assay. For PDE5, inclusion of either 100 μM 5′-GMP or cAMP is an excellent choice because neither compound interacts significantly with PDE5 allosteric cGMP-binding sites. By contrast, a 20-fold excess of Inosine-3′,5′-cyclic monophosphate (cIMP) competes effectively and specifically for the PDE5 allosteric cGMP-binding sites in the presence of PDE inhibitors; in the absence of PDE5 inhibitors, cIMP also interacts with the catalytic site. In crude systems, it is advisable to include 10 μM cAMP and/or 10 μM 8-bromo-cGMP to ablate cyclic nucleotide binding by other cyclic nucleotide-binding proteins such as PKA and PKG. 8-Bromo-cGMP interacts strongly with the allosteric cGMP-binding sites of PKG, but not with sites on PDE5.

3.2.5. Effect of PDE Inhibitors and Divalent Cations on Allosteric cGMP Binding

Occupation of the catalytic site of PDE5 by either cGMP, magnesium, or catalytic site inhibitors profoundly increases cGMP binding at the allosteric sites. Addition of PDE5 catalytic site inhibitors to the allosteric cGMP-binding assay not only prevents breakdown of [³H]cGMP but also apparently causes a conformational change in the protein that increases allosteric cGMP binding. In the absence of occupation of the catalytic site, there is little detectable allosteric cGMP binding by PDE5. When native PDE5 is exhaustively purified in the presence of EDTA, the enzyme lacks catalytic activity in the absence of added metal (**Fig. 4A**), but inhibitors such as IBMX, zaprinast, dipyridamole, papaverine, and sildenafil can still interact with the catalytic site to increase allosteric cGMP binding. Furthermore, in the absence of a PDE inhibitor, the addition of magnesium chloride increases cGMP hydrolysis at the catalytic site and concomitantly stimulates cGMP binding to the allosteric sites (**Fig. 4A,B**). This is most likely owing to increased occupation, albeit transient, of the catalytic site by the substrate cGMP, resulting in the conformational change that increases allosteric cGMP binding. As the cGMP substrate is depleted by hydrolysis, cGMP binding to the allosteric sites declines owing to the lowered concentration of cGMP (**Fig. 4B**).

More recently, a purification protocol that avoids exposure to EDTA has been used in purifying recombinant PDE5; this PDE5 retains significant catalytic activity even in the absence of added divalent cation (*[18]*; unpublished results). In this situation, cGMP hydrolysis during the course of the cGMP-binding assay becomes a significant problem. We routinely include PDE5 catalytic site inhibitors for two reasons: (1) when using enzyme that may retain endogenous metal, it is particularly important to use a concentration of catalytic site inhibitor that is sufficient to block catalysis and protect the cGMP substrate during the course of the cGMP-binding assay; and (2) interaction of a PDE5 inhibitor with the catalytic site evokes an apparent conformational change, which results in increased allosteric cGMP binding to the R domain. Substrate protection can be easily tested by terminating the reaction via boiling for 10 min or by adding a "stop mix" containing theophylline, EDTA, cAMP, and cGMP to reaction tubes followed by treatment with snake venom 5′-nucleotidase and filtration on QAE-Sephadex *(9)* (*see* **Note 3**). This allows one to measure the extent of cGMP hydrolyzed during the assay.

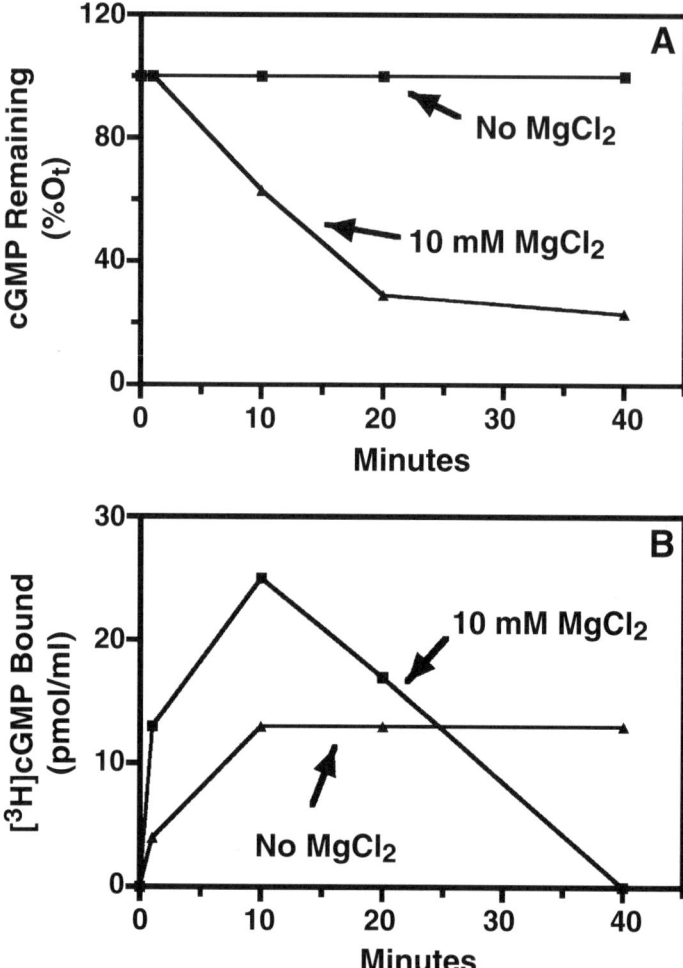

Fig. 4. Effect of magnesium on allosteric cGMP binding and cGMP hydrolytic activities of PDE5. Purified bovine PDE5 was incubated at 30°C with 10 μM [³H]cGMP in the absence or presence of 10 mM MgCl₂ as indicated. At different time points, aliquots of the incubation mixtures were removed for determination of the amount of cGMP hydrolyzed during the incubation (**A**) and another aliquot was removed for Millipore filtration at 4°C as described for the cGMP-binding assay using method A (**B**). The amount of [³H]cGMP that had been hydrolyzed at each time (A) was determined by removing duplicate aliquots at the same time points followed by the addition of 20 μL of a stop mix (*see* **Note 1**) in order to stop cGMP hydrolysis. These aliquots were then treated with 5′-nucleotidase for 10 min at 30°C as previously described. Data were corrected by subtraction of blank values obtained from processing parallel samples without PDE5.

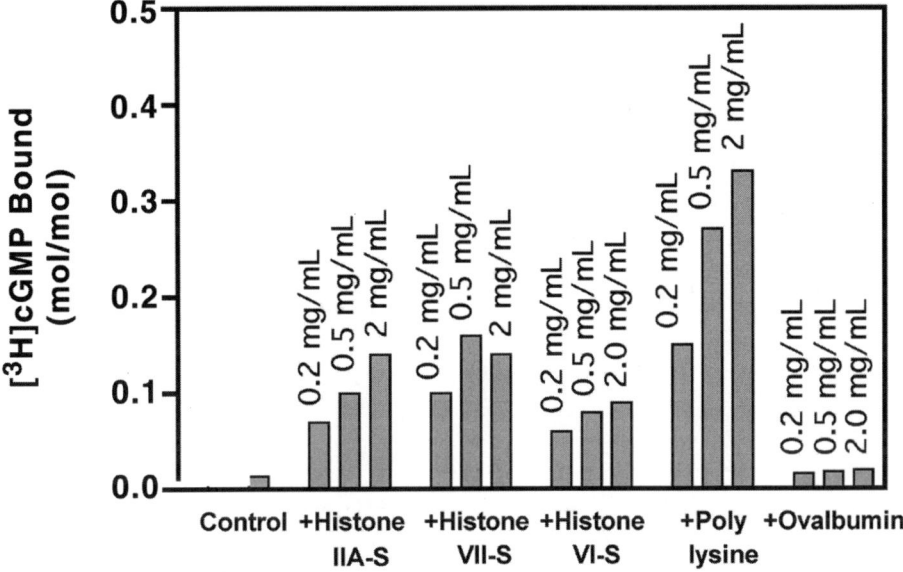

Fig. 5. Effect of various proteins on stimulation of PDE5 allosteric cGMP binding. Purified recombinant His-tagged bovine PDE5 diluted in KPM-BSA was used for these studies. The final concentration of the components in the [³H]cGMP-binding assay mixtures were as follows: 0.33 μM PDE5 (0.3 μg), 0.75 μM [³H]cGMP, 10 mM KPM, and 1 mM EDTA. Sildenafil (0.1 μM) was included in all samples except for the control. Where indicated, exogenous proteins were added to a final concentration of 0.2, 0.5, and 2 mg/mL. Following incubation for 1 h, the samples were filtered according to method A.

3.2.6. Effect of pH, Various Proteins, and Ionic Strength

Maximum allosteric cGMP binding to PDE5 has a broad pH range. Peak allosteric cGMP binding occurs between pH 7.8 and 9.0 *(19)*. Stimulation by IBMX is greater at pH 7.0–7.5 and diminishes with increasing pH values even though the catalytic rate is higher at these elevated pH conditions. This indicates that the effect of IBMX is not solely owing to substrate protection.

The addition of particular types of proteins can significantly improve the cGMP-binding activity of PDE5 (**Fig. 5**), but the mechanism for this is not fully understood. For allosteric cGMP binding to PDE5, the effects appear to be largely owing to increased recovery of PDE5 and the bound radioligand on the membrane owing to formation of a large complex between PDE5 and histone. Recovery has been determined by using PDE5 that has been phosphorylated using [³²P]adenosine triphosphate (ATP) and comparing the percentage of the

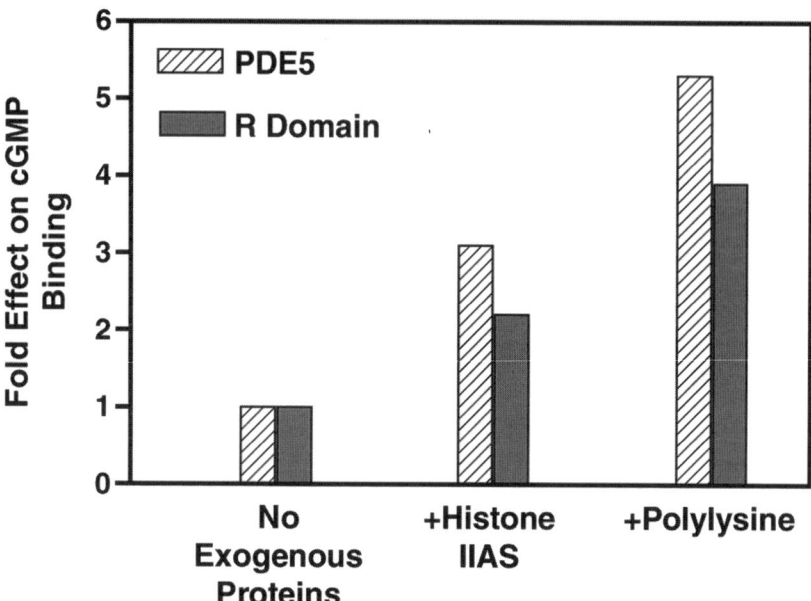

Fig. 6. Comparison of effect of histone IIA-S and polylysine on allosteric cGMP binding by full-length PDE5 and isolated R domain. (**A**) Purified recombinant His-tagged bovine PDE5 in 50 mM Tris, pH 7.5, and 1 mM EDTA was incubated with 30 mM DTT for 18 h at 4°C. The enzyme (0.012 µg [120 nM]) was then incubated for 1 h on ice in a reaction mixture containing 0.75 µM [^{3}H]cGMP, KP buffer (pH 6.8), and 1 mM EDTA in the presence and absence of histone IIA-S or poly-L-lysine at a final concentration of 2 mg/mL. Enzyme was also incubated with a combination of histone IIA-S (2 mg/mL) and varied concentrations of NaCl (30–450 mM). Reaction mixtures were then filtered using method A (**Fig. 3**). The purified His-tagged R domain of human PDE5 (Met1-Glu 540) (0.008 µg [120 nM]) was incubated as described for PDE5.

protein that is retained on the membrane with the total applied *(20)*. The effects of increasing concentrations of various histones, polylysine, and ovalbumin on cGMP binding by purified recombinant His-tagged bovine PDE5 are shown in **Fig. 5**. To date, we have found the following order of effectiveness for the following proteins: poly-L-lysine > histone VII-S > histone IIA-S > histone VI-S. Addition of ovalbumin, BSA, or phosvitin has no significant effects. The effectiveness of histone IIA-S or polylysine in increasing cGMP binding is comparable for the isolated His-tagged R domain of PDE5 and the recombinant holoenzyme (**Fig. 6**), but a cGMP-binding construct containing only the GAF a domain is inhibited by the addition of polylysine or histone IIA-S (not shown). This emphasizes, once again, the specific requirements of different proteins and even domains within the same protein for optimal binding.

On a sucrose density gradient (5–20%), recombinant PDE5 holoenzyme diluted in 10 mM KP, 25 mM β-mercaptoethanol containing BSA (0.17 mg/mL) sediments slightly faster than the internal standard, phosphorylase b (8.2 S). The sedimentation coefficient of PDE5 derived from the sucrose gradient is 9.1 S, which compares well with the value for native PDE5 of 9.35 S *(8)*. Combining PDE5 that has been diluted in buffer containing BSA with histone IIA-S (either 0.2 or 0.5 mg/mL) in the presence of 10 mM KP, pH 6.8, yields a solution that is slightly turbid, but the catalytic activity of the enzyme is not significantly affected (**Fig. 7A**). However, following centrifugation of the PDE5-BSA-histone mixture for 1 h at 115,000g at 4°C, almost all of the enzyme activity is lost from the supernatant (not shown), and a portion of the activity is recovered in the pellet following resuspension of the pellet in buffer (**Fig. 7B**). In samples containing only PDE5 and BSA, essentially all of the PDE5 activity remains in the supernatant and little is recovered in the pellet (**Fig. 7B**).

Likewise, following sucrose gradient centrifugation (5–20% gradient centrifuged for 18 h at 150,000g), PDE5 that is combined with both BSA and histone cannot be detected in fractions collected from the gradient. Rather, a small portion of the activity is found in the pellet (not shown). Despite the effect of low ionic strength on [^{3}H]sildenafil binding (*see* **Subheading 3.4.2.**) inclusion of 30 mM NaCl only slightly alters the distribution and recovery of the PDE5 in the pellet and the supernatant in either centrifugation (**Fig. 7B**). A modest portion of the PDE5 catalytic activity can be recovered when the pellet is resuspended in buffer, but the addition of 1% Triton to the pellet does not increase the recovery. In addition, when the PDE5-histone mixture is applied to a Sephacryl S300 column equilibrated in either 10 mM KP (pH 6.8); 25 mM β-mercaptoethanol; or 50 mM Tris-HCl (pH 7.5), 1 mM EDTA, the complex does not elute but appears to remain at the top of the column (not shown).

In a recent study, we demonstrated that the addition of histone IIA-S to the [^{3}H]sildenafil binding assay significantly increased retention of [^{32}P]PDE5 on a nitrocellulose and glass-fiber filter; histone IIA-S was also required to prevent adherence of [^{3}H]sildenafil to the tube surface *(9)*. For the allosteric [^{3}H]cGMP-binding assay, ligand adherence to the tube surface is not a problem, so the effects of histones on cGMP-binding are primarily related to improved retention of the enzyme on the filter owing to formation of a complex between PDE5 and histone. Formation of this complex occurs in the absence or presence of BSA. The two binding assays also differ in that cGMP binding by PDE5 progressively increases at concentrations of histones or polylysine from 0.1 to 2 mg/mL (**Fig. 5**); sildenafil binding is maximal at 0.2 mg/mL of histone IIA-S and decreases sharply at higher concentrations. As demonstrated by ultracentrifugation studies and by gel filtration chroma-

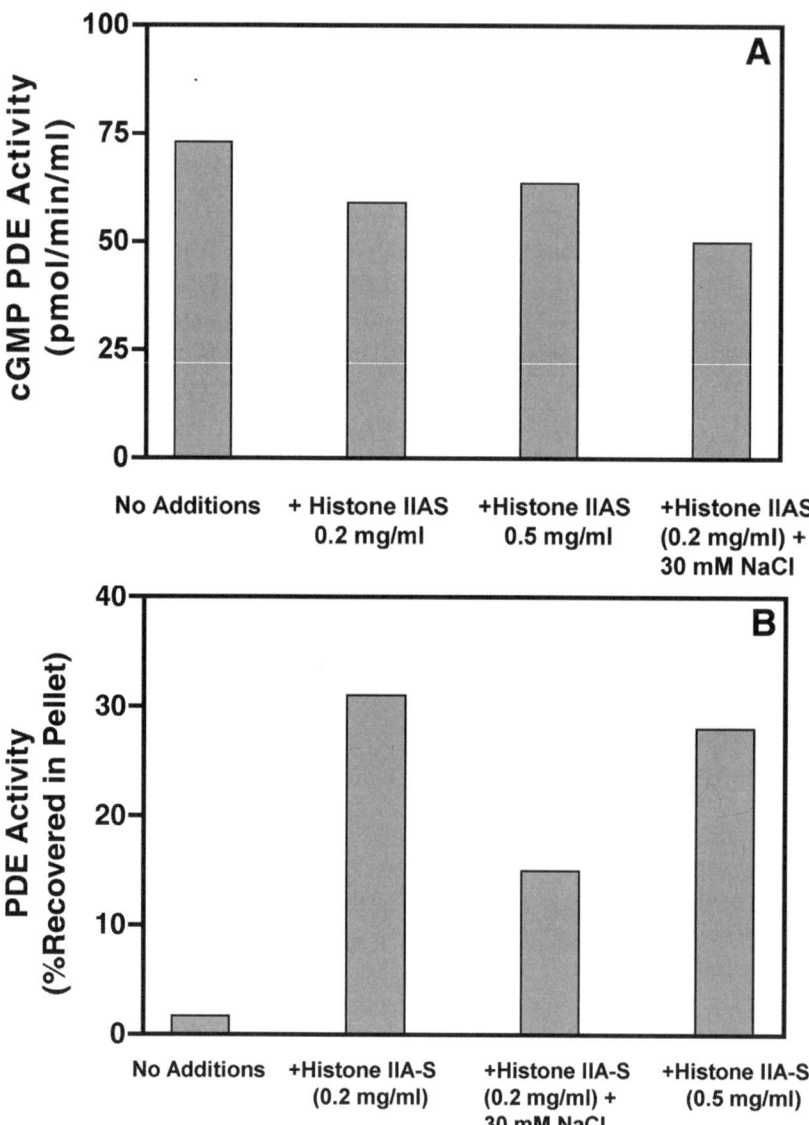

Fig. 7. Effect of histone and salt on PDE5 activity and solubility. Purified recombinant His-tagged bovine PDE5 was diluted in ice-cold KP, 25 mM β-mercaptoethanol, and BSA (1 mg/mL). (**A**) The diluted PDE5 (0.006 mg/mL [6 nM]) was then combined with histone IIA-S (either 0.17 or 0.43 mg/mL) or histone IIA-S (0.17 mg/mL) and 30 mM NaCl and assayed. The samples were next subjected to centrifugation at 115,000g for 1 h at 4°C, and the supernatant was removed and assayed. Essentially all PDE5 catalytic activity was lost from the supernatant in samples containing histone (not shown). (**B**) A small portion of the PDE5 activity was recovered in the pellet from each centrifuge tube following resuspension of the pellets in KPM containing BSA (1 mg/mL).

tography, PDE5 and histone IIA-S form a large complex over a broad range of histone concentrations (not shown). Allosteric cGMP binding by PDE5 in the presence of histone IIA-S is relatively insensitive to salt. Half-molar (0.5 M) NaCl inhibits binding by approx 33%, but 0.15 M salt has no effect (not shown).

3.3. Binding of [³H]PDE5 Inhibitors to Catalytic Site of PDE5

Both methods A and B have been used to study binding of radioligands to the catalytic site of PDE5. The protocol for method A is basically the same as described in **Subheading 3.1.** but with slight changes. We have developed a method for purifying and radiolabeling sildenafil from Viagra tablets *(10)*, and we have used it to study [³H]sildenafil binding to the catalytic site of PDE5. We have also synthesized tadalafil and obtained [³H]tadalafil via tritiation by Amersham.

3.3.1. Protocol for Method A

The general protocol for binding of [³H]PDE5 inhibitors to PDE5 and PDE11 is as follows:

1. Pre-wet a nitrocellulose filter by allowing 1 to 2 mL of filtration buffer to pass through. For this we use ice-cold KPT buffer.
2. Add the reaction mixture containing the radioligand alone (blank) or in combination with binding protein (we typically dilute PDE5 in KPM–1 mg/mL of BSA) in an appropriate ice-cold buffer and place 2 mL of this solution in a glass tube for a single reaction. The presence of certain proteins, such as histone IIA-S, prevents nonspecific binding of [³H]sildenafil to the surface of the tubes and increases filter retention of PDE5 and other PDEs.
3. Incubate the reaction mixture for an appropriate time (*see* **Note 4**).
4. Rapidly apply the reaction mixture onto the nitrocellulose filter by pipetting the mixture with a glass Pasteur pipet. Rinse the filter once with 3 mL of the prewetting buffer, and then remove the filter and place it in a vial box.
5. Dry the filter, place in a 6-mL scintillation vial containing 5 mL of nonaqueous scintillant, and then count in a scintillation counter.

3.3.2. Typical [³H]Sildenafil Binding Assay Using Method A

Add various concentrations of [³H]sildenafil to the tubes containing the ice-cold KPM-histone mixture. Next, start the reaction by the addition of purified His-tagged bovine PDE5 (80 µL of PDE5 diluted in KPM-BSA; 0.8 nM final PDE5 monomer concentration). This order of addition minimizes [³H]silde-nafil adsorption to the tube surface. Histone at 0.2 mg/mL has the optimum effect in blocking adherence. In a separate experiment, using [³²P]PDE5, we determined that histone increased the retention of PDE5 on the membranes, whereas NaCl concentrations of 0.02 *M* was strongly inhibitory on retention under the conditions of this particular assay (not shown).

Incubate the reaction mixtures in an ice-water bath for 20 min. For each individual filtration quickly add 200 µL of 25% Triton X-100 (2.2% final concentration) and then rapidly filter the samples (under vacuum) through a Millipore nitrocellulose filter (0.45 µm) that has been prewet with 1.5 mL of cold KPT. Inclusion of Triton X-100 in the prewash and wash buffers significantly lowers blank values for [³H]sildenafil and [³H]tadalafil binding. Rinse each tube once with 3 mL of the same buffer and filter. Remove the filter and place into slots of a scintillation vial box for drying and counting.

Using purified [³²P]PDE5 prepared by phosphorylation in the presence of cGMP, Mg[³²P]ATP, and the catalytic subunit of PKA followed by Sephadex G-25 chromatography, recovery of PDE5 by Millipore filters was determined to be 75% (not shown). Accordingly, [³H]sildenafil binding values were corrected for 25% loss of PDE5 through the filter. To correct for quenching of the radiolabel by the nitrocellulose filter, the [³H]sildenafil is spotted directly on the filter and counted as described in **Subheading 3.1.** above for [³H]cGMP.

3.3.3. Preparation of Method B Apparatus

This method uses the Brandel M24T-Cell space Harvester, which simultaneously filters 24 samples.

1. Wash the Brandel unit with a 50–70% ethanol solution. Rinse thoroughly with Millipore deionized water (~5 L) to remove the ethanol completely, as it can have deleterious effects on the assay.
2. Label three tube racks, each containing 24 glass tubes, Prewash, Wash, and Samples.
3. Place 1.5 mL of ice-cold KPT buffer into each tube of the Prewash rack, and add 6 mL of ice-cold KPT buffer to each tube of the Wash rack. Place both racks in an ice-water bath.

As with Method A, we have found that a low concentration of Triton X-100 in both the prewash and wash buffer significantly lowers the blank.

3.3.4. Typical [³H]Sildenafil Binding Assay Using Method B

1. Dilute the radioligand in an appropriate volume of KPM-His, transfer 2 mL of this solution to each tube in the Samples rack, and place the rack on ice. As with method A, we have found that certain proteins (e.g., histones) increase retention of ligand/enzyme complexes on the filter.
2. Dilute the PDE protein as described in method A and initiate the binding reaction by adding the PDE.
3. Incubate for approx 20–30 min on ice (*see* **Note 4**). Then place the glass filter in the filter apparatus, and close the apparatus. Next, remove the Prewash rack from the ice bath and place it on the sample rack holder as shown in **Fig. 8**.

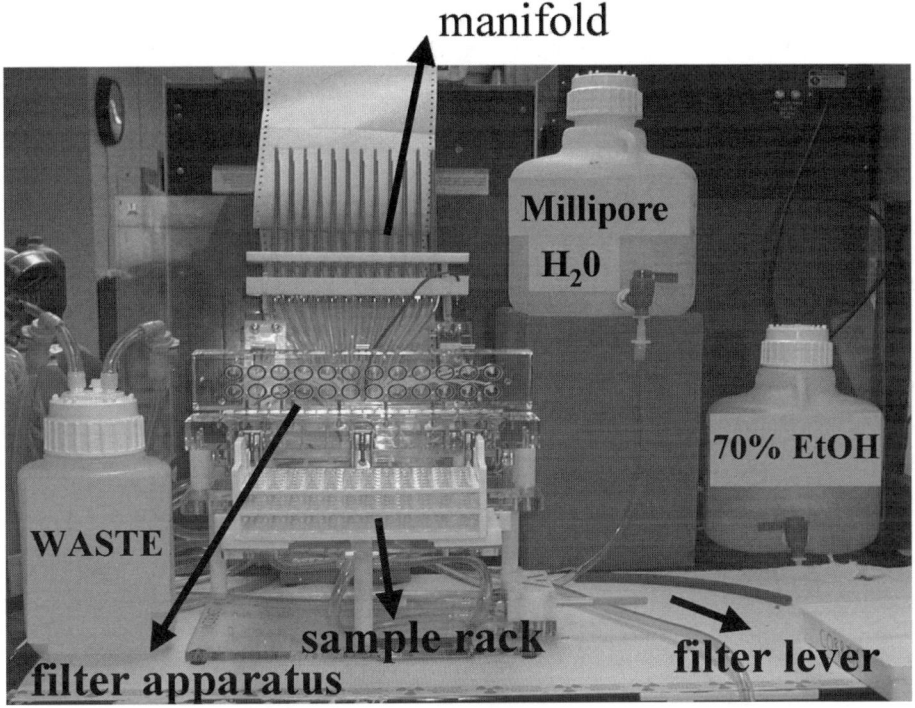

Fig. 8. Brandel multiple-well filtration apparatus used in method B.

4. Filter the Prewash rack by turning the vacuum source on and moving the filter lever to the Harvest position. After the prewash buffer has passed through the filter, move the filter lever back to the off position.
5. Place the Samples rack on the rack holder and immediately filter as described in **step 4**. Adjust the vacuum so that the filtering process is as fast as possible; this minimizes dissociation of the ligand from the enzyme.
6. Place the Wash rack on the rack holder and filter. Turn the vacuum source off, open the filter apparatus, remove the individual filter circles with forceps, and dry and count as described in **Subheading 3.3.1.**

3.4. Characteristics of [³H]PDE5 Inhibitors Binding to PDE5

This section deals with characterization of the effect of various conditions on the binding of tritiated PDE5 inhibitors to PDE5, focusing mainly on [³H]sildenafil.

3.4.1. Time Required to Reach Equilibrium

In studies using 3.6 nM PDE5 and subsaturating [³H]sildenafil, binding of the inhibitor to PDE5 reaches equilibrium within 1 min of incubation at 0–4°C. Using 30 nM [³H]sildenafil, binding is linear with increasing concentrations of

PDE5 up to 23 n*M* *(9)*. However, low concentrations of PDE5 (0.35 n*M*) require a longer time period (30–60 min) to reach equilibrium (unpublished results).

3.4.2. Effect of Ionic Strength

Binding of [³H]sildenafil to PDE5 requires low-ionic-strength conditions when using method A; 50 mm NaCl reduces [³H]sildenafil binding by approx 85% as assessed by method A *(9)*, primarily owing to decreased retention of PDE5 on the filter. A similar inhibition by 50 m*M* NaCl is observed for [³H]tadalafil binding to PDE5 (unpublished data). Increasing ionic strength has less effect using method B because 50 m*M* NaCl inhibits [³H]sildenafil binding by only 40% (not shown). Interestingly, preliminary data suggest that [³H]tadalafil binding to PDE11 is largely unaffected by NaCl; 50 mm NaCl inhibited [³H]tadalafil binding to PDE11 by only 10% using method B. In the absence of NaCl, the KP buffer used for the assay also was inhibitory at concentrations >10 m*M*. At 30 m*M* KP, [³H]sildenafil binding was inhibited by 90% compared with binding obtained with 10 m*M* KP, which yields the same values as 5 m*M* KP. Buffer conditions are likely to differ among various PDEs; thus, the investigator should determine the optimal buffer conditions for each protein and ligand being studied.

3.4.3. Effect of Temperature on [³H]Inhibitors Binding to PDE5

There is very little effect of temperature i.e. 4° vs 30°C on the K_D values of PDE5 for [³H]sildenafil or [³H]tadalafil but total binding is decreased substantially at 30°C (unpublished results). At 30°C the K_D values for some other PDE5 inhibitors are significantly lower (higher affinity) than at 4°C (unpublished results). This is true despite the close structural similarities of some of these compounds.

3.4.4. Effects of Exogenous Proteins, EDTA, and Nonionic Detergents

The effects of certain proteins and other substances on [³H]sildenafil binding to PDE5 are complex. As discussed earlier, histone in the assay helps to prevent [³H]sildenafil from sticking to the surface of the glass tubes and also increases retention of PDE5 on the filter. The resulting effect of histone IIA-S is to robustly stimulate recovery of [³H]sildenafil bound to PDE5, and the optimal concentration of histone IIA-S that produces these effects for PDE5 is 0.2 mg/mL. Also of note is that histone IIA-S does not significantly affect the catalytic activity of PDE5 (**Fig. 7A**), suggesting that the integrity of the catalytic site is maintained in the presence of histone IIA-S. At high concentrations of a purified PDE, or when using a crude preparation, the addition of exogenous proteins may have a reduced effect on the measurement of ligand binding. Using pure PDE5 (diluted in KPM-BSA before adding to reaction tubes), it was found that the presence of both BSA and histone is necessary for the efficient retention of counts. For example, in a binding assay containing 0.22 nM PDE5 and 0.2 mg/mL of histone IIA-S, retention of counts on the filter progressively increased up to 0.1 mg/mL of BSA and achieved a stoichiometry of approx 1 mol of [^{3}H]sildenafil/mol of PDE5 (unpublished results). BSA alone does not form a complex with PDE5 (**Fig. 7B**). Rather, this effect may occur through a PDE5-histone-BSA complex. The addition of 10 mM EDTA in the reaction mixture strongly inhibits [^{3}H]sildenafil binding to PDE5 (*9*), most likely owing to removal of endogenous metal(s) retained by the recombinant PDE5 during its purification. This finding further supports the interpretation that sildenafil is specific for the catalytic site. Furthermore, it suggests that divalent metal is not only necessary for PDE5 catalysis but also contributes to the overall integrity of the portion of the catalytic site that binds cGMP analogs such as sildenafil. However, this is not true for all cyclic nucleotide analogs or inhibitors because IBMX binds to the catalytic site and stimulates allosteric cGMP binding in PDE5 that has been exhaustively purified in the presence of EDTA.

The effects of nonionic detergents, such as Triton X-100, depend on the properties of the radioligand used. Low concentrations of Triton X-100 lower blanks when assaying [^{3}H]sildenafil binding to PDE5. However, for measuring [^{3}H]cGMP binding to PDE5, we have found that Triton X-100 actually increases the blank.

3.4.5. Specificity and Regulation of [^{3}H]Sildenafil Binding to PDE5 Catalytic Site

Specificity for [^{3}H]sildenafil binding to the catalytic site of PDE5 was examined by testing the effects of various compounds on [^{3}H]sildenafil binding (**Fig. 9**). Binding of [^{3}H]sildenafil to PDE5 was competed by unlabeled sildenafil, IBMX, T-0156 (PDE5 inhibitor, courtesy of Tanabe Seiyaku), and cGMP. cyclic AMP or 5′GMP, compounds that do not interact effectively with the PDE5 catalytic site, do not affect [^{3}H]sildenafil binding to PDE5. Furthermore, a recombinant construct containing the full R domain of PDE5 does not bind [^{3}H]sildenafil, although it binds [^{3}H]cGMP stoichiometrically. Taken together, these data indicate that sildenafil interacts specifically with the PDE5 catalytic site. Using this binding assay, the K_D values for [^{3}H]sildenafil and [^{3}H]tadalafil binding to PDE5 have been determined to be 3.8 and 1.8 nM, respectively, which compare well with the IC$_{50}$ values of these compounds for inhibition of PDE5 catalytic activity.

Regulation of ligand interaction with the PDE5 catalytic site can also be determined by this approach. As shown in **Fig. 10**, [^{3}H]sildenafil binding to PDE5 is stimulated up to fivefold by low micromolar concentrations of cGMP. We have proposed that this results from allosteric cGMP binding to the GAF

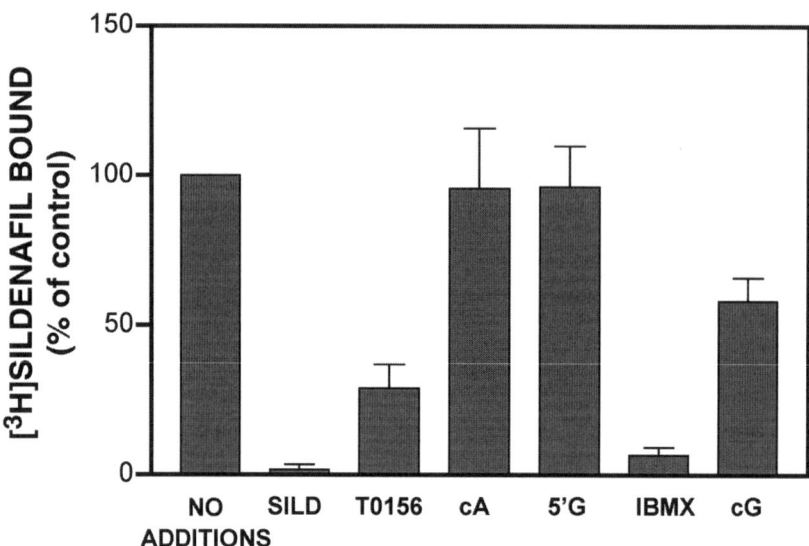

Fig. 9. Effects of nucleotides and unlabeled sildenafil on [^{3}H]sildenafil binding to PDE5. His-tagged bovine PDE5 (final concentration of 0.8 n*M*) was added to 2 mL of 10 m*M* KPM containing 0.2 mg/mL of histone IIA-S. [^{3}H]Sildenafil (final concentration of 6 n*M*) was added to the KPM-histone mixture before starting the reaction with the enzyme. Reactions were incubated in an ice-water bath for 20 min. To each, 200 μL of 25% Triton X-100 (2.2% final concentration) was added, and the samples were filtered according to method A. The Millipore nitrocellulose membranes had been prewet with 1.5 mL of cold KP containing 0.1% Triton X-100. Tubes were each rinsed with 3 mL of the same buffer and filtered. Binding units are based on the picomoles per milliliter of PDE5 added to the reaction. The concentrations of competing compounds were as follows: 0.083 μ*M* unlabeled sildenafil (SILD), 0.083 μ*M* T0156 (PDE5 inhibitor), 1.4 m*M* cAMP (cA), 1.4 m*M* 5'-GMP (5'G), 0.36 m*M* IBMX, and 1.4 m*M* cGMP (cG).

domains, which stimulates function of the C domain. A similar stimulation may be seen for other allosterically regulated proteins.

3.5. Advantages and Limitations of Methods A and B

Both methods A and B share a few limitations that are universal to vacuum-filtration methods. The first disadvantage is the possibility of variations in the pressure load of the house vacuum. We have not found this to be a significant factor, but this potential problem can be easily overcome by using a pump. The second disadvantage is that with vacuum-filtration methods, some dissociation of the radioligand-PDE complex invariably occurs during filtration. If filtration conditions are sufficiently rapid, dissociation is usually not a significant problem

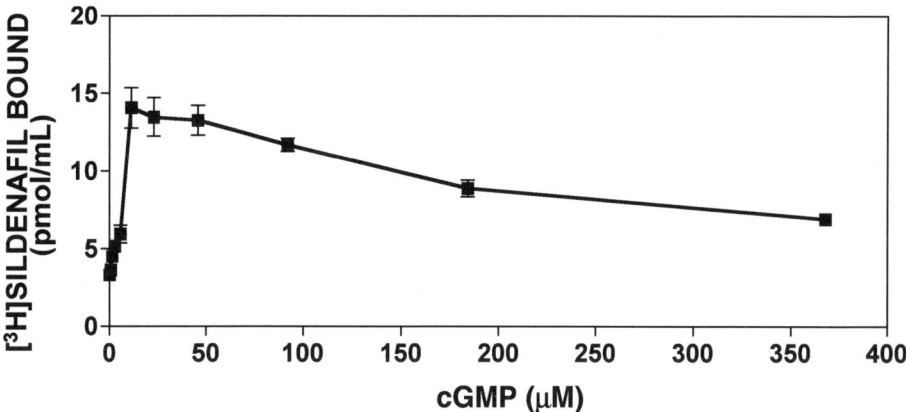

Fig. 10. cGMP binding to PDE5 allosteric sites increases [³H]sildenafil binding to the catalytic site of PDE5. Experiments were performed using method B and 5 nM [³H]sildenafil. The cGMP concentration in the assay mixture was varied between 0 and 368 µM, and [³H]sildenafil was 5 nM. The final concentration of PDE5 was 1.5 nM.

with radioligands of reasonably high affinity. Based on theoretical calculations, it has been suggested that the maximum allowable K_D for vacuum filtration is approx 10 nM. However, we have found the methods described herein to be effective with K_Ds ranging from 0.1 to 1000 nM and have obtained good stoichiometry throughout this range. A major limitation of binding assays in general is that they are stoichiometric in nature and thus are much less sensitive than catalytic assays. Despite these potential limitations, these methods are simple and fast and allow measurement of a number of important parameters (*see* **Subheading 1.1.**).

Methods A and B, although relatively similar, have some important differences with respect to their applications. Method A, a single-filter format, is more amenable to studies requiring multiple time points (e.g., dissociation and association kinetics). It is also much less expensive than method B and requires less space allocation. Method B can be used for such studies, but more time is required for filtration than with method A, and, thus, studies with very closely spaced time points are not possible. For studies with longer time-point intervals or single time-point studies, method B is well suited. In comparing [³H]sildenafil binding to PDE5 between method A and B the following advantages are noted for method B:

1. Nonspecific binding (blanks) is routinely twofold lower.
2. Recovery (percentage of PDE5 retained on the filter) approaches 100% for method B but is only approx 75% for method A.

3. Stoichiometry (mol of [³H]sildenafil/mol of PDE5) is roughly twofold higher using method B.
4. Variability between replicates is routinely lower using Method B.

Method A has the advantages that the reaction volume can be manipulated to fit the particular experiment and that lower quantities of enzyme and reagents are needed. Method B functions optimally at larger reaction volumes (about 2 mL). Thus, if only very limited amounts of protein or radioligand are available, method A is better suited. Using the same technique for calculating the recovery of PDE5 in method A, recovery was 97% using method B, and the blank was at least two times lower using method B. Maximum [³H]sildenafil binding stoichiometry was 0.31 and 0.61 mol/PDE5 subunit using methods A and B, respectively. This finding suggests that some bound [³H]sildenafil is lost during filtration, particularly with method A. For [³H]cGMP binding with method A, the blanks are typically very low, and recovery of protein varies depending on the construct being studied.

4. Notes

1. Although the Millipore filtration apparatus is sold as a set that includes the fritted disk base and the 15-mL funnel, we have found that the upper and lower portions of this device are not always well matched. Great care should be taken in optimizing the fit between the pieces, and it may be necessary to purchase several sets of the filtration device in order to find a combination that provides the best result. A good fit can usually be discerned by the symmetry of the wet circle after removal from the apparatus.
2. A weak vacuum increases the filtration time and increases the likelihood of significant ligand dissociation during the filtration process. The filtration should produce a well-formed circle. Irregular edges, resulting from poor reservoir-filter compatibility, should be avoided, because they may result in spurious background counts.
3. Stop mix is easily prepared by mixing 594 mg of theophylline, 25 mL of 200 m*M* EDTA (pH 6.8), 10 mL of 1 *M* Tris-HCl (pH 7.5), 440 mg of cGMP, and 370 mg of cAMP in a final volume of 50 mL.
4. The time of incubation required to reach binding equilibrium must be determined experimentally for each protein and set of conditions. This can depend on variables such as K_D between the radioligand and the enzyme, incubation temperature, and pH.

Acknowledgments

This work was supported by National Institutes of Health (NIH) research grants DK40029 and DK 58277, NIH training grant 5T32HL07752, American Heart Association (Southeast affiliate), and Pfizer.

References

1. Charbonneau, H. (1990) Structure-function relationships among cyclic nucleotide phosphodiesterases, in *Cyclic Nucleotide Phosphodiesterases: Structure, Regulation, and Drug Action* (Beavo, J. and Houslay, M. D., eds.), Wiley, New York, pp. 267–296.

2. McAllister-Lucas, L. M., Sonnenburg, W. K., Kadlecek, A., et al. (1993) The structure of a bovine lung cGMP-binding, cGMP-specific phosphodiesterase deduced from a cDNA clone. *J. Biol. Chem.* **268,** 22,863–22,873.

3. Burns, F., Zhao, A. Z., and Beavo, J. A. (1996) Cyclic nucleotide phosphodiesterases: gene complexity, regulation by phosphorylation, and physiological implications. *Adv. Pharmacol.* **36,** 29–48.

4. Francis, S. H. and Corbin, J. D. (1999) Cyclic nucleotide-dependent protein kinases: intracellular receptors for cAMP and cGMP action. *Crit. Rev. Clin. Lab. Sci.* **36,** 275–328.

5. Francis, S. H., Turko, I. V., and Corbin, J. D. (2001) Cyclic nucleotide phosphodiesterases: relating structure and function. *Prog. Nucleic Acid Res. Mol. Biol.* **65,** 1–52.

6. Aravind, L. and Ponting, C. P. (1997) The GAF domain: an evolutionary link between diverse phototransducing proteins. *Trends Biochem. Sci.* **22,** 458–459.

7. Francis, S. H., Lincoln, T. M., and Corbin, J. D. (1980) Characterization of a novel cGMP binding protein from rat lung. *J. Biol. Chem.* **255,** 620–626.

8. Thomas, M. K., Francis, S. H., and Corbin, J. D. (1990) Characterization of a purified bovine lung cGMP-binding cGMP phosphodiesterase. *J. Biol. Chem.* **265,** 14,964–14,970.

9. Corbin, J. D., Blount, M. A., Weeks, J. L. 2nd, Beasley, A., Kuhn, K. P., Ho, Y. S., Saidi, L. F., Hurley, J. H., Kotera, J., and Francis, S. H. (2003) [³H]Sildenafil binding to phosphodiesterase-5 is specific, kinetically heterogeneous, and stimulated by cGMP. *Mol. Pharmacol.* **63,** 1364–1372.

10. Francis, S. H., Sekhar, K. R., Rouse, A. B., Grimes, K. A., and Corbin, J. D. (2003) Single step isolation of siblenefil from commercially available Viagra tablet. *Int. J. Impot. Res.* **15,** 369–372.

11. Liu, L., Underwood, T., Li, H., Pamukcu, R., and Thompson, W. J. (2001) Specific cGMP binding by the cGMP binding domains of cGMP-binding cGMP-specific phosphodiesterase. *Cell. Signal.* **13,** 1–7.

12. Corbin, J. D. and Francis, S. H. (1999) Cyclic GMP phosphodiesterase-5: target of sildenafil. *J. Biol. Chem.* **274,** 13,729–13,732.

13. Francis, S. H., Thomas, M. K., and Corbin, J. D. (1990) Cyclic GMP-binding cyclic GMP-specific phosphodiesterase from lung, in *Cyclic Nucleotide Phosphodiesterases: Structure, Regulation, and Drug Action* (Beavo, J. and Houslay, M. D., eds.), John Wiley & Sons, New York, pp. 117–140.

14. Corbin, J. D. and Doskeland, S. O. (1983) Studies of two different intrachain cGMP-binding sites of cGMP-dependent protein kinase. *J. Biol. Chem.* **258,** 11,391–11,397.

15. Rannels, S. R. and Corbin, J. D. (1979) Characterization of small cAMP-binding fragments of cAMP-dependent protein kinases. *J. Biol. Chem.* **254,** 8605–8610.

16. Rannels, S. R. and Corbin, J. D. (1980) Two different intrachain cAMP binding sites of cAMP-dependent protein kinases. *J. Biol. Chem.* **255,** 7085–7088.

17. Francis, S. H., Noblett, B. D., Todd, B. W., Wells, J. N., and Corbin, J. D. (1988) Relaxation of vascular and tracheal smooth muscle by cyclic nucleotide analogs that preferentially activate purified cGMP-dependent protein kinase. *Mol. Pharmacol.* **34,** 506–517.

18. Turko, I. V., Haik, T. L., McAllister-Lucas, L. M., Burns, F., Francis, S. H., and Corbin, J. D. (1996) Identification of key amino acids in a conserved cGMP-binding site of cGMP-binding phosphodiesterases: a putative NKXnD motif for cGMP binding. *J. Biol. Chem.* **271,** 22,240–22,244.

19. Francis, S. H. (1985) Effectors of rat lung cGMP binding protein-phosphodiesterase. *Curr. Top. Cell. Regul.* **26,** 247–262.

20. Thomas, M. K., Francis, S. H., and Corbin, J. D. (1990b) Substrate- and kinase-directed regulation of phosphorylation of a cGMP-binding phosphodiesterase by cGMP. *J. Biol. Chem.* **265,** 14,971–14,978.

19

Analysis of Dimerization Determinants of PDE6 Catalytic Subunits

Khakim G. Muradov, Kimberly K. Boyd, and Nikolai O. Artemyev

Summary

An absolute majority of cyclic nucleotide phosphodiesterases (PDEs) form catalytic dimers. The structural determinants and functional significance of PDE dimerization are poorly understood. Furthermore, all known dimeric PDEs with the exception of retinal rod guanosine 3′,5′-cyclic-monophosphate PDE (PDE6) are homodimeric enzymes. Rod PDE6 is a catalytic heterodimer composed of α- and β-subunits. Gel filtration, sucrose gradient centrifugation, and immunoprecipitation are standard techniques used to study dimerization of proteins. We successfully applied these methods to investigate dimerization of chimeric proteins between PDE6αβ and PDE5, which allowed us to elucidate the structural basis for heterodimerization of rod PDE6. This chapter outlines approaches to the investigation of PDE6 dimerization that can be utilized in a broader analysis of PDE dimerization.

Key Words

Rod and cone PDE6; dimerization; guanosine 3′,5′-cyclic-monophosphate; gel filtration; sucrose gradient centrifugation; immunoprecipitation.

1. Introduction

Retinal guanosine 3′,5′-cyclic-monophosphate (cGMP) phosphodiesterases (PDEs) comprise the PDE6 family among 11 PDE families that have been recognized in mammalian tissues based on primary sequence, substrate selectivity, and regulation *(1,2)*. It has been long established that the vast majority of PDEs are homodimeric enzymes *(1,2)*. Still, the role of dimerization in enzyme function remains unknown. The monomeric catalytic domains of several PDEs have been isolated and shown to be catalytically active, suggesting that dimerization is not required for catalytic activity *(3,4)*. Similar to other dimeric PDEs, cone PDE6 catalytic dimer is composed of two identical PDEα′-subunits *(5)*. By contrast,

From: *Methods in Molecular Biology, vol. 307: Phosphodiesterase Methods and Protocols*
Edited by: C. Lugnier © Humana Press Inc., Totowa, NJ

rod PDE6 catalytic α- and β-subunits form a heterodimer, PDE6αβ *(6–8)*. Rod and cone PDE6 are the key phototransduction enzymes that regulate intracellular cGMP concentration on photoexcitation of photoreceptor cells *(9,10)*. The functional significance of heterodimerization of rod PDE in visual signaling is unclear. It may contribute to the kinetics of rod-specific photoresponses. However, a potential role of PDE6αβ heterodimers cannot be assessed in the absence of comparison with homodimeric PDE6 species αα and ββ. Such an analysis may become possible with identification of dimerization determinants of PDE6.

Our approach to investigate dimerization determinants of PDE6αβ was to construct chimeras between PDE6αβ and cGMP-specific PDE5 containing candidate dimerization regions of PDE6αβ *(11)*. PDE6 contains two N-terminal GAF domains (GAFa and GAFb) located upstream of the C-terminal catalytic domain. GAF domains have been recognized as a large family of domain homologs and are named for their presence in cGMP-regulated PDEs, adenylyl cyclases, and the *Escherichia coli* protein FhlA *(12)*. Besides PDE6 *(13)*, several other PDE families possess GAF domains, including cGMP-stimulated PDE (PDE2) and PDE5 *(14–16)*. Existing biochemical evidence indicates that dimerization of PDE2 and PDE5 occurs within the N-terminal parts of the molecules *(16,17)*. A recent crystal structure of PDE2A GAFa-GAFb domains revealed that the PDE2A regulatory region forms a dimer with the interface formed by the two GAFa domains *(18)*. Therefore, PDE6/PDE5 chimeras expressed for studying PDE6 dimerization have been designed to include various regions from the PDE6 GAFa/GAFb domains. These chimeras have been expressed in Sf9 cells in various combinations as His-, myc-, or flag-tagged proteins. We describe methods for the construction and expression of PDE6αβ/PDE5 chimeras and the application of gel filtration, sucrose gradient centrifugation, Western blotting, and immunoprecipitation for analysis of dimerization of these chimeras.

2. Materials.

1. pFastBacHTb vector (Life Technologies).
2. Restriction enzymes (New England Biolabs).
3. *E. coli* strains DH5α (Novagen) and DH10Bac (Gibco-BRL, Gaithersburg, MD) cells.
4. cGMP (Roche Molecular Biochemicals, Indianapolis, IN).
5. [^{3}H]cGMP (Amersham Pharmacia Biotech).
6. T4 DNA ligase (Roche Molecular Biochemicals).
7. Polymerase chain reaction (PCR) amplifier.
8. dNTP mix (25 m*M* of each) (Roche Molecular Biochemicals).
9. Agarose and agarose gel electrophoresis equipment.
10. AmpliTaq® DNA polymerase (Roche Molecular Biochemicals) and cloned Pfu DNA polymerase (Stratagene).

11. Qiagen Gel Extraction kit.
12. Luria-Bertani (LB)–agar dishes containing ampicillin (50 μg/mL).
13. Qiagen Miniprep kit.
14. Sf9 cells (Life Technologies).
15. TNM-FH insect medium for Sf9 cells (Sigma, St. Louis, MO).
16. Fetal bovine serum (HyClone).
17. Solution for washing Sf9 cells: 20 mM Tris-HCl (pH 8.0), containing 130 mM NaCl and 2 mM MgSO$_4$.
18. Cell lysis solution: 20 mM Tris-HCl buffer (pH 8.0) containing 2 mM MgSO$_4$.
19. Complete™ Mini protease inhibitor cocktail tablets (Roche Molecular Biochemicals).
20. 550 Sonic Dismembrator (Fisher).
21. Protein storing solution: 30 mM Tris-HCl buffer (pH 8.0) containing 130 mM NaCl, 2 mM MgSO$_4$, and 50% glycerin.
22. Equipment, reagents, and buffers for sodium dodecyl sulfate-polyacrylamide gel electrophoresis (SDS-PAGE).
23. Markers for SDS-PAGE "Dual Color Standards" (Bio-Rad, Hercules, CA).
24. High-performance liquid chromatography (HPLC) chromatography system.
25. Superose® 12 10/30 column (Amersham, Piscataway, NJ).
26. 30 mM Tris-HCl buffer (pH 8.0) containing 100 mM NaCl and 2 mM MgSO$_4$ for gel filtration.
27. Molecular weight protein standards for gel filtration and sucrose gradient centrifugation (Sigma): bovine thyroglobulin (670 kDa), horse ferritin (440 kDa), sweet potato β-amylase (200 kDa), rabbit aldolase (158 kDa), bovine serum albumin (BSA) (67 kDa), and chicken ovalbumin (45 kDa).
28. Enhanced chemiluminescence Western blotting detection reagents (Amersham).
29. Protein G–agarose beads (Sigma).
30. GradiFrac gradient former (Amersham).
31. Semidry blotting apparatus (Fisher).
32. Nitrocellulose membrane (Schleicher & Schuell).
33. Phosphate-buffered saline (PBS), pH 7.1.
34. PBS containing 0.1% Triton X-100 and 5% nonfat dry milk.
35. Monoclonal antipolyhistidine, M2 monoclonal antiflag, and monoclonal anti-c-myc (clone 9E10) antibodies, and antimouse antibodies conjugated to horseradish peroxidase (HRP) (Sigma).
36. 5–35% Sucrose density gradients in 30 mM Tris-HCl (pH 8.0) buffer containing 100 mM NaCl, 2 mM MgSO$_4$, and 4 mM 2-mercaptoethanol.

3. Methods

The following methods to study dimerization determinant of PDE6αβ are described: construction of PDE6αβ/PDE5 chimeras, expression of PDE6αβ/PDE5 chimeras in Sf9 cells, and assays to assess dimerization of the chimeric proteins (gel filtration, Western blotting, immunoprecipitation, and sucrose gradient centrifugation).

3.1. Construction of PDE6αβ/PDE5 Chimeras

3.1.1. Bac-to-Bac™ Baculovirus System for Expression of PDE6αβ/PDE5 Chimeras

Baculovirus expression systems have several important advantages for expression of PDE6αβ/PDE5 chimeras. They often provide high levels of heterologous gene expression. Moreover, simultaneous infection of insect cells with two or more viruses allows expression of heterooligomeric protein complexes. Most important, although an efficient functional expression system for the wild-type (WT) PDE6 has not yet been developed, PDE5 and cone PDE6αβ/PDE5 chimeras have been successfully expressed using a baculovirus/Sf9 cell system *(19,20)*. The Bac-to-Bac baculovirus expression system offers additional advantages allowing relatively rapid generation of recombinant viruses. Genes of interest, such as PDE cDNAs, are inserted into the multiple cloning site (MCS) of a pFastBac vector downstream of the baculovirus-specific promoter (**Fig. 1**). The plasmid represents a convenient template for mutagenesis of PDE6 and PDE5 cDNAs. A standard *E. coli* strain, DH5α, can be used for all manipulations during construction of PDE chimeras or mutants. Once the chimeric construct is prepared, the vector is used to transform DH10Bac *E. coli* cells that carry and propagate a baculovirus shuttle vector (bacmid). Site-specific transposition of the pFastBac vector into the bacmid produces the recombinant bacmid that allows a white/blue selection of colonies. Isolated recombinant bacmids are used to infect insect cells to generate recombinant viruses.

3.1.2. Approaches to Cloning PDE6ab/PDE5 Chimeras

Three vectors, pFastBacHTa, pFastBacHTb, and pFastBacHTc, are available for cloning of a gene of interest for expression using the Bac-to-Bac baculovirus system. The vectors contain sequences for the His6-tag, spacer region, and rTEV protease cleavage site, upstream of the MCS (**Fig. 1**). Restriction site *Rsr*II, upstream of the start codon and the His6-tag sequence, provides for convenient replacement of the His6-tag with other affinity tags such as a flag- or myc-tag. A DNA fragment can be amplified from the pFast-BacHTb template using a 5′ primer containing an *Rsr*II site and the tag sequence, and a 3′ primer containing one of the sites from the MCS, such as a *Bam*HI site. This fragment is then ligated with the large *Rsr*II/*Bam*HI fragment of pFastBacHTb to produce pFastBacflag or pFastBacmyc. The *Bam*HI and *Xho*I sites from the MCS are extremely useful for the construction of PDE6αβ/PDE5 chimeras because these sites are not found in cDNAs coding the PDE6αβ GAFa-GAFb domains and the PDE5 catalytic region, respectively. PDE6αβ GAFa/GAFb cDNA sequences can be amplified by PCR from a retinal cDNA library, and cDNA coding the C-terminal catalytic region of

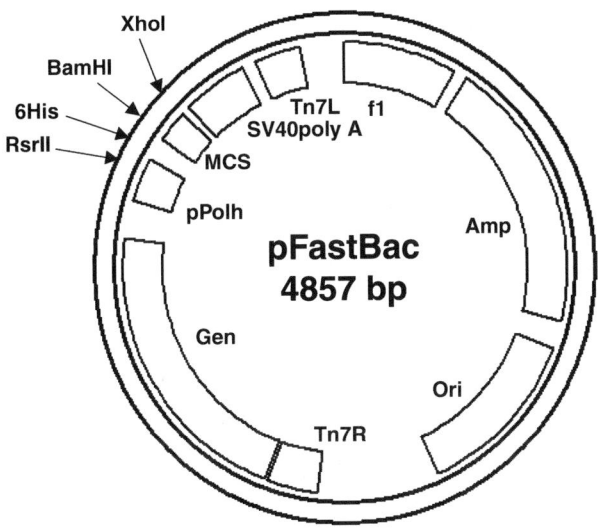

Fig. 1. pFastBac vector map and its features. pPolh, viral polyhedrin promoter; Amp and Gen, genes encoding resistance to ampicillin and gentamycin, respectively; Tn7R and Tn7L, right and left arms of the Tn7 transposition element, respectively. Arrows show the positions of the 6His tag and useful restriction sites.

PDE5 can be amplified using the pFastBacHTb-PDE5 template developed earlier *(20)*. It is important to select correctly a junction region where the PDE6αβ sequence is to be joined with the PDE5 sequence. Such junctions are preferred within regions of high homology between PDE6 and PDE5. Sequence alignment of PDE6α and PDE5 suggests one such junction between residues 1–443 of PDE6α and residues 506–865 of PDE5. The junction can be made using a *Hin*dIII restriction site. Accordingly, a chimera flag-α-α-5 (**Fig. 2**) is constructed by PCR amplification of DNA coding for PDE6α-1–443 with primers carrying the *Bam*HI and *Hin*dIII sites, and DNA coding for PDE5-506–865 is amplified with primers containing the *Hin*dIII and *Xho*I sites. The two PCR products are then ligated into the pFastBacflag vector using the *Bam*HI and *Xho*I sites. Analogously, chimera myc-β-β-5 (**Fig. 2**) is generated by ligation of the PCR-amplified PDE6β-1–441 and PDE5-506–865 DNAs into the *Bam*HI and *Xho*I sites of pFastBacmyc. PDE6αβ/PDE5 chimeras containing individual PDE6α/β GAFa or GAFb domains or their portions can be constructed by amplifying appropriate regions of the PDE6αβ-subunits with primers containing unique restriction sites. When unique restriction sites are not available at desired locations, first-round PCR products coding chimeric junctions can be extended to the nearest unique sites of the plasmid DNA in

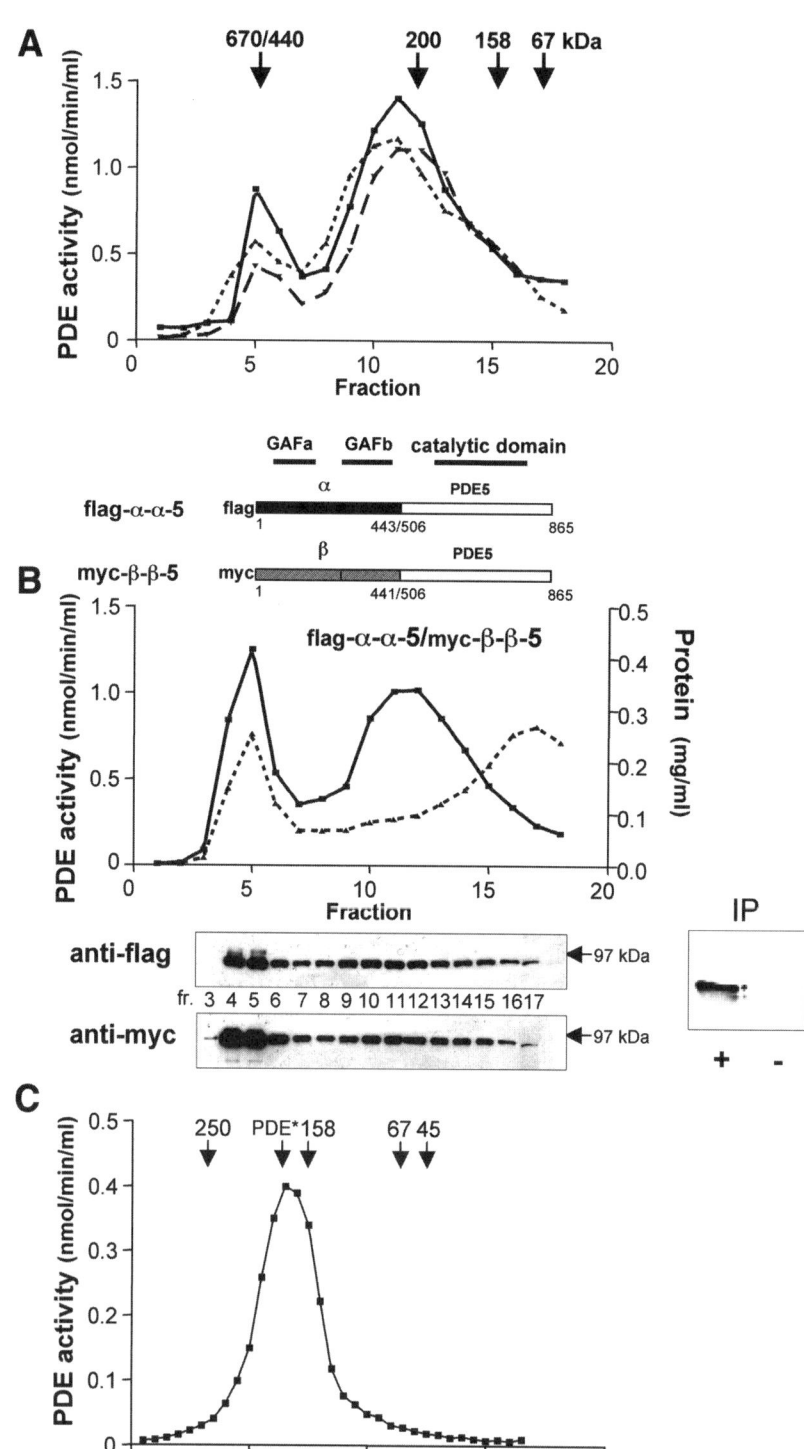

a second-round PCR amplification. An alternative approach to the construction of PDE6αβ/PDE5 chimeras may be used (*see* **Note 1**).

3.1.3. PCR-Based Cloning of PDE6αβ/PDE5 Chimeras

1. Design and order synthetic PCR primers carrying appropriate restriction sites or sequences coding the chimeric junctions. Primers coding chimeric junctions must have an 18- to 25-nucleotide sequence from each PDE subunit.
2. Prepare PCR reaction mixtures containing a template (10–50 ng of plasmid DNA or 100–200 ng of library DNA), 5′ and 3′ primer (1 μ*M* each), 0.2 m*M* dNTP mix, 5 μL of 10X PCR buffer, 5 U of Pfu DNA polymerase, and sterile distilled H_2O to a final volume of 50 μL.
3. Program a PCR amplifier and perform PCR amplifications as follows:
   a. First cycle: denaturing for 2 min at 95°C, annealing for 2 min at a temperature 10°C lower than the primer's melting temperature, and an extension at 72°C for a time depending on a size of amplified DNA (400-bp extension during 1 min).
   b. Subsequent 30 cycles: 50 s at 95°C, 50 s at annealing temperature, extension at 72°C for an appropriate time, and a final extension dwell at 72°C for 5–10 min.
4. Separate a PCR product on a 0.8–1.5% agarose gel and purify it using a standard Qiagen Gel Extraction kit. Incubate a purified PCR product with selected restriction enzymes for 2 to 3 h. If several restriction sites (enzymes) are suitable for

---

Fig. 2. *(opposite page)* Analysis of dimerization of PDE6αβ/PDE5 chimeras. **(A)** Gel filtration profiles of native rod holoPDE6, the wild-type PDE5 and α′-α′-5. Rod holoPDE6 (50 μg) (smaller dashed line) mixed with the dialyzed cell extract of non-infected Sf9 cells (100 μL), dialyzed cell extract of Sf9 cells (100,000*g*, 90 min) infected with virus for expression of PDE5 (50 μL) (solid line) and α′-α′-5 (larger dashed line) *(11)* were subjected to a fast protein liquid chromatography (FPLC) gel filtration on a Superose® 12 10/30 column (Amersham Pharmacia Biotech). Fractions of 0.4 mL were collected and analyzed for PDE activity. **(B)** Analysis of dimerization of PDE chimeras by gel filtration. Chimeras flag-α-α-5 and myc-β-β-5 were co-expressed in Sf9 cells and examined by gel filtration on a Superose® 12 column. Fractions of 0.4 mL were collected and analyzed for protein concentration (dashed line), PDE activity (solid line), and by Western blotting with anti-flag or anti-myc antibodies. **IP:** An aliquot (80 μL) of the dimeric PDE peak fraction 12 was incubated with (+) or without (–) anti-flag antibodies. Protein–antibody complexes were isolated using protein G-Agarose and analyzed by Western blotting with anti-myc antibodies. **(C)** Analysis of dimerization of PDE chimeras by sucrose gradient centrifugation. Combined peak PDE gel filtration fractions 10–13 were concentrated to a volume of 200 μL using YM-10 Microcon® devices (Millipore Corporation, Bedford, MA) and loaded onto the 5–35% sucrose density gradients. Following centrifugation for 24 h at 280,000*g* in a Beckman SW41 rotor, fractions of 300 μL were collected starting from the bottom of the tubes and analyzed for PDE activity. Arrows in **(A)** and **(C)** indicate elution or sedimentation of protein standards. PDE5 and holoPDE6 migrated at the same position of the gradient indicated as PDE*.

the mutagenesis, give the preference to the enzymes that are optimally active in the same restriction buffer. If simultaneous digestion is not possible, first cut a PCR fragment with one enzyme, precipitate using 0.7 vol of isopropanol, dissolve in HPLC-grade H_2O, and digest with a second enzyme.

5. Purify the digested PCR product using a 0.8–1.5% agarose gel and a Qiagen Gel Extraction kit. Ligate the purified PCR product to the appropriately digested purified pFastBac vector (~100 ng) at an insert to vector molar ratio of 3:1. Use 1 U of T4 DNA ligase and a 2-h incubation at 25°C or an overnight incubation at 16°C.

6. Transform DH5α *E. coli* cells with a ligation mixture containing 1–10 ng of DNA using standard procedures *(21)*. Plate the cells on LB-agar dishes containing ampicillin (50 µg/mL), and incubate overnight at 37°C.

7. Select single colonies and grow overnight at 37°C in LB medium containing ampicillin. Purify the plasmid DNA from 2 mL of the overnight cultures using a Qiagen Miniprep kit.

8. Check the plasmid DNA for correct insertion of the desired DNA fragment using restriction enzymes. Verify the DNA sequence of the construct by DNA sequencing.

3.2. Expression of PDE6αβ/PDE5 Chimeras in Sf9 Cells

1. Generate recombinant bacmids, transfect Sf9 cells, and carry out viral amplifications according to the manufacturer's recommendations (Life Technologies).

2. Infect the Sf9 cells with individual viruses or with a desired combination of viruses at a multiplicity of infection of 3–10 (*see* Note 2).

3. Pellet the Sf9 cells 48 h posttransfection. Wash the pellet with 20 mM Tris-HCl (pH 8.0) containing 130 mM NaCl and 2 mM MgSO$_4$. Store the cell pellet at –80°C or process immediately.

4. Resuspend the cells corresponding to 10mL culture in 3 mL of 20 mM Tris-HCl buffer (pH 8.0) containing 2 mM MgSO$_4$ and Complete Mini protease inhibitor cocktail (one-third tablet).

5. Disrupt the cells by sonication with two 10-s pulses using a microtip attached to a 550 Sonic Dismembrator (Fisher). Centrifuge the cell lysate at 100,000g for 90 min at 2°C.

6. Collect the supernatant and dialyze it against 30 mM Tris-HCl buffer (pH 8.0) containing 130 mM NaCl, 2 mM MgSO$_4$, and 50% glycerin.

7. Centrifuge the dialyzed lysates at 100,000g for 1 h at 2°C. Store the supernatant at –20°C.

3.3. Assays to Assess Dimerization of PDE6αβ/PDE5 Chimeras

3.3.1. Gel Filtration

1. Equilibrate a Superose 12 HR 10/30 column at 25°C with 30 mM Tris-HCl buffer (pH 8.0) containing 100 mM NaCl and 2 mM MgSO$_4$.

2. Calibrate the column with the following protein standards with known molecular masses and Stokes radii: bovine thyroglobulin (670 kDa, 85 Å), horse ferritin (440 kDa, 61 Å), sweet potato β-amylase (200 kDa), rabbit aldolase (158 kDa,

48.1 Å), BSA (67 kDa, 35.5 Å), and chicken ovalbumin (45 kDa, 30.5 Å). For calibration, individually inject 50–100 μL of solutions with standards (200–300 μg of protein) of the standards. Elute the proteins at 0.4 mL/min, and record the elution profiles using A_{280}.

3. Mix 50 μg of native bovine rod holoPDE6 with 100 μL of the Sf9 cell extract from noninfected culture and inject into Superose 12 HR 10/30. Elute the proteins at 0.4 mL/min and collect 0.4-mL fractions. Analyze the fractions for PDE activity using [^{3}H]cGMP as substrate *(22)*, and determine the protein concentration by the method of Bradford *(23)* (**Fig. 2B**).

4. Prepare dialyzed cell extract from Sf9 cells infected with virus for expression of the WT PDE. Carry out gel filtration of the PDE5 sample in **step 3** and analyze fractions for PDE activity and protein concentration (**Fig. 2A**).

5. Inject 50–100 μL of dialyzed cell extracts isolated from Sf9 cells infected with one or two selected viruses into a Superose 12 HR 10/30 column. Elute the proteins at 0.4 mL/min and collect 0.4-mL fractions. Analyze the fractions for PDE activity and protein concentration. Perform gel filtration analyses two times for each PDE chimera combination from at least two different preparations of Sf9 cell extracts (**Fig. 2B**).

6. Assay the fractions for the presence of chimeric PDE subunits by Western blot analysis with anti-His, antiflag, or anti-myc antibodies *(24)* (**Fig. 2B**).

7. When Western blot analysis demonstrates the presence of coexpressed chimeric PDE subunits in the same gel filtration fractions corresponding to the dimeric enzyme, analyze these fractions by immunoprecipitation (*see* IP in **Fig. 2**).

8. Calculate Stokes radii for chimeric PDEs from gel filtration data using the correlation of elution volumes with Stokes radius proposed by Porath *(25)*.

3.3.2. Western Blot Analysis

1. Load aliquots of gel filtration fractions onto a 10% SDS-polyacrylamide gel and separate the proteins by SDS-PAGE.

2. Transfer proteins onto a nitrocellulose membrane using a semidry blotting apparatus. Block the membranes with PBS (pH 7.1) containing 0.1% Triton X-100 and 5% nonfat dry milk for 1 h.

3. Incubate the membranes in the same solution with monoclonal antipolyhistidine, M2 monoclonal antiflag, or monoclonal anti-c-myc (clone 9E10) antibodies with the respective dilutions of 1:1500, 1:5000, and 1:5000. Wash the membranes four times with PBS containing 0.1% Triton X-100.

4. Incubate the membranes with HRP-conjugated secondary antibodies in PBS containing 0.1% Triton X-100 and 5% nonfat dry milk for 1 h. Wash the membranes four times with PBS containing 0.1% Triton X-100.

5. Rinse the membranes two times with PBS and visualize the blots with ECL Western blotting detection reagents (Amersham) (**Fig. 2B**).

3.3.3. Immunoprecipitation

1. Incubate aliquots of gel filtration fractions (50–100 μL) with or without antiflag or myc-antibodies (1 μL) for 30 min at 25°C.

2. Add 10 µL of a 50% suspension of protein G–agarose (Sigma) and incubate for 40 min at 25°C.

3. Wash the agarose beads four times with 300 µL of PBS and elute the bound proteins using an SDS-PAGE sample buffer.

4. Separate immunoprecipitated PDE complexes by SDS-PAGE in 10% gels, and analyze by Western blotting using anti-His or other appropriate antibodies (**Fig. 2B**).

3.3.4. Sucrose Gradient Centrifugation

1. Prepare sucrose density gradients (5–35%) in 30 mM Tris-HCl (pH 8.0) buffer containing 100 mM NaCl, 2 mM MgSO$_4$, and 4 mM 2-mercaptoethanol using 14 × 89 mm centrifugation tubes and a GradiFrac gradient former (Amersham).

2. Prepare solutions of protein standards with known sedimentation coefficients ($S_{20,w}$): bovine liver catalase (250 kDa, 11.3 S), rabbit aldolase (158 kDa, 7.3 S), BSA (67 kDa, 4.6 S), and chicken ovalbumin (45 kDa, 3.5 S).

3. Load protein standards (0.5–1.0 mg of protein in <200 µL) onto the gradients in 14 × 89 mm centrifugation tubes, and centrifuge the gradients for 24 h at 280,000g in a Beckman SW41 rotor at 4°C.

4. Collect 300-µL fractions from the bottom of the tubes. Analyze fractions for protein concentration.

5. Perform sucrose density centrifugations two times. Plot distances traveled by standards from meniscus vs the $S_{20,w}$ values of standards.

6. Load 200-µL aliquots from peak dimeric PDE gel filtration fractions onto the gradients in 14 × 89 mm centrifugation tubes, and centrifuge the gradients for 24 h at 280,000g in a Beckman SW41 rotor at 4°C.

7. Collect 300-µL fractions from the bottom of the tubes and analyze them for PDE activity (**Fig. 2C**).

8. Determine sedimentation coefficients ($S_{20,w}$) for chimeric PDEs using a linear plot of distances traveled by standards from meniscus vs the $S_{20,w}$ values of standards *(26)*.

9. Calculate the molecular weights of chimeric PDEs using the estimated sedimentation coefficients and the Stokes radii from the gel filtration data (*see* **Note 3**).

4. Notes

1. An alternative approach to construction of PDE6αβ/PDE5 chimeras is to carry out PCR amplification of the desired region from one PDE subunit using a 5′ primer containing a unique restriction site and a phosphorylated 3′ primer designed for blunt ligation. An appropriate region of the second PDE subunit is amplified with phosphorylated 5′ primer and 3′ primer containing a second unique restriction site. These two PCR products are cut with appropriate restriction enzymes and cloned into the plasmid.

2. Cultures should be infected in the middle of the logarithmic phase of growth. For suspension cultures, this corresponds to approx 1.5×10^6–2×10^6 cells/mL.

3. The molecular shape of a protein significantly influences its migration on a gel filtration column. Sucrose density centrifugation is an independent approach to estimate molecular weights of proteins *(26)*. The molecular weights of chimeric PDEs can be calculated using the estimated sedimentation coefficients and the Stokes radii from the gel filtration data using the following equation *(27)*:

$$M = (s \cdot N_A \cdot 6 \cdot \pi \cdot \eta \cdot r)/(1 - v \cdot \rho)$$

in which s is the sedimentation coefficient, N_A is Avogadro's number, η is the viscosity of the medium (0.01 g/[cm•s]), r is the Stokes radius, v is the partial specific volume of a protein (0.73 cm^3/g), and ρ is the density of the medium (1 g/cm^3).

Acknowledgment

This work was supported by National Institutes of Health grant EY-10843.

References

1. Beavo, J. A. (1995) Cyclic nucleotide phosphodiesterases: functional implications of multiple isoforms. *Physiol. Rev.* **75**, 725–748.
2. Francis, S. H., Turko, I. V., and Corbin, J. D. (2001) Cyclic nucleotide phosphodiesterases: relating structure and function. *Prog. Nucleic Acid Res. Mol. Biol.* **65**, 1–52.
3. Fink, T. L., Francis, S. H., Beasley, A., Grimes, K. A., and Corbin, J. D. (1999) Expression of an active, monomeric catalytic domain of the cGMP-binding cGMP-specific phosphodiesterase (PDE5). *J. Biol. Chem.* **274**, 34,613–34,620.
4. Richter, W. and Conti, M. (2002) Dimerization of the type 4 cAMP-specific phosphodiesterases is mediated by the upstream conserved regions (UCRs). *J. Biol. Chem.* **277**, 40,212–40,221.
5. Gillespie, P. G. and Beavo, J. A. (1988) Characterization of a bovine cone photoreceptor phosphodiesterase purified by cyclic GMP-sepharose chromatography. *J. Biol. Chem.* **263**, 8133–8141.
6. Baehr, W., Devlin, M. J., and Applebury, M. L. (1979) Isolation and characterization of cGMP phosphodiesterase from bovine rod outer segments. *J. Biol. Chem.* **254**, 11,669–11,677.
7. Fung, B. K.-K., Young, J. M., Yamane, H. K., and Griswold-Prenner, I. (1990) Subunit stoichiometry of retinal rod cGMP phosphodiesterase. *Biochemistry* **29**, 2657–2664.
8. Artemyev, N. O., Surendran, R., Lee, J. C., and Hamm, H. E. (1996) Subunit structure of rod cGMP-phosphodiesterase. *J. Biol. Chem.* **271**, 25,382–25,388.
9. Chabre, M. and Deterre, P. (1989) Molecular mechanism of visual transduction. *Eur. J. Biochem.* **179**, 255–266.
10. Yarfitz, S. and Hurley, J. B. (1994) Transduction mechanisms of vertebrate and invertebrate photoreceptors. *J. Biol. Chem.* **269**, 14,329–14,332.

11. Muradov, K. G., Boyd, K. K., Martinez, S. E, Beavo, J. A., and Artemyev, N. O. (2003) The GAFa domains of rod cGMP-phosphodiesterase 6 determine the selectivity of the enzyme dimerization. *J. Biol. Chem.* **278,** 10,594–10,601.

12. Aravind, L. and Ponting, C. P. (1997) The GAF domain: an evolutionary link between diverse phototransducing proteins. *Trends Biochem. Sci.* **22,** 458, 459.

13. Lipkin, V. M., Khramtsov, N. V., Vasilevskaya, N. V., Atabekova, K. G., Muradov, K. G., Li, T., Johnston, J. P., Volpp, K. J., and Applebury, M. L. (1990) β-subunit of bovine rod photoreceptor cGMP phosphodiesterase. *J. Biol. Chem.* **265,** 12,955–12,959.

14. McAllister-Lucas, L. M., Sonnenburg, W. K., Kadlecek, A., et al. (1993) The structure of a bovine lung cGMP-binding, cGMP-specific phosphodiesterase deduced from a cDNA clone. *J. Biol. Chem.* **268,** 22,863–22,873.

15. Charbonneau, H., Prusti, R. K., LeTrong, H., Sonnenburg, W. K., Mullaney, P. J., Walsh, K. A., and Beavo, J. A. (1990) Identification of a noncatalytic cGMP-binding domain conserved in both the cGMP-stimulated and photoreceptor cyclic nucleotide phosphodiesterases. *Proc. Natl. Acad. Sci. USA* **87,** 288–292.

16. Stroop, S. D. and Beavo, J. A. (1991) Structure and function studies of the cGMP-stimulated phosphodiesterase. *J. Biol. Chem.* **266,** 23,802–23,809.

17. Thomas, M. K., Francis, S. H., and Corbin, J. D. (1990) Characterization of a purified bovine lung cGMP-binding cGMP phosphodiesterase. *J. Biol. Chem.* **265,** 14,964–14,970.

18. Martinez, S. E., Wu, A. Y., Glavas, N. A., Tang, X. B., Turley, S., Hol, W. G., and Beavo, J. A. (2002) The two GAF domains in phosphodiesterase 2A have distinct roles in dimerization and in cGMP binding. *Proc. Natl. Acad. Sci. USA* **99,** 13,260–13,265.

19. Turko, I. V., Haik, T. L., McAllister-Lucas, L. M., Burns, F., Francis, S. H., and Corbin, J. D. (1996) Identification of key amino acids in a conserved cGMP-binding site of cGMP-binding phosphodiesterases: a putative NKXnD motif for cGMP binding. *J. Biol. Chem.* **271,** 22,240–22,244.

20. Granovsky, A. E., Natochin, M., McEntaffer, R., Haik, T. L., Franscis, S. H., Corbin, J. D., and Artemyev, N. O. (1998) Probing domain functions of chimeric PDE6α′/PDE5 cGMP-phosphodiesterase. *J. Biol. Chem.* **273,** 24,485–24,490.

21. Ausubel, F. M., Brent, R., Kingston, R. E., et al. (1993) *Escherichia Coli,* plasmids and bacteriophages, in, *Current Protocols in Molecular Biology* (Jansen, K., ed.), John Wiley & Sons, New York, pp. 1.8.1.–1.8.8.

22. Artemyev, N. O. Interactions between catalytic and inhibitory subunits of PDE6. in *Phosphodiesterase: Methods and Protocols* (Lugnier, C., ed.), Humana, Totowa, NJ, pp. 277–287.

23. Bradford, M. M. (1976) A rapid and sensitive method for the quantitation of microgram quantities of protein utilizing the principle of protein-dye binding. *Anal. Biochem.* **72,** 248–254.

24. Towbin, H., Staehelin, T., and Gordon, J. (1979) Electrophoretic transfer of proteins from polyacrylamide gels to nitrocellulose sheets: procedure and some applications. *Proc. Natl. Acad. Sci. USA* **76,** 4350–4354.

25. Porath, J. (1963) Some recently developed fractionation procedures and their application to peptide and protein hormones. *Pure Appl. Chem.* **6,** 233–244.
26. Martin, R. G. and Ames, B. N. (1961) A method for determining the sedimentation behavior of enzymes: application to protein mixtures. *J. Biol. Chem.* **236,** 1372–1379.
27. Siegel, L. M. and Monty, K. J. (1966) Determination of molecular weights and frictional ratios of proteins in impure systems by use of gel filtration and density gradient centrifugation: application to crude preparations of sulfite and hydroxylamine reductases. *Biochim. Biophys. Acta* **112,** 346–362.

20

Interactions Between Catalytic and Inhibitory Subunits of PDE6

Nikolai O. Artemyev

Summary

Rod and cone photoreceptor guanosine 3′5′-cyclic-monophosphate phosphodiesterases (PDEs) are classified into the PDE6 family of cyclic nucleotide PDEs. A unique feature of the PDE6 enzymes is the presence of inhibitory γ-subunits (Pγ). The inhibitory interaction between Pγ and the rod PDE6αβ catalytic subunits is critically important for understanding the mechanism of phototransduction. Recent insights into the molecular interface between Pγ and PDE6αβ have been achieved using mutagenesis of Pγ, fluorescence labeling, and crosslinking approaches.

Key Words

PDE6; guanosine 3′5′-cyclic-monophosphate; phototransduction; mutagenesis; polymerase chain reaction; fluorescence labeling; crosslinking.

1. Introduction

Rod and cone guanosine 3′,5′-cyclic-monophosphate (cGMP) phosphodi-esterases (PDEs) (PDE6 family) serve as the effector enzymes in the visual transduction cascade. The activity of PDE6 catalytic subunits (Pαβ) is controlled by two identical inhibitory γ-subunits (Pγ) (1,2). The Pγ-subunits are displaced from the enzyme catalytic core by the guanosine 5′-triphosphate-bound transducin-α molecules on light stimulation of photoreceptor cells. Insights into the inhibitory interaction between Pγ and the PDE6 catalytic subunits are essential for understanding the mechanisms of visual signaling. Pγ is a small protein (9.7 kDa, 87 amino acid residues) that can be readily expressed in *Escherichia coli* (3). Pγ mutants represent excellent tools to investigate the interaction of Pγ with the rod catalytic PDE6αβ-subunits. A Pγ residue at any position can be substituted by alanine or another residue to probe the role of individual Pγ residues. A cysteine-scanning mutagenesis is another valuable

From: *Methods in Molecular Biology, vol. 307: Phosphodiesterase Methods and Protocols*
Edited by: C. Lugnier © Humana Press Inc., Totowa, NJ

approach to investigate the Pγ/PDE6αβ interface using selective fluorescence labeling and modifications with crosslinking reagents.

This chapter describes approaches and methods for the construction, expression, and purification of Pγ mutants. Fluorescence labeling, fluorescence-binding assay, crosslinking, and PDE6 inhibition assays are described to illustrate the use of Pγ mutants to study the Pγ/PDE6αβ interaction. The fluorescence-binding assay is based on monitoring the change in fluorescence resulting from the binding of fluorescently labeled Pγ mutants to the Pαβ-subunits. The assay is particularly sensitive when Pγ is derivatized with 3-(bromoacetyl)-7-diethyl aminocoumarin (BC) *(4)*. For the crosslinking applications, 4-(*N*-maleimido) benzophenone (MBP) is selected because it was shown to be extremely effective in crosslinking specific Pγ mutants to PDE6αβ *(5)*. Standard methods to measure cGMP-PDE activity include the use of radiolabeled substrate [^{3}H]cGMP *(6)*, proton evolution *(7)*, and inorganic phosphate release assays *(8)*. A modification of the assay using [^{3}H]cGMP is described here as one of the most practical procedures to study PDE6αβ inhibition by Pγ or its mutants.

2. Materials

1. pET-11a-Pγ plasmid.
2. *E. coli* strains DH5α and BL21(DE3).
3. Oligonucleotide primers.
4. Restriction enzymes, T4 DNA ligase, Taq DNA polymerase, and shrimp alkaline phosphatase (SAP).
5. Polymerase chain reaction (PCR) amplifier and reagents for PCR reactions (dNTP mix, 10X PCR buffer).
6. Agarose and agarose gel electrophoresis equipment.
7. Isopropyl-β-D-thiogalactopyranoside (IPTG).
8. Luria-Bertani (LB) medium with ampicillin (1 L): 10 g of tryptone, 5 g of yeast extract, 5 g of NaCl, 80 μL of 12.5 *M* NaOH, 50 mg of ampicillin.
9. 2X TY medium with ampicillin (1 L): 16 g of tryptone, 10 g of yeast extract, 5 g of NaCl, 50 mg of ampicillin.
10. Cell lysis solution: 50 m*M* Tris-HCl buffer (pH 7.5) containing 20 m*M* NaCl, 5 m*M* EDTA, 1 m*M* dithiothreitol (DTT), and the protease inhibitors phenylmethylsulfonyl fluoride (PMSF) (1 m*M*) and pepstatin A (20 μg/mL).
11. Sonic dismembrator.
12. Equipment, reagents, and buffers for sodium dodecyl sulfate-polyacrylamide gel electrophoresis (SDS-PAGE).
13. Equipment, reagents, and buffers for Western blotting.
14. Liquid chromatography equipment and high-performance liquid chromatography (HPLC) chromatography system.
15. SP Fast Flow Sepharose (Amersham).
16. Solutions for chromatography on SP Fast Flow Sepharose:

    a. Solution A: 50 m*M* Tris-HCl buffer (pH 7.5) containing 20 m*M* NaCl, 5 m*M* EDTA, and 1 m*M* DTT.

    b. Solution B: 50 m*M* Tris-HCl buffer (pH 7.5) containing 400 m*M* NaCl, 5 m*M* EDTA, and 1 m*M* DTT.

17. Solution for labeling Pγ mutants with fluorescent probes: 20 m*M* HEPES buffer (pH 7.5).
18. Solution for separation of labeled Pγ mutants from excess fluorescent or crosslinking probes: 20 m*M* HEPES buffer (pH 7.5) containing 100 m*M* NaCl.
19. Solution for fluorescence binding assay: 20 m*M* HEPES buffer (pH 7.5) containing 100 m*M* NaCl and 4 m*M* MgCl$_2$.
20. C-4 reverse-phase HPLC (RP-HPLC) column for separation of proteins.
21. BC (Molecular Probes; Eugene, OR).
22. MBP (Molecular Probes).
23. PD-10 desalting columns (Amersham).
24. Fluorescence spectrophotometer.
25. AG1-X2 resin (Bio-Rad, Hercules, CA).
26. [^{3}H]cGMP (specific activity: 8.5 Ci/mmol) (Amersham).
27. Alkaline phosphatase (0.1 U) (Sigma Chemical Corporation, St. Louis, MO).
28. Liquid scintillation counter.

3. Methods

The following methods to study interactions between PDE6αβ and the Pγ-subunits are described: construction of Pγ mutants, expression and purification of Pγ mutants, and assays to probe the interaction between Pγ mutants and PDE6αβ.

3.1. Construction of Pγ Mutants

3.1.1. pET System for Expression of Pγ

The pET system originally developed by Studier and Moffatt *(9)* is the most efficient system for expression of Pγ in *E. coli*. The Pγ cDNA is cloned under the control of the T7 promoter, which is not recognized by *E. coli* RNA polymerase. Therefore, there is no expression of Pγ until the vector is transformed into an *E. coli* strain bearing the IPTG-inducible T7 polymerase gene, such as strain BL21(DE3). The selectivity and activity of T7 RNA polymerase lead to very high yields of Pγ expression using the pET system. The synthetic Pγ cDNA assembled by Brown and Stryer *(10)* represents a convenient template for site-directed mutagenesis of Pγ. For protein expression, this synthetic gene has been PCR amplified with primers containing *Nde*I and *Bam*HI sites and subcloned into the pET-11a vector (Novagen) *(11)*. The resulting pET-11a-Pγ vector has been widely used for Pγ mutagenesis *(11–13)*.

3.1.2. Approaches to Cloning of Pγ Mutants

Synthetic Pγ cDNA contains many restriction sites (**Fig. 1**). A significant number of these sites—*Nde*I, *Sma*I, *Kpn*I, *Nco*I, *Sac*I, and *Bam*HI—are unique in the pET-11a-Pγ vector (**Fig. 1**). These sites make a PCR-based mutagenesis of Pγ a method of choice to introduce mutations in proximity to one of these sites (*see* **Note 1**). For example, for cysteine-scanning mutagenesis, a single cysteine residue (Cys68) in the wild-type (WT) Pγ must be substituted first. A substitution of Cys68 for Ser can be made using the *Nco*I site and PCR-directed mutagenesis with a reverse primer containing an *Nco*I site and the mutation. A forward PCR primer may contain one of the upstream Pγ cDNA restriction sites, such as the *Nde*I site, that includes a start codon (**Fig. 1**). The *Nde*I/*Nco*I-digested PCR fragment is then inserted into the *Nde*I/*Nco*I fragment of the Pγ expression vector. The *Sma*I, *Kpn*I, *Nco*I, *Sac*I, and *Bam*HI sites allow the placement of residues practically at any position within the C-terminal portion of Pγ by PCR-directed mutagenesis (**Fig. 1**). A more complicated procedure is necessary to mutate residues near or within the polycationic region of Pγ, Pγ-24-45. The available restriction sites, BstEII, ApaI, and MluI, are not unique (**Fig. 1**). The pET-11a vector contains an additional copy of each site. A potential cloning strategy includes partial digestion of the pET-11a-Pγ vector with one of these enzymes. The linearized plasmid is isolated and subsequently cut at one of the unique restriction sites: *Nde*I, *Sma*I, *Kpn*I, *Nco*I, or *Bam*HI (**Fig. 1**). A plasmid DNA of appropriate size is separated and used for subcloning of a PCR-generated DNA fragment carrying a mutation and flanked with the chosen restriction sites. Alternatively, the Pγ cDNA can be subcloned from the pET-11a vector into a vector that does not contain *Bst*EII, *Apa*I, and *Mlu*I sites. Afterward, mutations can be generated by PCR-directed mutagenesis as just outlined.

3.1.3. PCR-Directed Mutagenesis of Pγ

1. Design and order synthetic PCR primers carrying mutations. Allow a mutant codon to be flanked by at least 10 unmodified bases.
2. Prepare PCR reaction mixtures containing 10–50 ng of pET-11a-Pγ vector as a template, 1 μ*M* each of 5′ and 3′ primer, 0.2 m*M* dNTP mix, 5 μL of 10X PCR buffer, 5 U of Taq DNA polymerase (Perkin-Elmer), and HPLC-grade H$_2$O to a final volume of 50 μL.
3. Program a PCR amplifier and perform PCR amplifications as follows: 2-min dwell at 95°C; 30 cycles of PCR at 94°C for 50 s, 56°C (or 10°C lower than the primer's melting temperature) for 50 s, and 72°C for 35 s.
4. Separate the PCR product on a 1.5% agarose gel and purify it using a standard Qiagen Gel Extraction kit. Sizes of PCR products for Pγ mutants usually are between 150 and 250 bp.

NdeI   XhoI   BstEII   ApaI
ca/tatgaatc/**tcgag**ccgccgaaagctgaaatccgttccgctacccgtgttatgggtggtcc**g/gttacc**ccgcgtaa**gggcc/c**gccg
  M N L   E   P   P   K   A   E   I   R   S   A   T   R   V   M   G   G   P   V   T   P   R K   G   P   P

                    MluI   SmaI
aaattcaaacagcgtcag**a/cgcgt**cagttcaaatccaaaccgccgaaaaaaggtgtccagggttttggtgacgatat**ccc/ggg**tatggaaggc

K F K Q R Q   T   R Q F K S   K P   P   K K G   V Q G F G   D D I   P   G M E G

     KpnI   NcoI   SacI   BamHI
ct**gggtac/c**gacatcaccgttatctg**c/ccatgg**gaggctttcaaccacctg**gagct/c**cacgaactggctcagtacggtatcatcta**g/gatcc

L G T   D I T   V I C   P W E   AFN H L E L   H E   L A Q Y G I   IStop

Fig. 1. Protein sequence of Pγ and corresponding sequence of synthetic cDNA. Restriction sites useful for construction of Pγ mutants are shown in bold. Unique restriction sites in the PEt-11a-Pγ plasmid are underlined.

5. Incubate a purified PCR product with selected restriction enzymes for 2 to 3 h. If several restriction sites (enzymes) are suitable for the mutagenesis, give the preference to the enzymes that are optimally active in the same restriction buffer. If simultaneous digestion is not possible, first cut a PCR fragment with one enzyme, precipitate using 0.7 vol of isopropanol, dissolve in HPLC-grade H_2O, and digest with a second enzyme.

6. Purify the digested PCR product using a 1.5% agarose gel and a Qiagen Gel Extraction kit. Ligate the purified PCR product to the appropriately digested and agarose gel–purified pET-11a-Pγ plasmid (~100 ng) at an insert-to-vector molar ratio of 3:1. Use 1 U of T4 DNA ligase and a 2-h incubation at 25°C or an overnight incubation at 16°C (*see* **Note 2**).

7. Transform DH5α *E. coli* cells with a ligation mixture containing 1–10 ng of DNA using standard procedures *(14)*. Plate the cells on LB-agar dishes containing ampicillin (50 μg/mL) and incubate overnight at 37°C.

8. Select single colonies and grow overnight at 37°C in LB medium containing ampicillin. Purify the plasmid DNA from 2 mL of the overnight cultures by standard methods, and identify colonies carrying desired mutations in the Pγ cDNA by DNA sequencing.

3.2. Expression of Pγ or Its Mutants

1. Transform BL21(DE3) *E. coli* cells with the WT or mutant pET-11a-Pγ plasmid using standard methods.

2. Grow BL21(DE3) cells carrying the pET-11a-Pγ plasmid or mutant plasmid at 37°C overnight in LB medium containing 50 μg/mL of ampicillin.

3. Dilute the overnight culture (1:100) with 2X TY medium containing 50 μg/mL of ampicillin, and continue to grow cells at 37°C until the OD_{600} reaches approx 0.8.

4. Reduce the incubation temperature to 30°C, and induce Pγ expression by the addition of 0.5 mM IPTG. Allow the induction to proceed for 3 to 4 h.

5. Following the induction, pellet the cells and wash the pellet two times with 50 mM Tris-HCl buffer (pH 7.5). Resuspend the pellet in 30 vol of 50 mM Tris-HCl buffer (pH 7.5) containing 20 mM NaCl, 5 mM EDTA, 1 mM DTT, and the protease inhibitors PMSF (1 mM) and pepstatin A (20 μg/mL).

6. Disrupt the cells by sonication with 30-s pulses for a total sonication time of approx 4 min. Centrifuge the cell lysate at 100,000g for 30 min and collect the supernatant. Keep the supernatant frozen at –20°C until further purification.

3.3. Purification of Pg or Its Mutants

1. Prepare a 5-mL column with SP Fast Flow Sepharose (Pharmacia), and equilibrate the column with 50 mM Tris-HCl buffer (pH 7.5) containing 20 mM NaCl, 5 mM EDTA, and 1 mM DTT.

2. Load a supernatant obtained from 1 L of bacterial culture (from **step 6** of **Subheading 3.2.1.**) onto the column. Elute bound proteins at a flow rate of 0.5 mL/min using 100 mL of a 20–400 mM NaCl gradient. Pγ elutes at approx 250 mM NaCl. Collect fractions of 2 to 3 mL and analyze for the presence of Pγ using a PDE6 inhibition assay or SDS-PAGE and Western blotting using standard procedures *(15)*.

3. Apply Pγ-containing fractions onto a reverse-phase protein C-4 HPLC column (Microsorb-Rainin or 214TP54-Vydac). Elute bound proteins at a flow rate of 0.5 mL/min with a 45-min 0–80% gradient of acetonitrile in 0.1% trifluoroacetic acid (TFA)/H$_2$O. Pγ elutes as a single major peak at approx 45% acetonitrile.

4. Speed-dry (lyophilize) HPLC fractions containing purified Pγ using a Speed Vac SC100 (Savant). Dissolve the lyophilized Pγ protein in 20 mM HEPES buffer (pH 7.5), determine the protein concentration, and store at –80°C until use (*see* **Note 3**). This procedure typically yields up to 20 mg of more than 95% pure Pγ or mutant/L of culture.

3.4. Assays to Probe Interaction Between Pγ or Its Mutants and PDE6 Catalytic Subunits

3.4.1. Labeling of Pγ Mutants With the Environmentally Sensitive Fluorescent Probe BC (see **Note 4**)

1. Prepare a fresh stock solution of BC (5 mM) by dissolving BC in N',N-dimethylformamide in a light-protected tube. Use $\varepsilon_{445} = 53,000$ for BC to confirm/adjust the concentration of the probe.

2. Prepare a working solution of Pγ or Pγ mutant (100–200 μM) containing a single Cys residue in 20 mM HEPES buffer (pH 7.5).

3. Add a threefold molar excess of BC from the stock solution to the working solution of Pγ or mutant. Incubate the mixture for 30 min at 25°C and stop the reaction by adding β-mercaptoethanol (5 mM final concentration). Pass the BC-labeled Pγ or mutant (~0.5 mL) through a desalting PD-10 column (Phar-

macia) equilibrated with 20 mM HEPES buffer (pH 7.5) containing 100 mM NaCl to remove excess BC.

4. Separate the BC-labeled Pγ from the unlabeled protein on a C-4 HPLC column (214TP54-Vydac) using a 0–80% gradient of acetonitrile/0.1% TFA as described in **Subheading 3.2.2., step 3**.

3.4.2. Fluorescence Assay of Binding Between BC-Labeled Pγ Mutants and P$\alpha\beta$

1. Prepare rod PDE6$\alpha\beta$ catalytic subunits using limited proteolysis of the holo-enzyme with trypsin and purification of trypsinized PDE (tPDE) as described previously *(16,17)* (*see* **Note 5**).
2. Mix BC-labeled Pγ mutant (10 nM final concentration) with 1 mL of 20 mM HEPES buffer (pH 7.5) containing 100 mM NaCl and 4 mM MgCl$_2$ in a cuvet for fluorescence measurements. Record a basal fluorescence of the sample with exci-tation at 445 nm and emission at 495 nm using a fluorescence spectrophotometer.
3. Add increasing concentrations of PDE6$\alpha\beta$ to the sample, and record the BC fluorescence after each addition (*see* **Note 6**).
4. Control for the specificity of BC-labeled Pγ mutant binding to PDE6$\alpha\beta$ by adding an excess of unlabeled Pγ at the end of each titration experiment. Excess Pγ should displace BC-labeled Pγ mutant from the complex with PDE6$\alpha\beta$, and the BC fluorescence should return to near-basal level (*see* **Note 7**).

3.4.3. Labeling of Pγ Mutants With the Photoreactive Crosslinking Probe MBP

1. Label Pγ or mutant immediately after RP-HPLC purification of a C-4 column (*see* **Subheading 3.2.2., step 3**) (*see* **Note 8**). Adjust the pH of the Pγ-containing HPLC fraction to 7.5 using 200 mM HEPES buffer (pH 8.4).
2. Carry out the following operations in dim light, and collect all fractions in light-protected tubes: Add a threefold excess of MBP from a 10 mM stock solution in acetonitrile to Pγ solution. Allow the reaction to proceed for 20 min at 25°C. Stop the reaction with the addition of 5 mM β-mercaptoethanol.
3. Separate the MBP-labeled Pγ mutants from free MBP on a PD-10 column (Pharmacia) equilibrated with 20 mM HEPES buffer (pH 7.5) containing 100 mM NaCl.
4. Determine the concentration of the MBP-labeled Pγ mutants using ε_{260} = 23,000 for MBP (*see* **Note 9**).

3.4.4. Crosslinking of Pγ or Its Mutants to PDE6$\alpha\beta$

1. Mix purified PDE (PDE6$\alpha\beta$) at a final concentration of 2 μM with 10 μM MBP-labeled Pγ or mutant in a polypropylene microcentrifuge tube.
2. Irradiate the sample for 4 min at a distance of 4 cm with a Transilluminator ultra-violet (UV) lamp (UVP). The polypropylene tube cuts off short-wavelength UV light (<300 nm), thus preventing protein damage. The longer-wavelength UV light

(300–350 nm), which is necessary for the excitation of benzophenone derivatives, passes through.

3. Analyze the efficiency and specificity of crosslinking between Pαβ and Pγ by SDS-PAGE and Western blotting using standard procedures *(15)*.

4. Purify and carry out proteolytic digestion of crosslinked products. Identify site(s) of crosslinking using appropriate mass spectrometric techniques (*see* **Note 10**).

3.4.5. PDE Activity Assay to Test Inhibition of PDE6αβ by Pγ or Its Mutants

1. Prepare a 20% slurry of AG1-X2 resin (Bio-Rad) by washing the resin three times with 0.5 M NaOH, another three times with 0.5 M HCl, followed by washing with HPLC-grade H_2O until the pH reaches 5.0.

2. Add PDE6αβ (0.05–0.1 nM) and varying concentrations of Pγ or mutant to the PDE reaction mixture (100 μL) containing 20 mM Tris-HCl (pH 7.8), 100 mM NaCl, 8 mM MgSO$_4$, and 0.2 mg/mL of bovine serum albumin.

3. Initiate the reaction with the addition of 5 μL of 100 μM cGMP containing 0.05 μCi of [^{3}H]cGMP (specific activity: 8.5 Ci/mmol) (Amersham). Allow the reaction to proceed for 10–20 min at 37°C, and terminate it by heating the samples for 2 min at 100°C.

4. Add alkaline phosphatase (0.1 U) (Sigma Chemical Corporation) to the mixture, and incubate for an additional 10 min.

5. Mix the reaction samples with 1 mL of a 20% slurry of AG1-X2 resin (Bio-Rad) and incubate for 5–10 min at room temperature. Spin the resin down for 1 min at 10,000g. Count 0.6-mL aliquots of the supernatants using a liquid scintillation counter.

4. Notes

1. A "cassette mutagenesis" strategy is another convenient approach for mutagenesis of Pγ. A relative proximity of many restriction sites in the synthetic Pγ gene facilitates the generation of mutations by using two annealed complementary mutant oligonucleotides with protruding ends that are compatible with selected restriction enzymes. This strategy has been successfully applied for Ala-scanning mutagenesis of the Pγ C-terminus *(11)*. The following procedure for the introduction of Pγ mutations by "cassette mutagenesis" can be used: Two complementary mutant oligonucleotides in equimolar amounts (0.2 nmol/10 μL) are mixed and incubated at 95°C for 1 min. Afterward, the mix is incubated at room temperature for 20 min. The pET-11a-Pγ vector is prepared by cutting with selected restriction enzymes, separation on an agarose gel, purification with a Qiagen Gel Extraction kit, and treatment with SAP. The oligonucleotide duplex and the vector are mixed at a molar ratio of 20:1 and ligated overnight at 25°C. The control ligation mix contains no duplex oligonucleotide and normally yields no colonies after transformation.

2. Typically, sizes of fragments excised from the pET-11a-Pγ plasmid are small (150–250 bp). Therefore, it is very difficult to separate a fully digested plasmid from an incompletely digested (linearized) plasmid. To avoid a high yield of

colonies with the original plasmid, the digested vector is additionally treated with SAP (1 U/0.5 µg of DNA) prior to ligation with a PCR fragment. In control experiments, the digested and SAP-treated vector is tested using ligation in the absence of the PCR fragment.

3. Perhaps owing to the fact that Pγ is a small basic polypeptide, some standard assays of protein concentration, such as the Bradford *(18)* assay, are extremely inaccurate in determining Pγ concentration. A simple and quite accurate way is to determine Pγ concentration from an absorbance spectrum of the HPLC-purified preparation using an extinction coefficient of 67,000 at 280 nm. A most accurate, but a time-consuming, method is to determine Pγ concentration on the basis of protein amino acid analysis.

4. Other environment- and conformation-sensitive thiol-reactive probes such as lucifer yellow vinyl sulfone, coumarin maleimide (7-diethylamino-3-[4′maleimidylphenyl]-4-methylcoumarin), and acrylodan (6-acryloyl-2-dimethylaminonaphthalene) from Molecular Probes (Eugene, OR) can be used for derivatization of Pγ and its mutants.

5. Tryptic proteolysis removes intrinsic Pγ-subunits and small farnesylated and geranyl-geranylated C-terminal fragments of Pα and Pβ, respectively. The time course for tryptic proteolysis is selected by monitoring the activation of PDE. The trypsinolysis of holoPDE should be stopped as soon as the PDE activity reaches a plateau. Additional control may include Western blotting of purified trypsinized PDE for the presence of the Pγ polypeptide.

6. The K_d values for BC-Pγ mutant binding to Pαβ are calculated by fitting the data to the following equation:

$$\frac{F}{F_o} = 1 + \frac{\left(\dfrac{F}{F_{omax}} - 1\right) \times X}{K_d + X} \tag{1}$$

in which F_o is the basal fluorescence of a BC labeled-Pγ mutant, F is the fluorescence after additions of tPDE, F/F_{omax} is the maximal relative increase in fluorescence, and X is the concentration of free tPDE.

7. Unlabeled Pγ mutants or competitive inhibitors of PDE such as zaprinast can be used to compete with BC labeled-Pγ mutants for the binding to Pαβ. This competition leads to dissociation of the BC labeled-Pγ mutant from Pαβ and a decrease in the BC fluorescence. IC_{50} values can be calculated by fitting the data to the one-site competition equation with variable slope:

$$\frac{F}{F_o} = 1 + \frac{\dfrac{F}{F_{omax}} - 1}{1 + 10^{(X - \log IC_{50}) \times H}} \tag{2}$$

in which X is the concentration of the competing ligand (Pγ mutant, zaprinast, cGMP analogs) and H is the Hill slope. Fitting of the data is performed with nonlinear least squares criteria using appropriate commercial software. The dissociation constants ($K_{1/2}$) for the ligand/Pαβ interaction can be calculated from the IC_{50} values using the Cheng and Prusoff *(19)* equation describing competitive displacement

$$K_{1/2} = IC_{50}/[1 + P\gamma mBC/K_d]$$

in which IC_{50} is the concentration of ligand that reduces the relative fluorescence increase by 50% (from **Eq. 2**), PγmBC is the total concentration of BC-labeled Pγ mutant, and K_d is the dissociation constant for the PγmBC/P$\alpha\beta$ complex (from **Eq. 1**). If the K_d and $K_{1/2}$ values are comparable to the total concentration of P$\alpha\beta$ in the assay, then equations derived by Linden *(20)* are appropriate for the calculation of $K_{1/2}$.

8. The labeling of Pγ and mutants at a Cys residue with MBP is more efficient in the presence of an organic solvent such as N',N-dimethylformamide or acetonitrile. However, high concentrations of solvent may lead to reduced selectivity of MBP toward Cys. Alternatively to using C-4 HPLC fractions for Pγ labeling, lyophilized Pγ protein can be dissolved in 20 mM HEPES buffer (pH 7.5) containing acetonitrile (40% [v/v]).

9. Typical efficiency of the MBP labeling of Pγ calculated from the absorption spectra of the unlabeled and labeled Pγ is 70–80%.

10. Specific strategies for purification, proteolytic digestion of Pγ-PDE6$\alpha\beta$ crosslinked products, and identification of crosslinked sites would depend on a position of the MBP photoprobe in the Pγ polypetide. Examples of successful strategies to identify Pγ-PDE6$\alpha\beta$ crosslinked sites can be found in **refs. 5** and *21*.

Acknowledgment

This work was supported by National Institutes of Health grant EY-10843.

References

1. Chabre, M. and Deterre, P. (1989) Molecular mechanism of visual transduction. *Eur. J. Biochem.* **179,** 255–266.
2. Yarfitz, S. and Hurley, J. B. (1994) Transduction mechanisms of vertebrate and invertebrate photoreceptors. *J. Biol. Chem.* **269,** 14,329–14,332.
3. Ovchinnikov, Y. A., Lipkin, V. M., Kumarev, V. P., Gubanov, V. V., Khramtsov, N. V., Akhmedov, N. B., Zagranichny, V. E., and Muradov, K. G. (1986) Cyclic GMP phosphodiesterase from cattle retina: amino acid sequence of the γ-subunit and nucleotide sequence of the corresponding cDNA. *FEBS Lett.* **204,** 288–292.
4. Granovsky, A. E., Natochin, M., and Artemyev, N. O. (1997) The γ-subunit of rod cGMP-phosphodiesterase blocks the enzyme catalytic site. *J. Biol. Chem.* **272,** 11,686–11,689.
5. Artemyev, N. O., Natochin, M., Busman, M., Schey, K. L., and Hamm, H. E. (1996) Mechanism of photoreceptor PDE inhibition by its γ-subunits. *Proc. Natl. Acad. Sci. USA* **93,** 5407–5412.
6. Thompson, W. J. and Appleman, M. M. (1971) Multiple cyclic nucleotide phosphodiesterase activities from rat brain. *Biochemistry* **10,** 311–316.

7. Liebman, P. A. and Evanczuk, A. T. (1982) Real time assay of rod disk membrane cGMP phosphodiesterase and its controller enzymes. *Methods Enzymol.* **81,** 532–542.
8. Gillespie, P. G. and Beavo, J. A. (1989) Inhibition and stimulation of photoreceptor phosphodiesterases by dipyridamole and M&B 22,948. *Mol. Pharmacol.* **36,** 773–781.
9. Studier, F. W. and Moffatt, B. A. (1986) Use of bacteriophage T7 RNA polymerase to direct selective high-level expression of cloned genes. *J. Mol. Biol.* **189,** 113–130.
10. Brown, R. L. and Stryer, L. (1989) Expression in bacteria of functional inhibitory subunit of retinal rod cGMP phosphodiesterase. *Proc. Natl. Acad. Sci. USA* **86,** 4922–4926.
11. Slepak, V. Z., Artemyev, N. O., Yun, Z. Dumke, C. L., Sabacan, L., Sondek, J., Hamm, H. E., Bownds, M. D., and Arshavsky, V. Y. (1995) An effector's site that stimulates G-protein GTPase in photoreceptors. *J. Biol. Chem.* **270,** 14,319–14,324.
12. Granovsky, A. E., McEntaffer, R., and Artemyev, N. O. (1998) Probing functional interfaces of rod PDE γ-subunit using scanning fluorescent labeling. *Cell Biochem. Biophys.* **28,** 115–133.
13. Granovsky, A. E. and Artemyev, N. O. (2001) A conformational switch in the inhibitory G-subunit of PDE6 upon the enzyme activation by transducin. *Biochemistry* **40,** 13,209–13,215.
14. Ausubel, F. M., Brent, R., Kingston, R. E., Moore, D. D., Seidman, J. G., Smith, J. A., and Struhl, K. (1993) *Escherichia Coli,* plasmids and bacteriophages, in, *Current Protocols in Molecular Biology* (Jansen, K., ed.), John Wiley & Sons, New York, pp. 1.8.1.–1.8.8.
15. Towbin, H., Staehelin, T., and Gordon, J. (1979) Electrophoretic transfer of proteins from polyacrylamide gels to nitrocellulose sheets: procedure and some applications. *Proc. Natl. Acad. Sci. USA* **76,** 4350–4354.
16. Hurley, J. B. and Stryer, L. (1982) Purification and characterization of the γ regulatory subunit of the cyclic GMP phosphodiesterase from retinal rod outer segments. *J. Biol. Chem.* **257,** 11,094–11,099.
17. Artemyev, N. O. and Hamm, H. E. (1992) Two-site high-affinity interaction between inhibitory and catalytic subunuts of rod cyclic GMP phosphodiesterase. *Biochem. J.* **283,** 273–279.
18. Bradford, M. M. (1976) A rapid and sensitive method for the quantitation of microgram quantities of protein utilizing the principle of protein-dye binding. *Anal. Biochem.* **72,** 248–254.
19. Cheng, Y.-C. and Prusoff, W. H. (1973) Relationship between the inhibition constant (K1) and the concentration of inhibitor which causes 50 per cent inhibition (I50) of an enzymatic reaction. *Biochem. Pharmacol.* **22,** 3099–3108.
20. Linden, J. (1982) Calculating the dissociation constant of an unlabeled compound from the concentration required to displace radiolabel binding by 50%. *J. Cyclic Nuc. Res.* **8,** 163–172.
21. Muradov, K. G., Granovsky, A. E., and Artemyev, N. O. (2002) Direct interaction of the inhibitory γ-subunit of rod cGMP phosphodiesterase (PDE6) with the PDE6 GAF domains. *Biochemistry* **41,** 3884–3890.

21

Purification, Reconstitution on Lipid Vesicles, and Assays of PDE6 and Its Activator G Protein, Transducin

Theodore G. Wensel, Feng He, and Justine A. Malinski

Summary

PDE6 in rod and cone photoreceptors is the principal effector of phototransduction. It is kept at a very low activity level in the dark, and in the light it is strongly activated by the guanosine 5′-triphosphate-bound form of the α-subunit of the G protein, transducin. Both transducin and PDE6 are peripheral proteins, and understanding both their interactions with one another and the roles of lipids in their function requires reconstituting purified proteins on the surfaces of defined lipid bilayers. We describe here methods for purifying the proteins, reconstituting them with vesicles, and assaying catalytic activity and binding.

Key Words

G proteins; phosphodiesterase; membrane reconstitution; cyclic nucleotides; phototransduction; retina; vision.

1. Introduction

The guanosine 3′,5′-cyclic-monophosphate (cGMP) phosphodiesterase (PDE) of vertebrate vision, PDE6, is a heterotetrameric membrane protein found on disk membranes of rod and cone photoreceptors. It is regulated by binding of a G protein, transducin ($G_{\alpha t}$) in its guanosine 5′-triphosphate (GTP) form. Formation of this activated GTP-bound form of $G_{\alpha t}$ is the pivotal reaction catalyzed by rhodopsin, the primary photon receptor, after its photoconversion to metarhodopsin II (R*). Although the transducin-PDE6 couple seems to be unique to photoreceptor cells, there are many features that this complex shares with other G protein-effector systems, so it can serve as a model for this very common type of signal transduction mechanism.

Both transducin and PDE6 are peripheral membrane proteins with covalently attached lipids *(1)*. $G_{\alpha t}$ has one of four fatty acids (C12:0, C14:0, C14:1, C14:2)

From: *Methods in Molecular Biology, vol. 307: Phosphodiesterase Methods and Protocols*
Edited by: C. Lugnier © Humana Press Inc., Totowa, NJ

attached to its N-terminal glycine in an amide linkage *(2–4)*. The initial translation products of the catalytic subunits of PDE6 in rods, PDE6α and PDE6β, have CAAX boxes at their carboxyl termini. These are substrates for cellular enzymes that remove the last three amino acids by proteolysis, attach either a 15-carbon farnesyl (PDE6α) or a 20-carbon geranylgeranyl (PDE6β) isoprenyl group to the cysteine via a thioether linkage *(5)*, and convert the C-terminal carboxyl group into a methyl ester. These hydrophobic modifications are thought to help keep the proteins tethered to the disk membranes on which phototransduction occurs.

The lipid surface on which PDE6 activation occurs is very important for the interaction with $G_{\alpha t}$ *(6,7)*. Although some activation of PDE6 by transducin can be observed in solution, it is very weak compared with what is observed on membranes, requiring a large molar excess of activated $G_{\alpha t}$ in order for substantial PDE6 activation to be observed. By contrast, on membranes under optimal conditions, a nearly stoichiometric binding of $G_{\alpha t}$ with PDE6 can be observed when both are present at nanomolar concentrations. The lipid head group has a strong influence on the strength of the activation, and the side chains of the phospholipids may play a role as well.

To probe the interactions of $G_{\alpha t}$ with PDE6 without interference from endogenous membrane proteins in rod outer segment (ROS) disk membranes, and to define the effects of lipid composition on the protein–protein interactions, it is highly useful to purify the proteins and reconstitute them on the surface of lipid vesicles with well-defined composition and structure. We provide here methods for preparing $G_{\alpha t}$ and PDE6 in highly purified form; for preparing and characterizing well-defined large unilamellar vesicles; for reconstituting $G_{\alpha t}$ in, and PDE6 on, their surface; and for assaying protein binding to, and enzymatic activity on, the vesicle surfaces.

2. Materials

2.1. Biological Materials

2.1.1. Bovine ROSs

ROSs are conveniently purified from commercially available frozen dark-dissected cattle retina. Papermaster and Dreyer *(8,9)* provide an excellent protocol for preparing them.

2.1.2. Lipids

Phospholipids and other membrane lipids of high quality are available in both powder and solution (chloroform or some other organic solvent) from Avanti Polar Lipids. Both synthetic and natural lipids can be used in the reconstitution procedures described here. Other suppliers of biochemicals also have phospholipid (e.g., Sigma, St. Louis, MO), including some not available from Avanti. Lipids should be stored at –80°C under argon.

2.1.3. Nucleotides

1. cGMP (Sigma G-6129; sodium salt, molecular weight = 367.2 g/mol).
   a. To prepare a 100 mM cGMP solution, dissolve the solid cGMP in Milli-Q water with vortexing. Adjust the pH to 8.0 with 0.1 M NaOH.
   b. Filter through a 0.22-µm syringe nitrocellulose filter.
   c. Use absorbance at 254 nm to determine the concentration of cGMP. Measure the absorbance at 254 nm in a quartz cuvet and use Beer's law to determine the concentration of cGMP: GMP ε_{254} = 12,950 M^{-1} cm^{-1}.
   d. Make an aliquot (20–50 µL) of cGMP and store at –20°C.
2. 100 mM GTP (Sigma G5884; lithium salt, formular weight [FW] = 523.2 g/mol) or GTPγS (Sigma G8634 tetralithium salt, FW = 563 g/mol) stock solution to extract transducin from ROSs. Dissolve GTP or GTPγS in Milli-Q water, adjust the pH with NaOH to 7.2 (*Caution:* if the pH has not been adjusted in the solution of GTP, it will denature or inactivate transducin owing to low pH during the GTP/low-salt wash steps), filter through a 0.2-µm nitrocellulose filter, and then determine GTP concentration by spectrophotometry using an extinction coefficient of 13,700 M^{-1} cm^{-1}.
3. Guanosine 5'-diphosphate (GDP) (Sigma G7127; sodium salt, FW = 443.2 g/mol) stock: This is made up in the same way as GTP.

2.1.4. Proteins

1. Trypsin (Sigma T1426; TPCK treated, Type XII): Prepare small aliquots of 10 mg/mL of trypsin with filtered Milli-Q water. Store the aliquots at –20°C.
2. Soybean trypsin inhibitor (Sigma T9003; Type I-S from soybean): Prepare small aliquots of 50 mg/mL of soybean trypsin inhibitor with filtered Milli-Q water. Its molecular weight is about the same as that of trypsin, so 10X by mass excess over trypsin mass is nearly a 10X molar ratio also. Store the aliquots at –20°C.

2.2. Buffers

2.2.1. General Considerations for All Buffers

All buffers are filtered through filters with 0.2-µm pores. The filters must be made of nitrocellulose to maximize protein binding. All buffers (except pH assay buffer) should contain, in addition to the ingredients detailed in **Subheadings 2.2.2.–2.2.8.**, the protease inhibitor phenylmethylsulfonyl fluoride (PMSF) and dithiothreitol (DTT). DTT is normally added from a 1 M stock just before filtering and using, and PMSF is added in solid form after filtering (except for high-performance liquid chromatography [HPLC] buffers, in which case the buffer with added solid PMSF is allowed to stir for 30 min just before filtration and use). PMSF is poorly soluble in water and hydrolyzes fairly rapidly; a few suspended crystals of PMSF thus serve as a reservoir of intact PMSF, slowly releasing it into solution as the pool in the solution hydrolyzes. Unless otherwise indicated, buffers are stored and used at 4°C. Buffers used for chromatography are degassed under vacuum, and this is especially important if buffers are used at a higher temperature than the one at which they were stored. Vacuum filtration through 0.2-μm filters is a very efficient degassing method.

2.2.2. Buffers for Extraction of Proteins From ROSs

1. Moderate-salt buffer: 10 mM 3-(N-Morpholino)propane sulfonic acid 4-Morpholino propane sulfonic acid (MOPS), pH 7.4, 30 mM NaCl, 60 mM KCl, 2 mM MgCl$_2$. Prepare 1.5 L of moderate-salt buffer for three washes (65 mL for each Ti45 ultracentrifuge bottle, six bottles for each spin, and a total of three washes). Another 1 L is needed the day of transducin extraction.
2. Low-salt buffer: 5 mM Tris-HCl, pH 7.2, 0.5 mM MgCl$_2$. Prepare 1 L for two washes of PDE on the day it is extracted, and an additional 1.5 L for low-salt wash and two low-salt/GTP transducin extractions for the day transducin is extracted.

2.2.3. Phosphate Buffers for Hydroxylapatite Chromatography

1. 1 L of 30 mM sodium phosphate buffer, pH 7.2, 50 mM NaCl.
2. 500 mL of 80 mM sodium phosphate buffer, pH 7.2.
3. 200 mL of 150 mL of sodium phosphate buffer, pH 7.2.
4. 200 mL of 300 mL of sodium phosphate buffer, pH 7.2.

2.2.4. Buffers for DEAE HPLC Chromatography of PDE6

Prepare 500 mL each of buffer A and buffer B:

1. Buffer A: 20 mM Tris-HCl, pH 7.4.
2. Buffer B: 20 mM Tris-HCl, pH 7.4, 1 M NaCl.

2.2.5. Buffer for HPLC Gel Filtration of PDE6

500 mL of 30 mM sodium phosphate buffer, pH 7.2, 300 mM NaCl.

2.2.6. Buffers for Transducin Chromatography on Hexylagarose

1. 1 L of buffer A$_0$: 5 mM Tris-HCl, pH 7.4, 0.5 mM MgCl$_2$.
2. 1 L of buffer B$_0$: 10 mM MOPS, pH 7.4, 2 mM MgCl$_2$.
3. 500 mL of buffer B$_{75}$: 10 mM MOPS, pH 7.4, 2 mM MgCl$_2$, 75 mM KCl.
4. 500 mL of buffer B$_{300}$: 10 mM MOPS, pH 7.4, 2 mM MgCl$_2$, 300 mM KCl.
5. 500 mL of buffer B$_{500}$: 10 mM MOPS, pH 7.4, 2 mM MgCl$_2$, 500 mM KCl.

2.2.7. Buffers for Separation of Transducin Subunits on Blue Sepharose

1. 1 L of buffer KCl$_0$: 10 mM MOPS, pH 7.4, 5 mM MgCl$_2$, 0.1 mM EDTA.
2. 500 mL of buffer KCl$_{100}$: 10 mM MOPS, pH 7.4, 100 mM KCl, 5 mM MgCl$_2$, 0.1 mM EDTA.

3. 500 mL of buffer KCl$_{750}$: 10 m*M* MOPS, pH 7.4, 750 m*M* KCl, 5 m*M* MgCl$_2$, 0.1 m*M* EDTA.

2.2.8. Buffers for PDE Activity Assays and for Preparation and Reconstitution of Vesicles

Two different buffer systems are used, depending on the total PDE activity and concentration of cGMP in the experiments. DTT and PMSF are omitted from buffers to be used only in assays. They should be included, as for other buffers (*see* **Subheading 2.2.**), when used for preparative purposes, such as the preparation of vesicles. In addition, for the preparation of sucrose-loaded vesicles, the buffer should be supplemented with 170 m*M* sucrose during the vesicle freeze-thaw and extrusion cycles.

1. MOPS buffer: This is routinely used for pH assay. There is not much buffering capacity at pH 8.0 because it is far from its pK_a (MOPS buffer pK_a = 7.2), so the assay is very sensitive to pH changes owing to cGMP hydrolysis. 5X pH assay MOPS buffer (40 mL): 100 m*M* MOPS, pH 8.0, 750 m*M* KCl, 10 m*M* MgCl$_2$. Filter with a 0.2 µ*M* nitrocellulose membrane, and then aliquot to 1 mL/tube and store at –20°C.
2. Tris-HCl buffer: This is used only for high PDE activity or high concentrations of cGMP. Tris pK_a = 8.1. When less buffering capacity is needed, a lower Tris concentration buffer is used. We use 10 m*M* Tris-HCl (final; 5X is 50 m*M*) for PDE activity <25 µ*M*-s^{-1} and 25 mM Tris-HCl (final, 5X is 125 m*M*) for PDE activity in the range of 25–160 µ*M*-s^{-1}.

2.3. Chromatography and Filtration Media

1. Nitrocellulose filters: Syringe filters, 0.2 µm nitrocellulose (other materials should not be used); 45-mm, 0.2-µm, or 0.45-µm nitrocellulose filters and vacuum filtering apparatus for filtering large buffer volumes.
2. Hydroxylapaptite (HAP): HAP Bio-Gel HTP Gel (cat. no. 130-0420, Bio-Rad, Hercules, CA). Each milliliter of packed HAP will bind about 5 mg of total protein, so we routinely use a 30- to 40-mL packed HAP column for 300–450 retinas' worth (two to three preparations of ROSs) of extracts. To prepare the column, add 1 part of HAP powder to 6 parts of phosphate buffer (30 m*M* sodium phosphate buffer, pH 7.2), swirl gently, then allow to settle for 5–10 min. Decant the supernatant containing fines. Repeat at least twice until most of the fines are removed. Degas and then pour into a 2.5 cm i.d. × 10 cm column.
3. DEAE HPLC column: The Protein-Pak™ DEAE-5PW is an anion (diethylaminoethyl)-exchange column from Waters (cat. no. WAT088044). The recommended flow rate is 0.5–1 mL/min (*do not exceed 1.2 mL/min*), and the maximum backpressure is 500 psi.
   a. Do not exceed 20% organic content in the mobile phase.
   b. Do not use sodium azide, sodium dodecyl sulfate (SDS), or any anionic detergents in the mobile phase.

c. Do not change the flow rate faster than increments of 0.5 mL/min.

d. Do not freeze the column.

e. Other ion-exchange columns also work for PDE purification, including Mono Q, AP, or QMA.

4. HPLC gel filtration column. Bio-Sil® SEC 250-5 is a size-exclusion column from Bio-Rad. The bed volume of this column is 14 mL, so the maximum sample volume is 0.7 mL (1–5% of bed volume is recommended). The operating flow rate is 1 mL/min and maximum pressure is 1500 psi. One can also use other compatible size-exclusion columns if available.

5. Gel filtration standards: Bio-Rad gel filtration standard (cat. no. 151-1901):

   a. Thyroglobulin (5 mg, 670 kDa).

   b. Bovine gamma globulin (5 mg, 158 kDa).

   c. Chicken ovalbumin (5 mg, 44 kDa).

   d. Equine myoglobin (2.5 mg, 17 kDa).

   e. Vitamin B12 (0.5 mg, 1.35 kDa).

   Add 0.5 mL of Milli-Q water to the lyophilized standard, swirl gently to mix, and allow the vial to stand for 2 to 3 min, the final 36-mg components/mL. Centrifuge the standard solution at 16,000g for 15 min at 20°C before application to remove any fine particulates.

6. Hexylagarose (cat. no. H-1882; Sigma): This comes as a suspension in 500 mM NaCl. The capacity of the packed hexylagarose is about 10–12 mg/mL of total proteins based on binding of bovine serum albumin (BSA), so we routinely use a 10- to 15-mL column for two preparations of ROSs.

   a. Degas, pour, and equilibrate the column. The flow rate for this column usually is 0.5 to 0.6 mL/min.

   b. Hexylagarose contains 0.5 M NaCl; ten column volumes of buffer A$_0$ is necessary to completely equilibrate the column in zero NaCl.

7. Reactive Blue 2 Sepharose CL-6B (Sigma): Each gram of resin can make 5 mL of the column. We routinely use 20–30 mL of column to isolate Gt-subunits from the Gt extract of two ROS preparations. At room temperature, swell the resin in buffer KCl$_0$, decant the fines, degas the media under vacuum, and pour the column. Then transfer the column to a cold room and equilibrate the column with 5 column vol of buffer KCl$_0$. If a used column is available, regenerate the column with 10 column vol of 10 mM MOPS (pH 7.4), 2 M KCl or with 6 M urea (in most cases, using 2 M KCl in MOPS buffer is sufficient); then equilibrate the column with 10 column vol of buffer KCl$_0$.

8. Polycarbonate filters for vesicle extrusion and tubes for freeze-thaw:

   a. Poretics® polycarbonate filters, 0.1 µm (cat. no. K01CP02500, Osmonics).

   b. Corex® centrifuge tubes, for drying, hydrating, freezing, and thawing lipids.

2.4. Assay Reagents

2.4.1. Phosphate Assay Reagents

1. 0.1 N H$_2$SO$_4$: concentrated.

2. 70% HClO$_4$, as supplied.

3. Ammonium molybdate ($[NH_4]_6Mo_7O_{24}$) (Sigma A1343, tetrahydrate, FW = 1235), 2.5% (w/v) solution.
4. Ascorbic acid (Sigma A0278), 10% (w/v) free acid form. This solution should be less than 7-wk old, stored at 4°C, and nearly colorless.
5. Phosphate standard (Sigma 6661-9).

2.4.2. Protein Assay Reagent

This is used for the Coomassie blue binding assay of Bradford *(10)*. The reagent is available premixed with instructions from Pierce or Bio-Rad.

2.5. Apparatus

2.5.1. Spectrophotometer

An ultraviolet (UV)/visible spectrophotometer is needed for determining phosphate, nucleotide, and protein concentrations. Use plastic cuvets for protein Bradford assays and phosphate assays, and quartz cuvets for measuring the absorbance of nucleotide solutions.

2.5.2. Ultracentrifuge, Rotor, and Bottles

1. Beckman L7-55 ultracentrifuge (or equivalent).
2. Type 45-Ti rotor and bottles.
3. TL-100 or equivalent tabletop ultracentrifuge.
4. TLA-100.3 fixed-angle ultracentrifuge rotor or equivalent.
5. Tubes for TLA-100.3, both polycarbonate tubes and microcentrifuge-style polyallomer tubes with adapters.

2.5.3. Syringes and Pipetors

1. Standard air displacement adjustable volume micropipeptors.
2. Glass/stainless steel/Teflon microsyringes (e.g., Hamilton). Standard biological airdisplacement pipets (e.g., Pipetman, Eppendorf) can be used for aqueous solutions but must not be used for organic solvents. The solvent degrades the plastic materials and the high vapor pressure interferes with accurate volume measurement.
3. Positive displacement micropipetors or glass microcapillary pipets for strong acids; these can also be used for organic solvents.

2.5.4. Mixers and Homogenizers

1. Potter-Elvehjem (glass vessel, round-end Teflon pestle) homogenizer, for washing ROS membranes during extraction of PDE and transducin.
2. Glass rod with rubber policeman on the end, for resuspending ROS membranes.

2.5.5. Chromatography and Electrophoresis

1. Standard setup for low-pressure chromatography: peristaltic pump, tubing, fraction collector. A UV detector is useful but not essential.

2. HPLC or fast protein liquid chromatogrpahy (FPLC) system: Either standard stainless-steel systems, or glass-Teflon, or titanium-based systems can be used. To avoid heavy metal contamination do not use porous stainless steel "sinkers" on the intake ends of the solvent supply tubes in the reservoirs. For standard HPLC systems, use seals designed for use with aqueous salt solutions, and thoroughly clean out all parts of the system before and after use with HPLC-grade water to avoid contact of salt solutions with organic phases.

3. Sodium dodecyl sulfate-polyacrylamide gel electrophoresis (SDS/PAGE): A standard setup for protein electrophoresis, either "minigel" or standard size, is essential for checking the purity of protein fractions and assaying protein binding to vesicles.

2.5.6. Vesicle Extrusion Device

Nitrogen pressure extruder supplied by Lipex Biomembranes: This requires an N2 tank and a high-pressure regulator. See the manufacturer's instructions for details of high-pressure connections.

2.5.7. pH Assay Assembly

1. Microtiter plate (e.g., Dynatech 001-012-9205; order from PGC 05-6124-24).
2. Magnetic stir plate: If the plate is warm to the touch, a thermostated plate must be placed between the stir plate and the microtiter plate to ensure constant temperature. We use one scavenged from an old spectrophotometer with tubes for water inlet and outlet connected to a circulating water bath.
3. Flea-sized magnetic stirring bars (cat. no. 37119-0005, Bel-Art products).
4. Microelectrode: MI410 Combination pH electrode (provided by Microelectrodes).
   a. When ordering the electrode, make sure the connectors match those on the pH meter, e.g., BNC.
   b. Check the electrode before the experiment.
   c. Remove the sleeve from the electrode fill hole; check that the outer reference chamber is full to just below the fill hole. If it is not at the proper level, it must be refilled with 3 M KCl/saturated AgCl solution (provided by Microelectrodes).
   d. Calibrate the electrode with pH 7.0 and 10.0 standards prior to the experiment.
5. pH meter with voltage output.
6. Chart recorder and/or analog-to-digital converter in PC.

3. Methods

3.1. Extraction of PDE and Transducin From Bovine ROSs

3.1.1. General Strategy

Bleached ROSs are sequentially washed with buffers varying in ionic strength to produce extracts that are highly enriched in either PDE6 or transducin *(11)*. This protocol is written for ROSs prepared from 300 to 450 (two to three preparations of ROSs) retinas, but it is easily scaled up or down.

1. Wash three times with moderate-salt buffer to remove the soluble and not tightly bound proteins.
2. Wash two times with low-salt buffer to extract PDE from the ROS membrane.
3. Freeze the membranes overnight (or longer) while the PDE extract is processed (alternatively, if enough personnel are available, one person can extract the transducin while another processes the PDE).
4. Wash two additional times with moderate-salt and once with low-salt buffer to remove proteins released by the freeze-thaw cycle (can be skipped if transducin is extracted immediately without freezing the membranes).
5. Wash two times with low-salt buffer containing GTP or GTPγS to extract transducin from the ROS membrane.

The total time from thawing the ROSs to finishing the second GTP/low-salt wash is about 7–9 h if both PDE and transducin are extracted in 1 d, and it is important to process the PDE immediately. Therefore, in general, if it is not necessary to extract both in one day, the pellet after the second low-salt wash is frozen at –80°C, and when transducin extract is needed, the low-salt-washed pellets must be washed twice with moderate-salt buffer and once with low-salt buffer before the GTP/low-salt buffer is used to extract transducin, because contaminating proteins from ROSs are released from the freeze/thaw step.

3.1.2. General Cautions

PDE6 is exceptionally sensitive to proteolysis. Therefore, it is very important that any surface that will touch the protein-containing extract be kept clean, cold, and covered. Wear gloves all the time during the preparation to protect the extracts from personal proteases. Filter all the buffers with a 0.2-μm nitrocellulose membrane. Transducin is extremely sensitive to oxidation; DTT or other reducing reagents are important for extraction and purification of transducin.

3.1.3. Selective Extractions of Proteins From ROSs

1. Thaw and bleach the ROSs on ice (remove aluminum foil in room light). Occasional inverting to mix the ROSs will ensure that all of the rhodopsin is activated to R* and help the ROSs thaw quickly.
2. While the ROSs are thawing, turn on the ultracentrifuge and the vacuum, and prechill all the Ti45 bottles, homogenizer, and a 500-mL beaker on ice.
3. Carry out the first isotonic wash (moderate-salt wash) as follows:
   a. Combine all ROSs (two to three preparations) in a 500-mL beaker, and dilute the ROSs to a final volume of about 390 mL with moderate-salt buffer; add solid PMSF to the diluted ROSs, and mix with a rubber policeman.
   b. Add about 60 mL of diluted ROSs to the homogenizer each time; homogenize the ROSs with a Teflon pestle until no macroscopic particles are visible. Because proteins denature at air-water interfaces, avoid foaming and minimize the number of times the pestle is pulled completely out of the solution during

homogenization. Keep the homogenizer surrounded by ice, but do not let any ice contaminate the sample.

c. Transfer the homogenized ROSs to six Ti45 bottles. Make sure the Ti45 bottles are well balanced. Clean the outside of the bottles with a paper towel, and then load into a Ti45 rotor. The Ti45 bottles must be filled to the lip (point of narrowing). If partially full Ti45 bottles are spun, there is a high probability that they will collapse.

d. Centrifuge with the Ti45 rotor at 150,000g for 20 min at 4°C. Watch the speed of the rotor until it comes up to 150,000g before leaving the centrifuge.

e. After centrifugation, remove the supernatant from the Ti45 bottles with a plastic tube (iv tubing) attached to a 60-mL syringe. Pool and store the supernatants from each spin on ice until the assays are complete to make sure they do not contain PDE or transducin by SDS-PAGE or pH assay. Then discard the supernatant.

4. Carry out the second isotonic wash as follows:
   a. Add moderate-salt buffer to the pellet (final volume of 65 mL/tube), use the rubber policeman to resuspend the pellet, and then place in the homogenizer.
   b. Homogenize as for the first wash.
   c. Spin again in the Ti45 bottles at 150,000g for 20 min at 4°C.
   d. Save all the supernatants and keep at 4°C until the assays are completed.

5. For the third isotonic wash, repeat **steps 4a–d**.

6. Carry out the first hypotonic wash, to extract PDE from the ROSs, as follows:
   a. Add low-salt buffer to each pellet in a final volume up to 65 mL, use the rubber policeman to resuspend the pellet, and then put into the homogenizer.
   b. Homogenize as in **step 3b**.
   c. Spin again in the Ti45 bottles at 150,000g for 40 min at 4°C, because pelleting is much less efficient in low salt. Take care when removing the supernatant to avoid contamination with the pellet.
   d. Pool the first hypotonic wash containing 75–90% of total PDE in a 500-mL bottle, and add fresh DTT (2 mM) and solid PMSF to the extract.
   e. Measure the PDE activity in the two isotonic washes (*see* **Subheading 3.7.2.**) and the first hypotonic wash. Determine the amount of PDE in the hypotonic wash (should be about 1 to 2 mg/300 retinas), and start loading it to the HAP column.

7. Carry out the second hypotonic wash as follows:
   a. Repeat **steps 6a–e**.
   b. Pool the second wash in another 500-mL bottle, and add solid PMSF and fresh DTT to the PDE extract. Because the second one only contains 10–25% of total PDE, it is not recommended that the first and second washes be combined until one checks the PDE activity or runs a gel.
   c. PDE extract can be further concentrated for use as a crude extract or for chromatographic purification as described in **Subheading 3.2.2.** If the PDE is going to be purified the next day, keep the extract on ice in a cold room or cold box

after adding DTT and solid PMSF. If the PDE cannot be purified PDE immediately, try to concentrate the extract (the extract should be subjected to ultracentrifugation first, to remove traces of membrane that will clog ultrafiltration filters in concentrators); add glycerol (40% [v/v]), fresh DTT, and solid PMSF to the crude PDE; and then keep it at –20°C for a short time (a couple of weeks). Long-term storage is not recommended because the crude extract may still be contaminated with proteases.

8. Carry out the first GTP/low-salt wash, to extract transducin from ROSs, as follows:
   a. Add low-salt buffer to each pellet with a final volume to 65 mL, use the rubber policeman to resuspend the pellet, and then transfer to the homogenizer.
   b. Homogenize as for **steps 6a–e**.
   c. Transfer the homogenized ROSs to Ti45 bottles, add 65 μL of 100 mM GTP (pH 7.2) to each bottle (final: 100 μM GTP), and mix by inverting several times. Then balance the bottles and load into the rotor. Timing is important in this step, because transducin will hydrolyze GTP to GDP and rebind to R* at a rate of approx 2/min. In our typical prep, R* is diluted to approx 5–10 μM and transducin is approx 0.25–0.5 μM, so it will take 100–200 min to hydrolyze 100 μM GTP.
   d. Spin in the Ti45 rotor at 150,000g for 40 min at 4°C.
   e. Pool the first GTP/hypotonic wash that contains most of the transducin in a 500-mL bottle, and add fresh DTT and solid PMSF to the extract.
   f. Concentrate the transducin with an Amicon pressure concentrator at 4°C. After concentration, either one can add fresh DTT to 2 mM with solid PMSF and 40% (v/v) glycerol and then store at –20°C, or one can purify further by chromatography as described in **Subheading 3.3.**

9. Carry out the second GTP/low-salt wash as follows:
   a. Repeat **steps 8a–f**.
   b. The second GTP/low-salt wash usually contains little or no transducin. Check the second wash on SDS-PAGE, and if it contains sufficient amounts of transducin, the second wash can be pooled with the first one and concentrated together.
   c. Resuspend the final pellets in a minimum volume of the low-salt buffer, pool them in a 50-mL conical tube, and store at –80°C for further purification of other ROS proteins (e.g., RGS9-1, R9AP, PKC).

10. Save a 1-mL aliquot of each supernatant for use in gel samples (100–200 μL of each supernatant + 100–200 μL of 10% trichloroacetic acid [TCA]), PDE activity pH assay (25–50 μL), and Bradford assay (100–200 μL).

3.2. Purification of PDE6 From Low-Salt Extracts of ROSs

3.2.1. Overview

This protocol is for use with extracts from ROSs equal to 300–450 retinas (two to three preparations of ROSs). PDE isolated from low-salt washes can be sequentially purified in HAP *(7)*, DEAE, and gel filtration columns *(12)*, depending on the requirements of the experiments. HAP chromatography is particularly useful for separating PDE6 from HSP-90, arrestin, and other contaminating ROS proteins. The DEAE HPLC column is a weak anion-exchange column, and elution is with an ionic strength gradient. It removes proteins not removed by HAP but will not separate PDE from HSP-90. High-resolution DEAE HPLC will separate PDEαβ from PDEαβγγ. After sequential HAP and DEAE chromatography, the purity of PDE is usually more than 90%. If even more pure PDE is required, PDE can be further purified by gel filtration.

3.2.2. HAP Chromatography of PDE6 Using Step Elution

All the purification steps should be performed in a cold room (4°C) to stabilize PDE activity and minimize protease activity. This step-elution method is routinely used to purify PDE from ROS low-salt extract, but a gradient method also can be used.

1. Equilibrate 40-mL of packed HAP column with 400 mL of 30 mM sodium phosphate (NaPi), pH 7.2. The flow rate for the HAP column usually is 0.5–1.5 mL/min under gravity. If one needs to use the pump, determine the flow rate of HAP column under gravity first, and do not exceed the flow rate under gravity with the pump.
2. Add 300 mM sodium phosphate, pH 7.2, to PDE extract to make the final concentration of NaPi 30 mM. Add a few crystals of PMSF.
3. Load the PDE to HAP column under gravity and save the flowthrough. It is recommended that the SAFETY-LOOP be set up and the PDE be loaded overnight on the day of PDE extraction. The PDE extract is about 390 mL for each wash and the flow rate in the HAP column is 1 mL/min, so one needs at least 7 h to complete the loading. Begin to load the first low-salt extract as soon as possible, even while the second one is in progress.
4. Wash the column with 300–400 mL (about 10 column vol) of 80 mM NaPi (pH 7.2), 1 mM DTT, and PMSF, and collect and save the flowthrough. Most of the contaminating protein in this fraction is arrestin; occasionally, a little PDE will elute in this fraction too.
5. Elute PDE with 150 mM sodium phosphate (pH 7.2), 1 mM DTT, and PMSF. Collect 5 mL/fraction with the fraction collector. During the elution, monitor the PDE by an UV monitor, a mini Bradford assay (add 100 µL of 1X Bradford reagent and 15–20 µL of each fraction to each well in a 96-well plate), or PDE activity (2–10 µL for each fraction).
6. Elute (optional) with 300 mM sodium phosphate. This step has very little PDE but most of the HSP-90.
7. Run a 10% acrylamide SDS-PAGE: 0.5 mL for loading flow through (FT) and wash FT fractions and 20–100 µL of each elution fraction. After running the gel, the HAP can be discarded, and it is not recommended that the column be regen-

erated and reused again, because HAP crystals can gradually break down to fines that block the column.

8. Pool and concentrate the PDE (from **step 5**) based on the gel or PDE activity. If this is the only column one is going to use to purify the PDE, make the concentrate 40% in glycerol, add a few crystals of PMSF, and store at –20°C. Otherwise, during the concentration, change the buffer to 20 mM Tris-HCl (pH 7.4), 100 mM NaCl, and then purify in a DEAE-5PW column.

3.2.3. DEAE HPLC Chromatography of PDE6

1. Connect a Protein-Pak™ DEAE-5PW column to an HPLC or FPLC system, and set the maximum pressure at 400 psi and the flow rate to 1 mL/min. Set up the instrument software for injection, washing, and gradient elution according to the conditions listed in **steps 4–6**. HPLC can be carried out at room temperature, although operation at 4°C probably helps the protein stability, and fractions should be covered and placed on ice immediately after elution.
2. Equilibrate the DEAE column with 10% buffer B and 90% buffer A (i.e., starting condition is 100 mM NaCl) for 20 min at a flow rate of 1 mL/min.
3. Spin the concentrated PDE (from **step 5** in **Subheading 3.2.2.**) at 82,000g for 20 min at 4°C using a TL100 centrifuge to remove the insoluble proteins.
4. Inject the PDE into the sample loop and column. Save the flowthrough.
5. Wash the column with 10% buffer B and 90% buffer A for 10 min. Save the wash fraction.
6. Elute the PDE from the DEAE column with a gradient from 10% buffer B to 100% buffer B in 50 min. Collect fractions of 0.5–1 mL.
7. Check the purity of the PDE by analyzing 10 µL of each fraction by 10% acrylamide SDS-PAGE.
8. Pool the pure PDE fractions and dialyze against 20 mM Tris-HCl (pH 7.4), 2 mM MgCl$_2$, 1 mM DTT, and PMSF at 4°C.
9. Concentrate the PDE with Centri-prep 50 or another concentrator. If this is the last column to be used to purify PDE, make the concentrate 40% in glycerol, add a few crystals of PMSF, and store at –20°C. Otherwise, during the concentration, change the buffer to 30 mM sodium phosphate buffer, pH 7.2, and 300 mM NaCl for gel filtration column.

3.2.4. Gel Filtration Chromatography of PDE6

1. Set the maximum pressure at 1300 psi and the flow rate to 1 mL/min in HPLC. *See* **Subheading 3.2.3.1.** for comments on temperature.
2. Equilibrate the gel filtration column with 30 mM NaPi (pH 7.2), 300 mM NaCl for >30 min at a flow rate of 1 mL/min.
3. Test the column with gel filtration standard at least twice. Inject 20 µL of gel filtration standard onto the column, run HPLC at 1 mL/min, and record the retention times.
4. Spin the concentrated PDE (from DEAE fractions) at 82,000g for 20 min at 4°C with a TL-100 ultracentrifuge.

5. Inject the supernatant of PDE onto the column and elute the PDE with equilibration buffer (30 m*M* sodium phosphate buffer, pH 7.2, and 300 m*M* NaCl) at a flow rate of 1 mL/min. Collect 0.5- to 1-mL fractions.
6. Check the purity of the PDE by SDS-PAGE; load 10 μL for each fraction to 10% acrylamide gel. It usually is very pure (about 99%), as evidenced by a lack of detectable contaminating proteins on the gel.
7. Pool the pure fractions, concentrate and dialyze the PDE, add glycerol up to 40% (v/v) with a few crystals of PMSF, and store at –20°C.
8. After purification, wash the size-exclusion column extensively, using at least 10 column vol, with the equilibration buffer to remove small-molecule contamination. Then rinse with several column volumes of HPLC-grade water, and replace the mobile phase with 0.05% sodium azide or 5% methanol in water.

3.3. Purification of Transducin by Hexylagarose Chromatography

3.3.1. Overview

This protocol is for use with extracts from our standard 300 retinas (two preparations of ROSs). It is easily scaled up or down. Transducin is extracted from low-salt-stripped ROSs with low-salt buffer and GTP and then purified on hexylagarose using steps of increasing NaCl (*11,13*). In general, multiple washes with low-salt buffer, before stripping with low-salt buffer and GTP, will eliminate 95% of the proteins and the crude transducin extract will already be very clean. The hexylagarose column primarily concentrates transducin and cleans up lipids and a few contaminating proteins. If highly pure transducin is required, transducin can be further purified with a DEAE column (*see* **Subheading 3.2.3.**). All the purification steps should be performed in a cold room (4°C) to stabilize transducin activity and minimize protease activity.

3.3.2. Hexylagarose Chromatography of Gt

1. Hexylagarose comes as a suspension in 500 m*M* NaCl. The capacity of the packed hexylagarose is about 10–12 mg/mL of total proteins based on binding of BSA, so we routinely use a 10- to 15-mL column for two preparations of ROSs. Degas the resin under vacuum, and then pour and equilibrate the column. The flow rate for this column usually is 0.5 to 0.6 mL/min. Ten column volumes of buffer A_0 are necessary in order to completely equilibrate the column in zero NaCl.
2. Equilibrate the column with 10 column vol of buffer A_0. If one needs to use the pump, determine the flow rate of hexylagarose column under gravity first and do not exceed the flow rate under gravity with the pump.
3. Add a shake of PMSF and fresh 1 m*M* DTT to the GTP/low-salt extract (*see* **Subheading 3.1.3.8.**) and load the extract to the column under gravity or with the pump turned on. After the first extract is available, start to load the extract, which contains more than 90% transducin, onto the column; there is no need to wait and mix with the second extract. It is recommended that a "safety-loop" be set up in the column and that the Gt extract be loaded overnight starting immediately on extraction of Gt. Save the flowthrough.

4. During the loading, look at the column; occasionally, if an extract has been contaminated with ROSs, the column will be orange. If this happens, do not reuse the column.

5. Wash the column with 10 column vol of buffer B_0, and save the flowthrough, because Gt occasionally does not bind to the column owing to a problem with either the column or the buffer.

6. Wash the column with 30 mL of buffer B_{75}, and collect 3-mL fractions with a fraction collector (using a 12×75 mm plastic tube). It is very important to monitor these fractions carefully, because sometimes Gt will start to bleed off in this step.

7. Elute the Gt with buffer B_{300}, and collect 3-mL/fractions. Gt usually elutes in fractions 8–10 in this step.

   a. On some occasions, two peaks may be seen at this step, the first of which contains a contaminating protein that runs just above Gt on SDS-PAGE and the second of which contains most of the Gt. If this happens, remove the buffer "head" carefully before changing the step buffer.

   b. Sometimes Gαt and Gβγt are separated in different fractions. If this happens, monitor the fractions carefully and pool fractions of Gαt and Gβγt together.

8. Run a 12% acrylamide SDS-PAGE; load 20–100 µL for each fraction on the gel (precipitate the protein with TCA prior to adding SDS-PAGE sample application buffer to obtain a manageable volume). Always run a purified holo-Gt standard on the gel and run the gel long enough to separate the α and β transducin subunits.

9. Pool the Gt based on the gel and dialyze the Gt against the low-salt buffer (10 mM Tris-HCl, pH 7.4; 2 mM MgCl$_2$; 1 mM DTT; and solid PMSF). Do not concentrate the Gt before dialysis because it tends to precipitate in high- or moderate-salt buffer during the concentration.

10. Concentrate the Gt with a concentrator and make the concentrate 40% in glycerol; add a few crystals of PMSF, 2 mM DTT, and 50 µM GDP; and store at –20°C.

11. Wash the column with buffer B_{500}. This step usually does not elute a significant amount of Gt but can clean up and regenerate the column. Finally, wash the column with 10 column vol of buffer B_0 containing 5 mM EDTA and 0.05% NaN$_3$.

12. If the Gt is going to be purified further with a DEAE column, after dialysis and concentration, add NaCl (100 mM final concentration) to the Gt. Spin at 82,000g for 20 min at 4°C with a TLA-100.3 tabletop ultracentrifuge rotor. The supernatant of the Gt is now ready to load onto a DEAE column in HPLC, using essentially the same procedure as described for PDE (*see* **Subheading 3.2.3.**).

3.4. Separation of Gαt- and Gβγt-Subunits by Blue Sepharose Chromatography

3.4.1. Overview

Transducin subunits bind tightly to Cibacron blue dye (Reactive Blue 2) on blue Sepharose columns and are individually eluted with "steps" of increasing concentrations of KCl, with Gβγt eluting before Gαt *(14)*. This protocol *(7,15)* is used to separate the Gt-subunits from the Gt extract of two pooled preparations of ROSs, and can be used to separate either GTPγS-bound or GDP-bound Gαt from Gβγt-subunits.

3.4.2. Purification of Gt-Subunits

1. Reactive Blue 2 Sepharose CL-6B (Sigma): Each gram of resin can make 5 mL of the column. We routinely use 20–30 mL of column to isolate Gt-subunits from the Gt extract of two ROS preparations. At room temperature, swell the resin in buffer KCl_0, decant the fines, degas the media under vacuum, and pour the column. Then transfer the column to a cold room and equilibrate the column with five-column vol of buffer KCl_0.
2. If a used column is available, regenerate the column with 10 column vol of 10 mM MOPS (pH 7.4), 2 M KCl or with 6 M urea (in most cases, using 2 M KCl in MOPS buffer is sufficient). Then equilibrate the column with 10 column vol of buffer KCl_0.
3. Prepare the extract for Blue Sepharose column as follows:
   a. To separate GDP-bound Gαt from Gβγt-subunits, use either a concentrated low-salt/GTP-stripped extract from ROSs, or hexylagarose-purified Gt. In the latter case, the Gt must be dialyzed against buffer KCl_0 before being applied to the blue sepharose column. Add GDP (50 µM final concentration, pH 7.4) to the Gt just prior to loading onto the column.
   b. To obtain activated Gαt (GTPγS-Gαt), it is recommended that the Gt be stripped from the ROSs with GTPγS/low-salt buffer directly. Otherwise, Gαt can be activated with R*. Briefly, mix the Gt with urea-washed ROSs containing rhodopsin at a ratio of 1 rhodopsin for 100 Gt in the presence of GTPγS in a darkroom. Then incubate the mixture on ice under room light for 5–10 min to convert R into R* and activate the Gt. ROS membranes can be removed by centrifuging at 150,000g at 4°C for 40 min in low-salt buffer. The supernatant should contain GTPγS-Gαt and Gβγt.
   c. Measure the extract volume. We found that cleaner and more efficient separation of transducin subunits occurs when the total volume of extract is approximately equivalent to 1 column vol. Therefore, it is helpful to adjust the volume by concentrating or adding buffer. Loading a large volume of Gt extract is not recommended.
4. If the column has not been equilibrated, equilibrate it with 10 column vol of buffer KCl_0 under gravity. If the pump must be used, determine the flow rate of the blue Sepharose column under gravity first and do not exceed the flow rate under gravity with the pump, or the beads will be crushed and the flow obstructed.
5. Add 50 µM GDP to the concentrated Gt extract in order to isolate GDP-Gαt from Gβγt. It is not necessary to add GDP to the GTPγS-activated Gt extract.
6. Adjust the concentration of $MgCl_2$ to 5 mM in the Gt extract, because transducin binds much better to the column in 5 mM $MgCl_2$.
7. Let the equilibration buffer run down to the top of the column bed or remove the "head" of the buffer (to avoid diluting the extract), load the extract onto the

column with a pipet, and run the column by gravity. Collect and save the flowthrough.

8. After the sample has been completely loaded and allowed to drain to the top of the resin bead, add a head of buffer KCl_0 to the column (but without disturbing the media), and then wash the column with 100–150 mL of buffer KCl_0 (3–5 column vol).

9. Elute the Gβγt with buffer KCl_{100}. Remove the head of the buffer KCl_0, add a new head of buffer with buffer KCl_{100}, and then elute the Gβγt with 4 to 5 column vol of buffer KCl_{100}. Collect 3-mL fractions in 12 × 75 mm plastic tubes. The Gβγt should begin to elute following one-third of the column volume in the buffer KCl_{100} elution step. Check the last fraction with a Coomassie-Blue binding *(10)* assay to determine whether a detectable amount of protein is still coming off after elution with 5 column vol, and, if so, use more buffer KCl_{100} to elute the proteins until no more protein can be detected in the eluent.

10. Elute the Gαt with buffer KCl_{750}. Remove the head of the buffer KCl_{100}, add a new head of buffer with buffer KCl_{750}, and then elute the Gαt with 4 to 5 column vol of buffer KCl_{750}. Collect 3 mL fractions in 12 × 75 mm plastic tubes.

11. Run a 12% acrylamide SDS-PAGE gel to determine which fractions to pool. Load 20–100 μL/fraction onto the gel. Include a holo-Gt as a standard so that each subunit can be properly identified. Often a little Gβγt bleeds out with the KCl_0 wash, and some Gαt bleeds out at the end of the KCl_{100} elution.

12. Pool the fractions of pure Gβγt and Gαt separately. Dialyze the pooled fractions with low-salt buffer (10 m*M* Tris-HCl, pH 7.4, 2 m*M* $MgCl_2$, 1 m*M* DTT, and solid PMSF). Do not concentrate Gt-subunits before dialysis, to avoid losses owing to precipitaton.

13. After dialysis, concentrate Gt-subunits with Amicon or Centricon ultrafiltration devices. Then make the concentrate 40% in glycerol, add a few crystals of PMSF, 2 m*M* DTT, and 50 μ*M* GDP, and store at −20°C.

14. Clean the column with 10 column vol of a buffer containing 10 m*M* MOPS (pH 7.4), 2 *M* KCl, and 6 *M* urea, then wash with 10 column vol of buffer KCl_0; and, finally, store the column in buffer KCl_0 supplemented with 1 m*M* EDTA and 0.05% NaN_3. Seal the column top and bottom with the caps and store at 4°C. The column can be reused many times.

15. $G_{\alpha t}$-GTPγS prepared using this procedure is pure enough for most experiments. However, if highly pure $G_{\alpha t}$-GTPγS is required—e.g., if high concentrations will be used so that small traces of PDE become a concern—$G_{\alpha t}$-GTPγS can be further purified by DEAE HPLC chromatography and/or gel filtration using essentially the same procedures as those described for PDE6 in **Subheadings 3.2.3. and 3.2.4.**).

3.5. Preparation of Lipid Vesicles

3.5.1. Overview

This procedure *(16,17)* results in a preparation of vesicles that predominantly have a single bilayer, and a narrow size distribution in the hundred-nanometer range, in addition to a well-defined lipid composition. Thus, it is possible to know with some accuracy the total lipid surface available for binding proteins, and to eliminate the degree of curvature of the membrane surface as an uncontrolled variable. Fairly homogeneous small unilamellar vesicles can also be prepared by sonication, but their small radius of curvature is not optimal for reconstitution of the $G_{\alpha t}$complex *(7)*. The basic idea is to form a thin layer of lipid by drying, hydrate it uniformly by multiple freeze-thaw cycles, and then to force it multiple times through well-defined submicron pores to produce uniform vesicles. This last step relies on a commercially available extrusion device using high-pressure nitrogen, but *see* **Note 1**.

3.5.2. Preparation of Vesicles

1. Clean a Corex tube and gastight (e.g., Hamilton) syringes with chloroform.
2. Remove the desired volume of each lipid solution from stock, under argon, and place in the tube. Cover the stocks with an argon blank, reseal, and store at –80°C. For example, to make 20 mg of vesicles containing 40:39:20:1 (molar ratio) phosphocholine (PC):phosphoethanolamine (PE):phosphoserine (PE):rhodamine-labeled PE, remove (403-2 µL of 20 mg/mL PC, 372 µL of 200mg/mL PE, 207-7 µL of 20 mg/mL PS, and 34.2 µL of 10 mg/mL rhodamine PE. The rhodamine-labeled PE is included to facilitate separation of the supernatant and pellet in sedimentation experiments and is optional if such procedures are not to be carried out.
3. Dry lipid to a thin homogeneous film using a gentle stream of argon or nitrogen in a fume hood. Holding the tube in one hand, tilt it so that the lipid solution spreads up the sides of the tube. While rotating the tube, direct a very gentle stream of gas into the tube. It is very important that the lipid surface be smooth, with no trapped solvent. If the surface after drying does not look smooth, redissolve in a small amount (200–500 µL) of chloroform and carefully repeat the drying procedure.
4. Extensively clean the syringes with chloroform immediately after using.
5. Place the tube in a vacuum chamber (e.g., a vacuum desiccator or a stoppered side-arm flask), and dry under vacuum for at least 2 h (it is convenient to keep under vacuum overnight, but no longer, to ensure complete removal of solvent. Longer periods under vacuum can lead to contamination with oil from the vacuum system).
6. Add 1–10 mL of buffer, depending on the final concentration desired. Remember that some lipid will be lost during subsequent procedures. For PDE assays, a good buffer is pH assay buffer *(see* **Subheading 2.2.8.**). It should be supplemented with 170 m*M* sucrose if sedimentation experiments are planned *(17)*. The lipids should be covered with an argon blanket and the tube covered with parafilm. Vortex the sample for 1-min intervals every 5 min over a 15- to 30-min period, until the surface film is no longer apparent and the suspension has a uniform "chalky" appearance. The lipids are now hydrated in large multilamellar

structures, but with uneven distribution of aqueous solution throughout the lamellae.

7. Subject the lipids to six freeze-thaw cycles. These involve immersing in liquid nitrogen (or, if unavailable, a dry ice–acetone slurry) until frozen solid, then placing in water until thawed. This procedure enhances the distribution of aqueous solution within the lamellae. To prevent contamination from acetone or water, cover the tubes with Parafilm pierced with two or three tiny holes.

8. Extrude the lipids through submicron pores in two stacked polycarbonate filters to produce uniformly sized unilamellar vesicles. This procedure is most conveniently carried out using a nitrogen pressure extruder supplied by Lipex Biomembranes (*see* **Note 1**). The directions supplied by the manufacturer should be followed carefully. The filters used must be polycarbonate filters supplied by Osmonics. These have much more uniform size than normal commercial filters. They are available in a range of pore sizes, but 0.1-µm pores are recommended, because they produce predominantly unilamellar vesicles with a narrow and stable size distribution. Load the lipid suspension into the sample compartment, and then seal. Apply pressure until all the liquid is eluted into a collection tube. Relieve the pressure, and load the sample again, until 10 extrusion cycles have been carried out. At this point, the lipid suspension should no longer be "chalky" but, rather, should be transparent with an opalescent appearance.

9. To remove unwanted solutes (e.g., sucrose) from the external solution, dialysis can be used. Alternatively, if the vesicles are sucrose loaded, they may be diluted (at least fivefold) with the desired buffer and pelleted by ultracentrifugation before resuspending in the desired buffer. Spinning for 60 min at $10,000g$ in a polycarbonate tube using a TLA-100.3 rotor is sufficient. Gently but thoroughly resuspend the pellet in the desired final buffer. Longer or harder spinning may collapse some vesicles and release trapped solution. To prepare vesicles for the binding assay described in **Subheading 3.6.**, after resuspension, spin the vesicles again at $52,000g$ in microfuge-style TL-100 tubes (with adapters) for 20 min. The vesicles collected from the pellet at this step can be relied on to be pelleted by the assay spin at $86,000g$.

10. To accurately measure the amount of phospholipid collected in the vesicles after the procedures, it is necessary to carry out an assay of total phosphate; this step is critical for comparing the effects of different lipid mixtures. This is a colorimetric assay based on the formation of molybdate blue *(18)*. Sensitivity is in the range of 5–100 nmol/assay sample (e.g., 5–100 µL of 1 mM, 0.76 mg/mL egg PC). Samples and phosphorous standards covering the expected range of phosphate are prepared in triplicate by pipetting 1–100 µL into 13×100 mm glass tubes washed with phosphate-free detergent and rinsed with Milli-Q water. To each sample add 20 µL of 70% $HClO_4$ and 20 µL of concentrated H_2SO_4 using microcapillary glass pipets or positive displacement pipets or syringes. Add enough water to each sample so that the total volume is now 140 µL. Vortex thoroughly. Heat overnight in a fume hood at 110–120°C, or heat for 3 h at 150°C. Use safety goggles or a face shield to protect against hot acid splashes. Samples char and then become clear when

hydrolysis is complete. Add 500 μL of water and 400 μL of the following freshly prepared solution: 2 vol of Milli-Q water, 1 vol of 0.1 N H$_2$SO$_4$, 1 vol of 2.5% (w/v) (NH$_4$)$_6$Mo$_7$O$_{24}$. Add 10 μL of 10% (w/v) ascorbic acid, free acid form; that solution should be less than 7 wk old, stored at 4°C, and colorless. Heat tubes for 60 min at 37°C. Measure the absorbance at 820 nm. Standards should give a linear response in the range of 5–65 nmol of phosphorus, and the phosphorus content of the lipid samples is determined by comparison to standards. Correct the results for any lipids lacking phosphate or containing extra phosphates (e.g., phosphoinositides) to determine the total lipid content.

3.6. Vesicle-Binding Assays

3.6.1. Overview

Binding assays are facilitated by loading sucrose (170 mM) during the vesicle formation step *(17)*. The increased density allows the vesicles to sediment in an ultracentrifuge under conditions in which proteins remain in the supernatant fraction. The assay is also facilitated by incorporating a dye-labeled lipid (e.g., lissamine-rhodamine B sulfonyl-PE) into the vesicles during preparation to check for completeness of pelleting. Vesicles and proteins to be reconstituted should be exchanged into the proper buffers for the binding assays beforehand using the methods described in **Subheading 3.5.2.9.** or using gel filtration.

3.6.2. Procedure

Proteins (e.g., G$_{\alpha t}$-GTPγS and PDE6) are mixed with vesicles in microfuge-style polyallomer tubes for the TLA-100.3 or similar fixed-angle tabletop ultracentrifuge rotor and spun at 86,000g for 30 min. Typical concentrations in a final volume of 200 μL are 0–5 nM G$_{\alpha t}$-GTPγS, 1 nM PDE6, and 0–15 μM (total phospholipid) sucrose-loaded vesicles. The sucrose-loaded vesicles under many conditions do not form a tight pellet. A procedure for separating the supernatant and pellet that does not rely on flawless manual technique is to remove only the top 150 μL of the sample. A 150-μL volume of buffer is added to the pellet plus 50 μL of the supernatant, and these "75% supernatant" and "pellet plus 25% of supernatant" fractions are analyzed to determine the amount of specific proteins in each and the tracer fluorescent lipid is analyzed by measuring the fluorescence of 10-μL aliquots diluted to 200 μL in 1% SDS.

Protein quantification depends on the proteins to be analyzed. If antibodies are available, standard SDS-PAGE immunoblotting procedures with standards from the same stocks used to prepare the samples can be used to determine the fraction of protein in the supernatant and pellet fractions. If the G$_{\alpha t}$-GTPγS is prepared using GTPγ^{35}S, then the radioactivity can easily be measured by scintillation counting. Total PDE in each fraction can be determined by measuring PDE activity using the assay described in **Subheading 3.7.2.** after trypsin activation. If fairly large amounts of proteins are used (at least 200 ng/ protein subunit/sample), Coomassie staining of gels followed by densitometry can be used to quantify each protein.

3.7. Assays of PDE Activity

3.7.1. Overview

The assay described here, derived from a method developed by Liebman and Evanczuk *(19)*, is based on pH measurements over a range in which proton release increases linearly with cGMP hydrolysis. It is very convenient, provides activity readouts within a few minutes, and allows multiple additions to be made to a single sample and the activity recorded after each one. The disadvantages are that it is not tremendously sensitive (at least 100 μL of a few nanomolar PDE6 is required) and that each sample takes several minutes. Thus, if one wants to assay a large number of samples, each with much lower levels of PDE6 activity, then assays based on monitoring hydrolysis of radiolabeled cGMP are preferable.

The general principle of the assay is that the pK_a of 5'-GMP is lower than the pK_a of cGMP, so that hydrolysis of cGMP in a pH range of 7.6–8.2 liberates nearly one proton per cGMP hydrolyzed, and in this narrow pH range, pH decreases in a nearly linear fashion with protons released. A standard pH meter records this as a potential change, and this changing voltage can be recorded as a good approximation to cGMP hydrolysis using a chart recorder or a computer with an analog-to-digital converter card. With time as the *x*-axis, and cGMP hydrolysis as the *y*-axis, the slope at any point (as analyzed by computer or simply by using a ruler to draw a straight line on the chart) is proportional to the rate of hydrolysis. The proportionality constant is determined by multiplying the slope in (arbitrary *y* units/s) times the total *y* excursion, in the same units, resulting from hydrolysis of a known concentration of cGMP (e.g., 2 m*M*).

The assay is conveniently carried out in wells of 96-well microtiter plates containing magnetic stir fleas that just fit the wells. Loss of protein to surfaces is minimized by having the assay solution contain 0.1% (w/v) ovalbumin, or by coating the wells with such a solution before use. Microelectrode (these are "micro" by conventional pH electrode standards, but enormous by comparison with those used for cellular recording) volumes in the range of 100–200 μL can be used routinely. Use of a 5X buffer stock allows other components to be added as needed, with water making up the remainder of the volume, so that the buffer capacity does not vary significantly. In the same sample, one can sequentially assay basal activity, activity stimulated by one or more concentrations of an activator such as Gαt-GTPγS or a vesicle suspension, and maximal activity stimulated by trypsin. Trypsin rapidly degrades the inhibitory PDEγ-subunit and thus gives rise to the highest observable level of PDE6 activity, which can be used to normalize the G protein–stimulated activity.

3.7.2. pH Assay for PDE6 Catalytic Activity

1. Clean the electrode, microtiter plate, and stir flea with 0.1 N HCl, and then wash with Milli-Q water.
2. Condition the electrode with pH assay buffer, pH 8.0, at least 30 min before planning to start the assay. Start the recorder and observe the drift and noise level. Measuring PDE basal activity accurately requires a flat baseline. To minimize drift, do the following:
   a. Be sure to clean up traces of trypsin in the electrode between each assay with 0.1 N HCl and washing with Milli-Q water.
   b. Keep the Milli-Q water and cGMP aliquot at room temperature.
   c. Make sure that the assay solution covers the reference junction, about 1.5 mm from the tip of the electrode.
   d. Maintain a constant ionic strength.
3. Check the recorder range with solutions prepared from standards (pH 6.0 and 8.0 standards), adjust to zero, and vary the calibration knobs as necessary. The voltage output from the pH meter can be connected to a computer analog-to-digital converter card as an alternative or supplement to the chart recorder.
4. Calculate the volumes and concentrations of all planned components in the assay; we routinely use a 200-μL volume in each assay that contains a final 1X pH assay buffer: 2 mM cGMP, with 0.5–20 nM PDE and transducin, and at least a 1000-fold excess of total phospholipids over PDE. The final trypsin concentration is typically 0.1 mg/mL.
5. Put a clean stir flea into the microtiter well and start the stirring motor. Add 40 μL of 5X pH assay buffer and an appropriate volume of Milli-Q water (calculated to give the desired 200-μL volume after the other components are added) to the stir bar–containing well. Lower the electrode to the solution. The tip of the electrode should not touch the stir bar or the side of the well.
6. After the pH is stable at 8.0, start the chart recorder.
7. Add the assay components in the desired order, labeling the chart paper after each addition with a line pointing to the position (time) of the addition.
8. Between assays, wash the electrode with Milli-Q water, 0.1 N HCl, and then Milli-Q water.
9. At the end of the assay, extensively wash everything used with Milli-Q water and 0.1 N HCl, and then wash with Milli-Q water. Store the electrode probe in pH 4.0 standard.
10. Calculate PDE activity. The calculations are based on the assumption that the concentration of cGMP is much greater than the K_m (PDE K_m for cGMP is ~10 μM), the initial rate of cGMP hydrolysis is at its maximum, and it is directly proportional to the amount of PDE present in the assay. Raw data from a typical

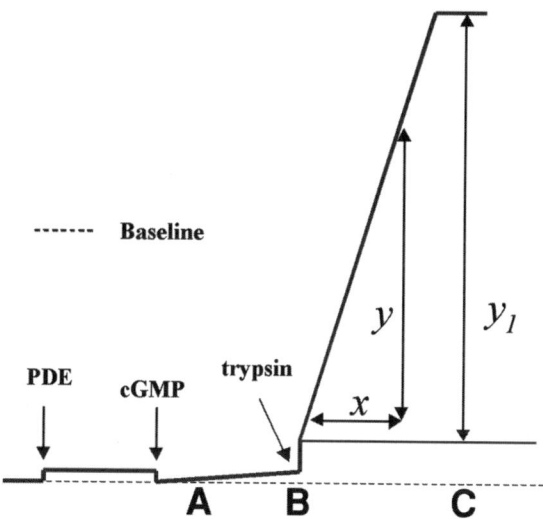

Fig. 1. **(A)** The data is the slope of the line (rise over run; y distance [cm]/x distance [cm]) that represents the initial velocity, change in pH/time, following each addition. **(B)** This is proportionality constant between y excursion and cGMP hydrolysis, based on the change in the $y1$ direction after complete hydrolysis of 2mM cGMP. It does not include the change owing to trypsin solution alone. **(C)** The speed of the chart; set up typically as 1 cm/60 s.

PDE assay are shown in **Fig. 1**. Use the following equation to determine PDE activity:

$$\text{PDE activity } (\mu M \text{ s}^{-1}) = \frac{y \text{ cm}}{x \text{ cm}} \bullet \frac{2000 \ \mu M \text{ CGMP}}{y_1 \text{ cm}} \bullet \frac{1 \text{ cm}}{60 \text{ s}}$$

The data in the first part of the equation represent the slope of the line ("rise over run"; y distance [cm]/x distance [cm]) that represents the initial velocity, change in pH/time, following each addition. The second part of the equation represents the proportionality constant between y excursion and cGMP hydrolysis, based on the change in the $y1$ direction after complete hydrolysis of 2 mM cGMP. It does not include the change owing to trypsin solution alone. The last part of the equation represents the speed of the chart, set up typically as 1 cm/60 s.

4. Note

1. For vesicle extrusion (*see* **Subheading 3.5.2.**), Avanti supplies a manual extruder in which the pressure is supplied via hand pressure on microsyringes. In our experience, an earlier version of this extruder works, but at the expense of multiple

broken microsyringe barrels, and it is not recommended. A more recent version includes an extruder stand/stabilizer block that may alleviate this problem and is less expensive than the Lipex device. Another alternative if an HPLC system is available involves the use of a high-pressure stainless steel filter holder supplied by Millipore for use with 2.5-cm filter disks. It can be connected to the injection port of an HPLC system in place of a column, with the outlet directed into a collection tube instead of the detector. The sample is injected via the system's injection loop and extruded using the same buffer used for hydrating the lipids, supplied by the system's solvent delivery pump. The flow rate setting should be initially set to 0.1 mL/min and slowly increased until the lipid suspension begins to elute from the filter holder. A pressure limit of 800 psi should be used. As soon as the injected volume is eluted the flow is stopped, and the eluted lipids are reinjected. A total of 10 cycles of extrusion is sufficient. Note that the volume unavoidably increases a little each time, and this dilution must be taken into account.

References

1. DeMar, J. C. Jr., Rundle, D. R., Wensel, T. G., and Anderson, R. E. (1999) Heterogeneous N-terminal acylation of retinal proteins. *Prog. Lipid Res.* **38(1),** 49–90.
2. Kokame, K., Fukada, Y., Yoshizawa, T., Takao, T., and Shimonishi, Y. (1992) Lipid modification at the N terminus of photoreceptor G-protein α-subunit. *Nature* **359,** 749–752.
3. Neubert, T. A., Johnson, R. S., Hurley, J. B., and Walsh, K. A. (1992) The rod transducin alpha subunit amino terminus is heterogeneously fatty acylated. *J. Biol. Chem.* **267,** 18,274–18,277.
4. Yang, Z. and Wensel, T. G. (1992) N-myristoylation of the rod outer segment G protein, transducin, in cultured retinas. *J. Biol. Chem.* **267(32),** 23,197–23,201.
5. Anant, J. S., Ong, O. C., Xie, H. Y., Clarke, S. O., Brien, P. J., and Fung, B. K. (1992) In vivo differential prenylation of retinal cyclic GMP phosphodiesterase catalytic subunits. *J. Biol. Chem.* **267,** 687–690.
6. Catty, P., Pfister, C., Bruckert, F., and Deterre, P. (1992) The cGMP phosphodiesterase-transducin complex of retinal rods: membrane binding and subunits interactions. *J. Biol. Chem.* **267(27),** 19,489–19,493.
7. Malinski, J. A. and Wensel, T. G. (1992) Membrane stimulation of cGMP phosphodiesterase activation by transducin: comparison of phospholipid bilayers to rod outer segment membranes. *Biochemistry* **31(39),** 9502–9512.
8. Papermaster, D. S. and Dreyer, W. J. (1974) Rhodopsin content in the outer segment membranes of bovine and frog retinal rods. *Biochemistry* **13(11),** 2438–2444.
9. Papermaster, D. S. (1982) Preparation of retinal rod outer segments. *Methods Enzymol.* **81,** 48–52.
10. Bradford, M. M. (1976) A rapid and sensitive method for the quantitation of microgram quantities of protein utilizing the principle of protein-dye binding. *Anal. Biochem.* **72,** 248–254.

11. Fung, B. K., Hurley, J. B., and Stryer, L. (1981) Flow of information in the light-triggered cyclic nucleotide cascade of vision. *Proc. Natl. Acad. Sci. USA* **78(1)**, 152–156.

12. Baehr, W., Devlin, M. J., and Applebury, M. L. (1979) Isolation and characterization of cGMP phosphodiesterase from bovine rod outer segments. *J. Biol. Chem.* **254(22)**, 11,699–11,707.

13. Ramdas, L., Disher, R. M., and Wensel, T. G. (1991) Nucleotide exchange and cGMP phosphodiesterase activation by pertussis toxin inactivated transducin. *Biochemistry* **30(50)**, 11,637–11,645.

14. Yamazaki, A., Tatsumi, M., and Bitensky, M. W. (1988) Purification of rod outer segment GTP-binding protein subunits and cGMP phosphodiesterase by single-step column chromatography. *Methods Enzymol.* **159**, 702–710.

15. Kleuss, C., Pallast, M., Brendel, S., Rosenthal, W., and Schultz, G. (1987) Resolution of transducin subunits by chromatography on blue sepharose. *J. Chromatogr.* **407**, 281–289.

16. Hope, M. J., Bally, M.B., Webb, G., and Cullis, P. R. (1985) Production of large unilamellar vesicles by a rapid extrusion procedure: characterization of size distribution, trapped volume and ability to maintain a membrane potential. *Biochim. Biophys. Acta* **812**, 55–65.

17. Kim, J., Blackshear, P. J., Johnson, J. D., and McLaughlin, S. (1994). Phosphorylation reverses the membrane association of peptides that correspond to the basic domains of MARCKS and neuromodulin. *Biophys. J.* **67(1)**, 227–237.

18. Chen, P. S., Toribara, T. Y., and Warner, H. (1956) Microdetermination of phosphorous. *Anal. Chem.* **28**, 1756–1758.

19. Liebman, P. A. and Evanczuk, A. T. (1982) Real time assay of rod disk membrane cGMP phosphodiesterase and its controller enzymes. *Methods Enzymol.* **81**, 532–542.

Index